# MATERIAL MATTERS

HERBERT PRESS
Bloomsbury Publishing Plc
50 Bedford Square, London, WC1B 3DP, UK
Bloomsbury Publishing Ireland Limited,
29 Earlsfort Terrace, Dublin 2, D02 AY28, Ireland



First published in Great Britain 2026



A catalogue record for this book is available from the British Library

Library of Congress Cataloguing-in-Publication data has been applied for

ISBN: HB: 978-1-7899-4230-9; ePub: 978-1-7899-4229-3; ePDF: 978-1-7899-4232-3

2 4 6 8 10 9 7 5 3 1

Designed and typeset by Five Twentyfive Limited
Printed and bound in China by C&C Offset Printing Co., Ltd.

MIX
Paper | Supporting responsible forestry
FSC® C008047
FSC
www.fsc.org

Bloomsbury Publishing Plc makes every effort to ensure that the papers used in the manufacture of our books are natural, recyclable products made from wood grown in well-managed forests. Our manufacturing processes conform to the environmental regulations of the country of origin.

To find out more about our authors and books
visit www.bloomsbury.com and sign up for our newsletters

For product safety related questions contact productsafety@bloomsbury.com

# MATERIAL MATTERS

## An Artist's Guide to Understanding Ceramic Materials

**Linda Bloomfield and Linda Swanson**

HERBERT PRESS
LONDON • OXFORD • NEW YORK • NEW DELHI • SYDNEY

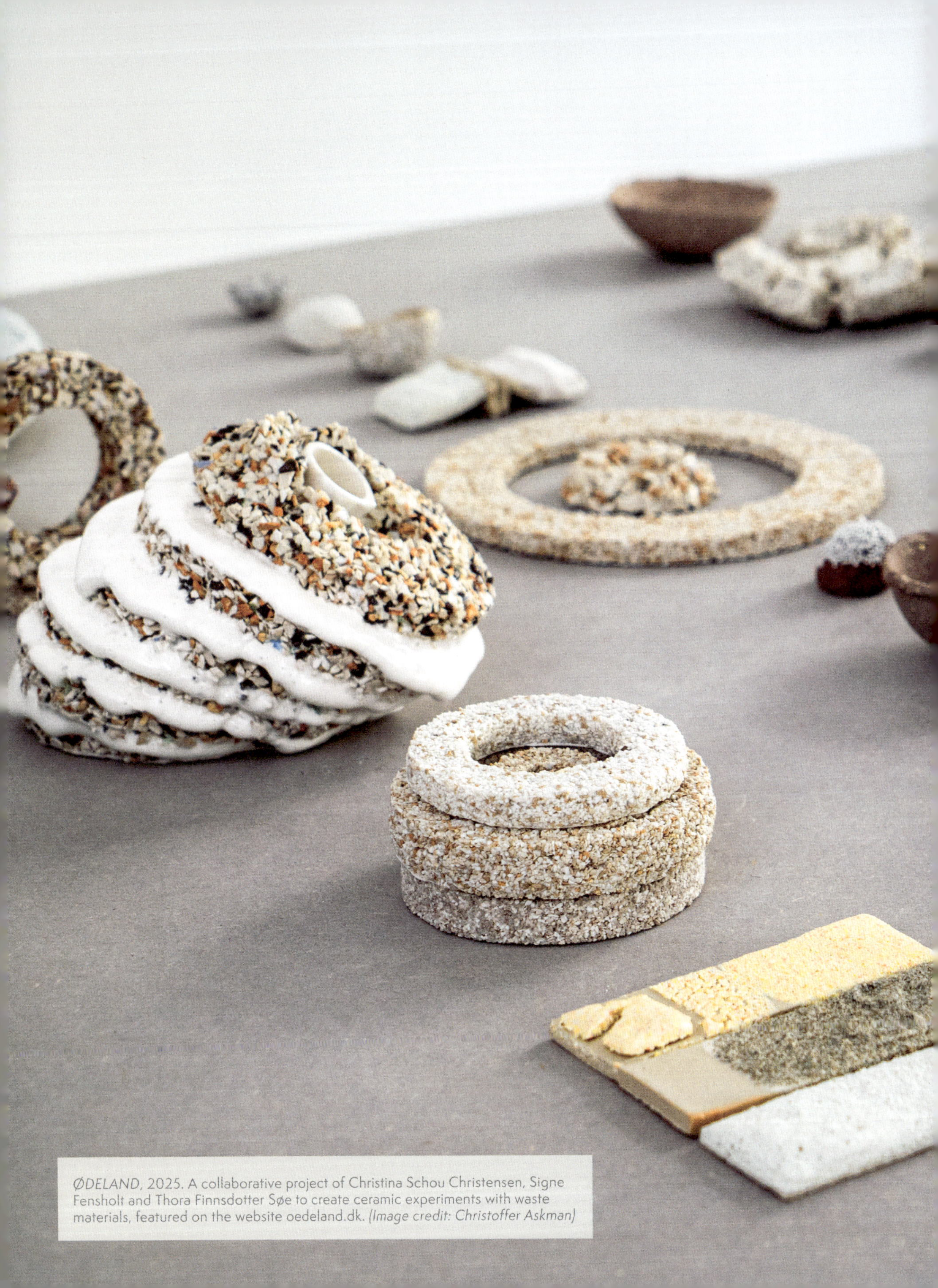

*ØDELAND*, 2025. A collaborative project of Christina Schou Christensen, Signe Fensholt and Thora Finnsdotter Søe to create ceramic experiments with waste materials, featured on the website oedeland.dk. *(Image credit: Christoffer Askman)*

# CONTENTS

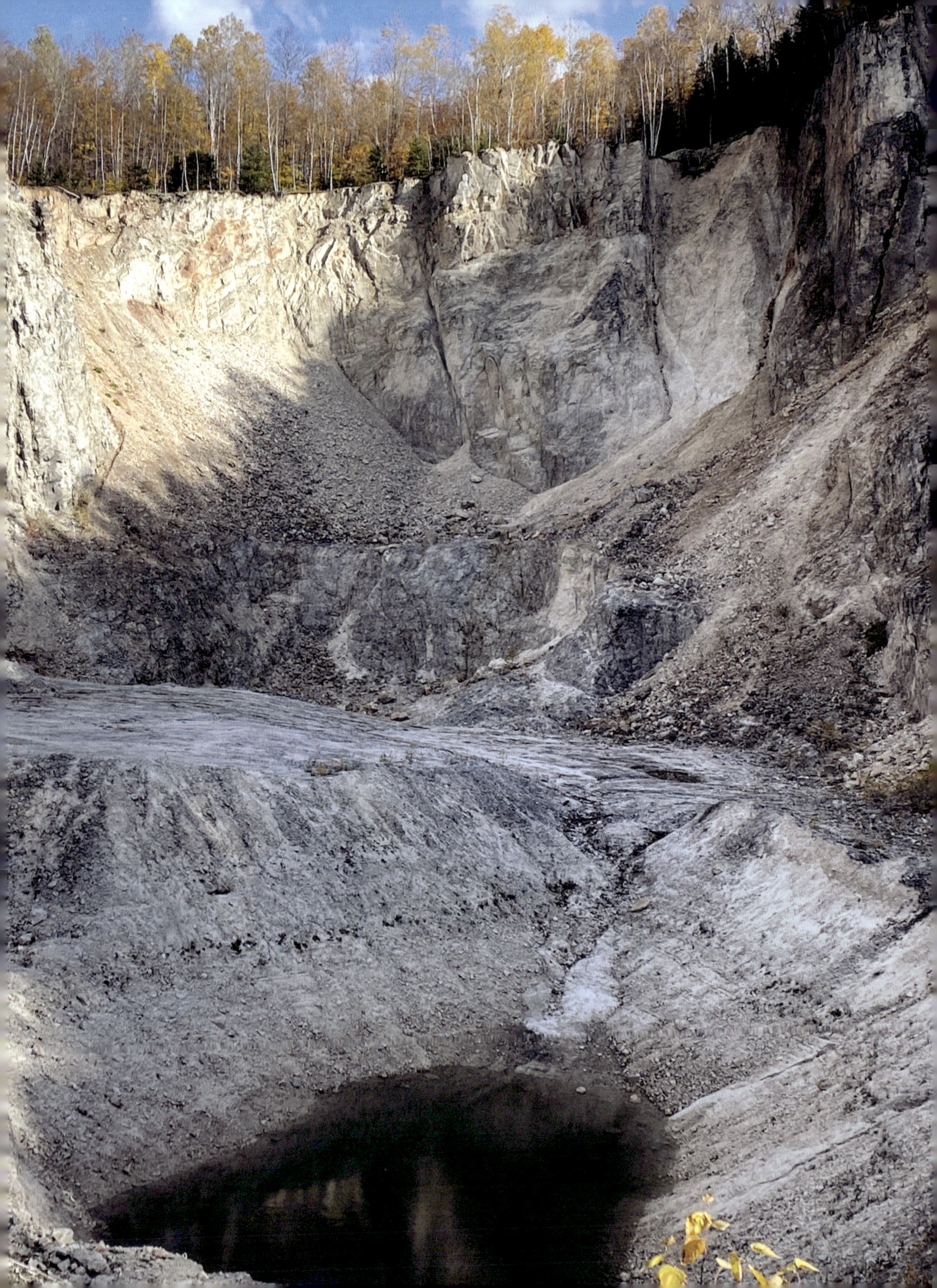

# INTRODUCTION

*Material Matters: An Artist's Guide to Understanding Ceramic Materials* provides an overview of the properties and composition of ceramic materials. It is aimed at providing a basic framework that explains how ceramic materials behave and interact in simplified terms. While this foundational knowledge is useful for understanding how to use materials, it is also crucial for developing more sustainable approaches in ceramics.

Many cultures around the world have sustained the production of ceramics with local resources for centuries. Most ceramic production today, including small-scale studios in Europe and North America, rely on industrially refined clay and glaze ingredients. Yet the long distances these materials travel and their links to global mining operations often have serious environmental and social consequences. The current environmental crisis demands that we reevaluate all aspects of how we make ceramics, and it is only by understanding our own materials and processes that we can begin to address it. Learning how to better use, source and recycle these materials will surely be essential in the future of ceramics.

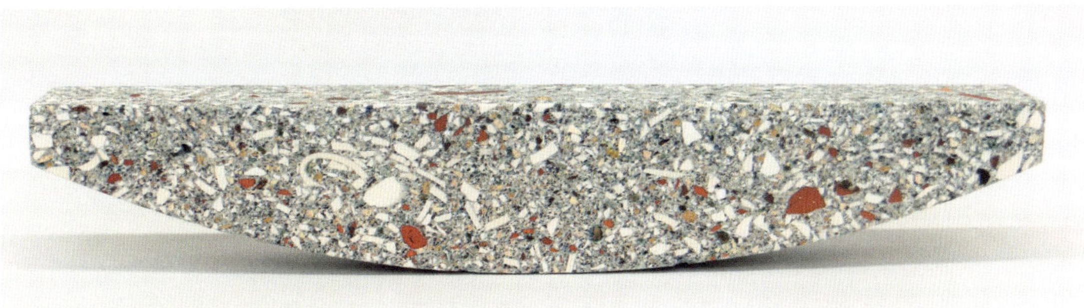

**Opposite:** Closed quartz quarry in Canada

**Above:** David Binns, *Shard Vessel*, 2017. Formed of recycled crushed ceramic material aggregate with additional clay and glaze materials that have been fired, cut, then polished after firing. *(Image credit: David Binns)*

Both the pursuit of a particular result and the delight of an unexpected one are enriched by the ability to posit how and why things happen. While ceramic materials may vary considerably due to their natural origins, some general principles can help us to strategize how to use and substitute them. While testing is necessary, knowing more about materials and how they function in clays and glazes leads to a level of efficiency that helps in saving effort as well as resources.

A challenge in ceramics arises from the fact that most materials occur as a combination of ceramic oxides rather than as singular sources. For example, clay is comprised of two oxides: silica ($SiO_2$) and alumina ($Al_2O_3$). Understanding the composition of ceramic materials in terms of their elemental makeup can provide some significant information about a recipe even before it is mixed or fired. Fortunately, there are only 13 ceramic oxides that comprise most base glazes. The role of each oxide is presented in this text with its natural and commercial sources.

There are five basic types of materials in ceramics: clay, flux, glass former, filler and colourant. Clay bodies and glazes are made from these same ingredients in varying proportions:

| | CLAY BODIES | GLAZES |
|---|---|---|
| **Clay** | High | Low |
| **Flux** | Low | High |
| **Glass former** | High | High |
| **Filler** | Some | None |
| **Colourant** | Low | Low |

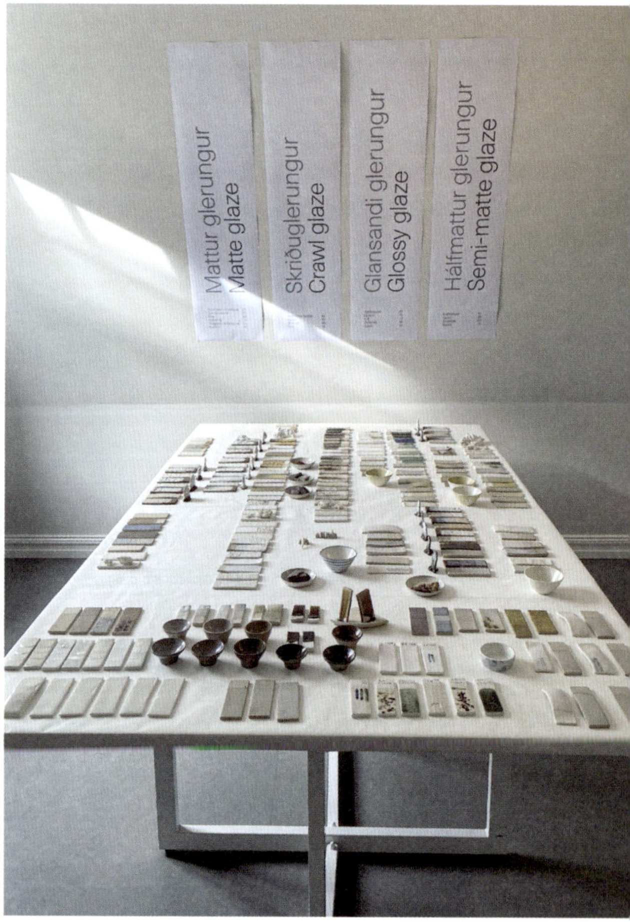


**Right:** *Glaze Dialogue* is an international collective of artists and designers working on developing glazes with local and recycled materials to broaden a public dialogue on ceramics and sustainability.

**Opposite:** Tamsin van Essen, *Psoriasis, Medical Heirlooms* Series, 2007. Normally considered a glaze fault, this shivering glaze offers a poignant reference to skin conditions. *(Image credit: Tamsin van Essen)*

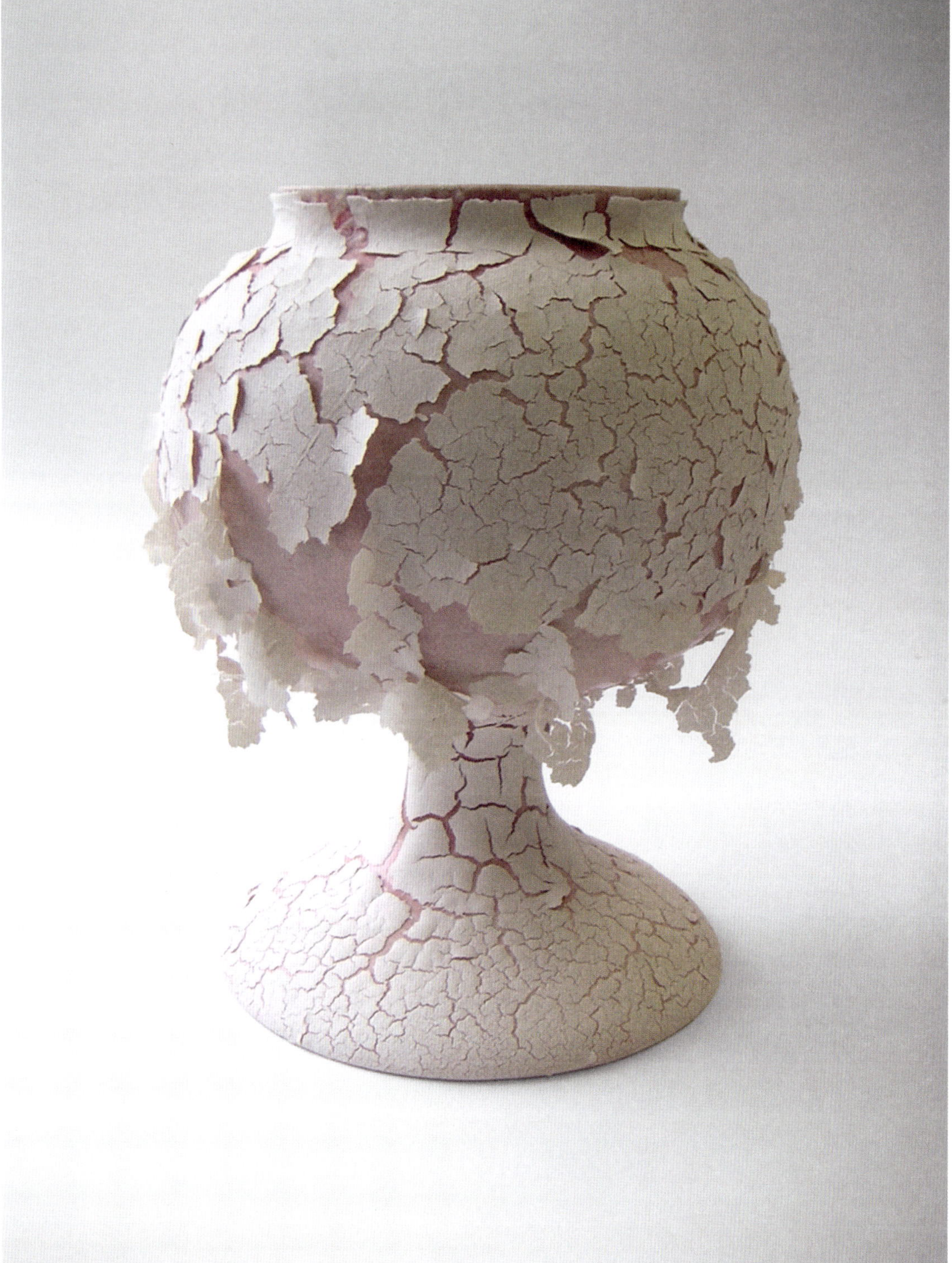

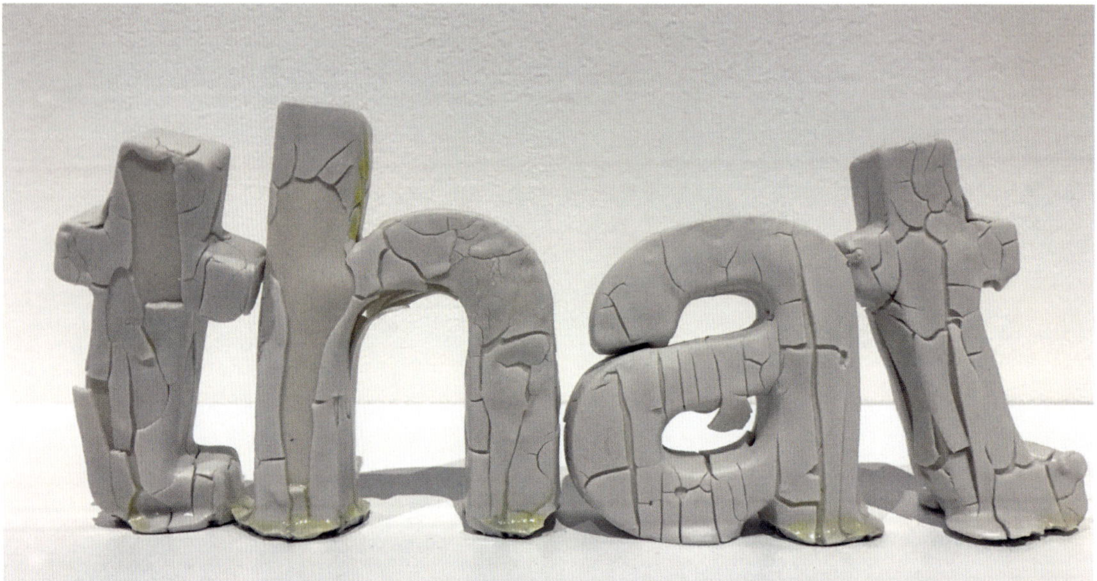


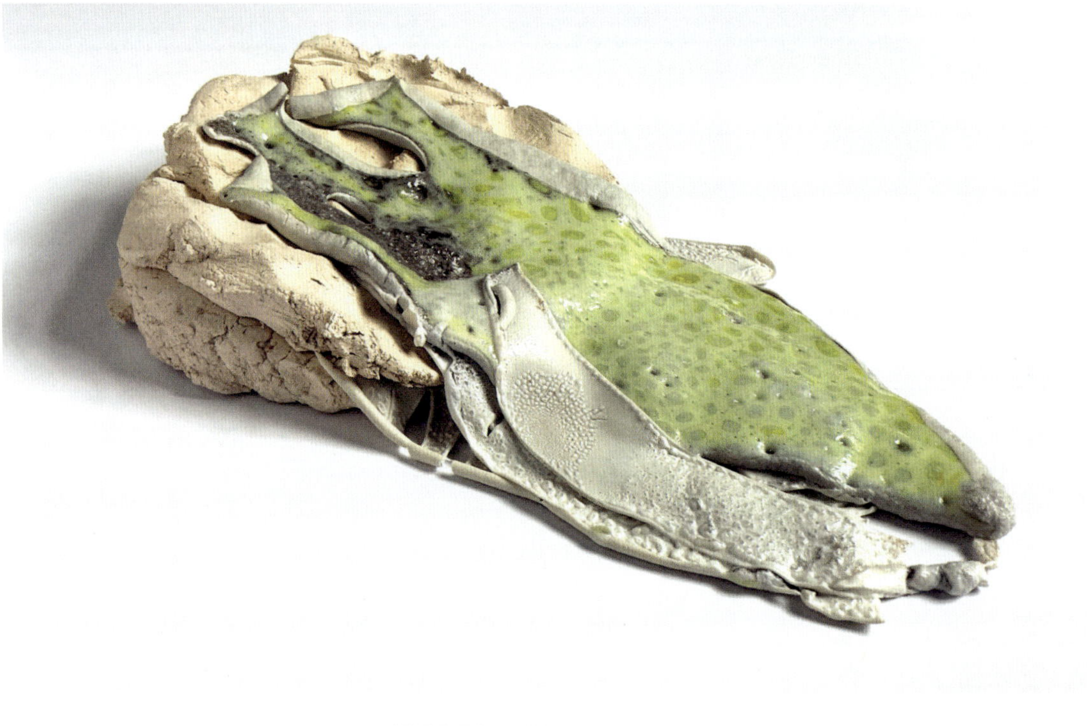

**Above top:** Thomas Müller, *Where'd you get that banana?*, 2017. High clay content causes cracking which adds dimensionality to this surface. *(Image credit: Thomas Müller)*

**Above bottom:** Annouk Thys, *Scape*, 2019. Recycled clay and glazes adopted from other artists are recomposed with an element of chance to create a daring new form. *(Image credit: Annouk Thys)*

Ceramic faults tend to occur when these materials are used in other proportions. For example, clay bodies usually contain low amounts of flux. A clay body mixed with very high amounts of flux may melt and lose its form. Likewise, adding more clay to a glaze can cause the glaze to stiffen, melt less or crawl. Understanding how to gauge the amounts of these five basic types of materials is important for modifying and creating recipes or substituting natural and recycled materials in them.

Knowledge of ceramic materials and processes establishes the means for artists and designers to become more inventive. Even with a limited number of materials in ceramics, there are endless possibilities in their combinations. What is truly wondrous about ceramics is that there will always be something more to discover.

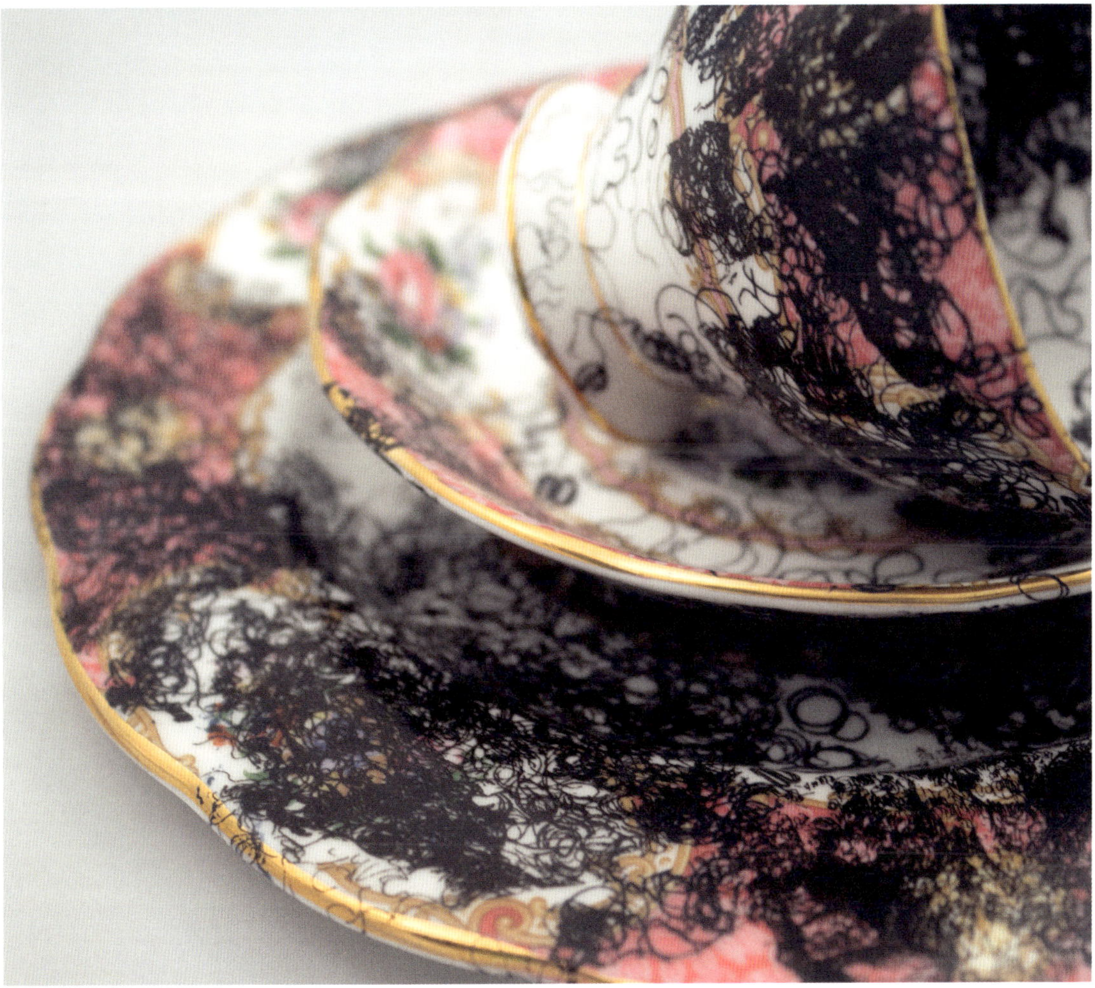

**Above:** Sharon Norwood, *Hair Matters*, 2018. Upcycled ceramics are re-envisioned with meanings reimagined and revised through the application of overglaze ceramic decals. *(Image credit: Sharon Norwood)*

## AUTHORS

We met during a residency at the Guldagergaard International Ceramic Research Center in Denmark. Working alongside each other, we discovered our common desire to make ceramics easier to understand. Our different backgrounds and approaches to the medium proved to be complementary for creating this simplified text on ceramic materials.

Linda Bloomfield originally trained as a scientist and currently works as an author and studio potter in the UK. She has written numerous books on ceramics including *Colour in Glazes, Special Effect Glazes* and *Science for Potters*, among others. Linda also teaches workshops on glaze development and regularly contributes articles to *Ceramics Monthly* and *Ceramic Review*.

Linda Swanson is an educator and artist who creates installations of raw materials and kiln-fired ceramics which have been exhibited internationally. She teaches ceramics at Concordia University in Montreal, Canada, where her current research interests include the development of unconventional ceramic materials and sustainability.

**Below:** Linda Bloomfield, *Lichen Forms in Woods*, 2021. Thrown porcelain forms with lichen-effect glazes. *(Image credit: Henry Bloomfield)*

**Opposite:** Linda Swanson, *Oculus*, 2018. Time-based work with bentonite clay in process of swelling to eventually fill the glass vessel. *(Image credit: Gaëtane Girard)*

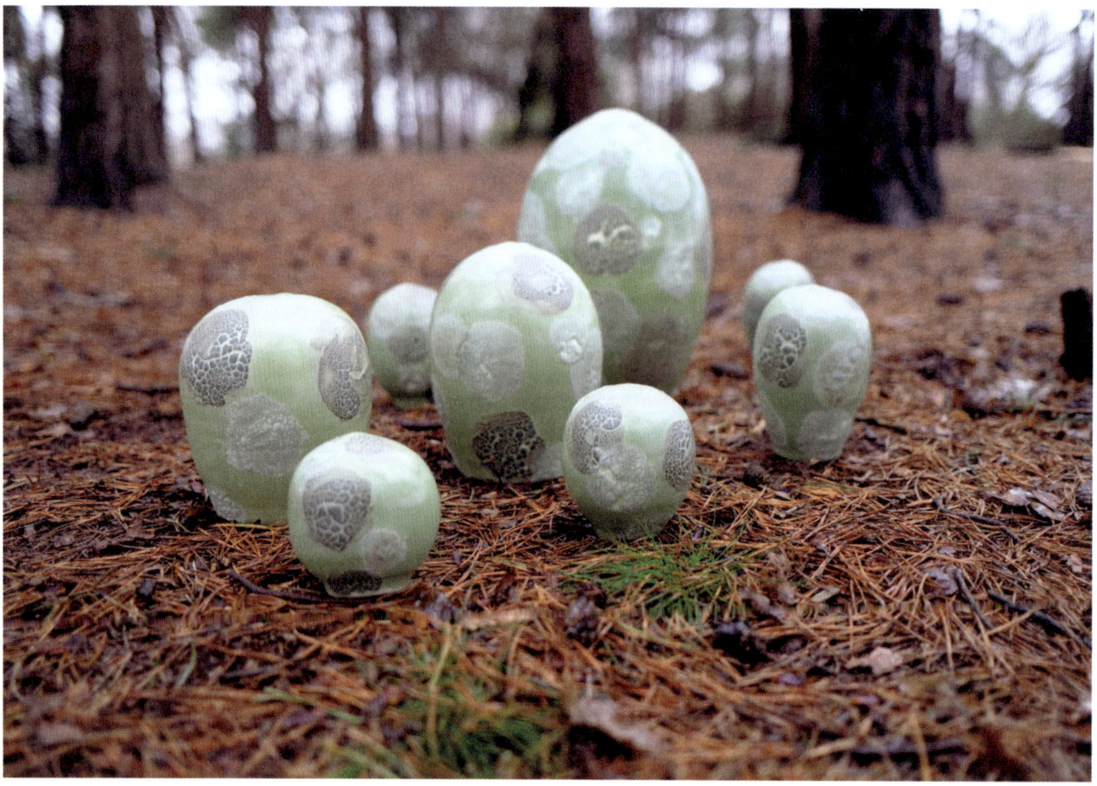

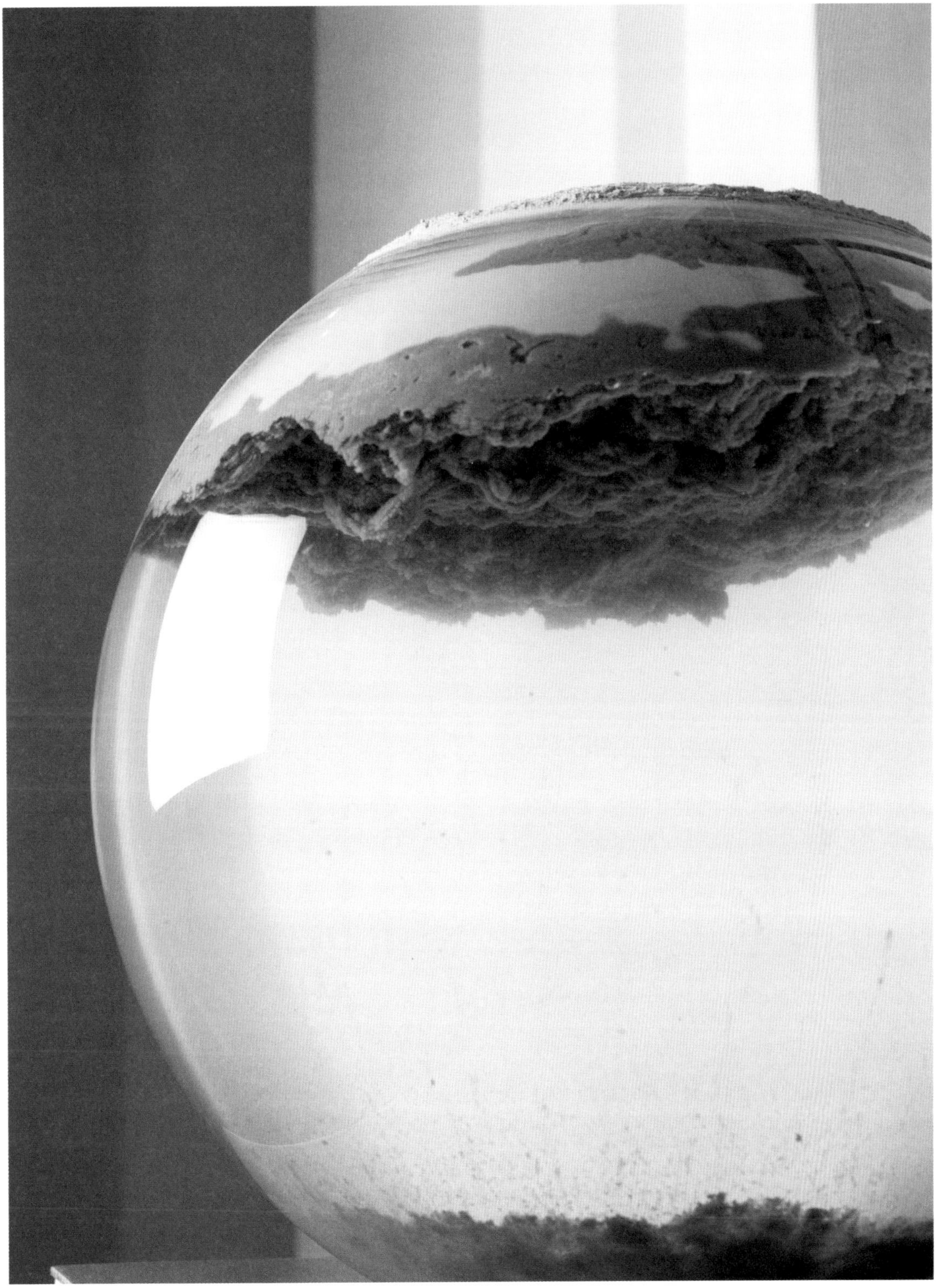

CHAPTER ONE

# GEOLOGY AND THE ORIGINS OF CLAY

## ROCKS

Clay is a unique natural material that cannot be made artificially or synthesized. The special properties inherent to clay come from the way it is formed through natural weathering processes. Clay comes from rocks that have eroded and broken down into clay minerals that comprise about 60% of the earth's surface. While clay is abundant, different types of clay and their properties vary widely depending on how and where it is formed. The local geology of any particular place can influence the type, quality and kind of ceramic materials found there.

Ceramic materials are made from minerals which are found in rocks. Some materials are simply crushed rocks, while clay is formed through erosion and weathering. The minerals used in ceramics come from all three of the basic types of rocks: igneous, sedimentary and metamorphic.

| | |
|---|---|
| **IGNEOUS ROCKS** | formed from molten magma that has cooled and crystallized |
| **SEDIMENTARY ROCKS** | formed under pressure from deposited layers of sediment |
| **METAMORPHIC ROCKS** | formed of other rocks that are changed by heat, pressure and/or hydrothermal processes |

**Opposite:** Magdolene Dykstra, *Polyanthroponemia*, 2020. Site-specific installation of unfired clay and mixed media. *(Image credit: Toni Hafkenscheid)*

**Overleaf:** Aneta Regel, *Untitled Vessel*, 2023. Crushed rocks are incorporated into the clay body and protrude from the form in unpredictable ways. *(Image credit: Sylvain Deleu & Matthew Booth)*

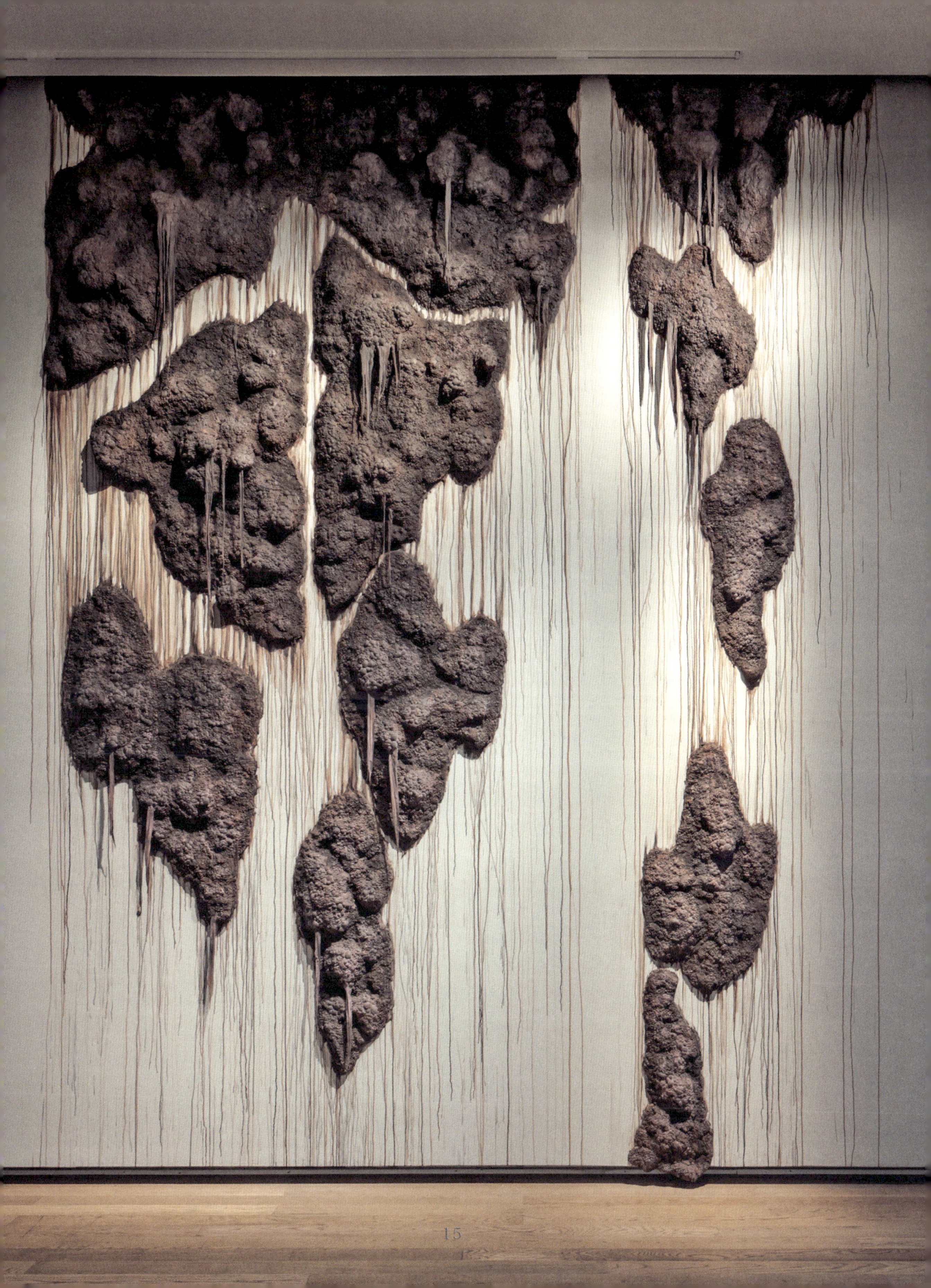

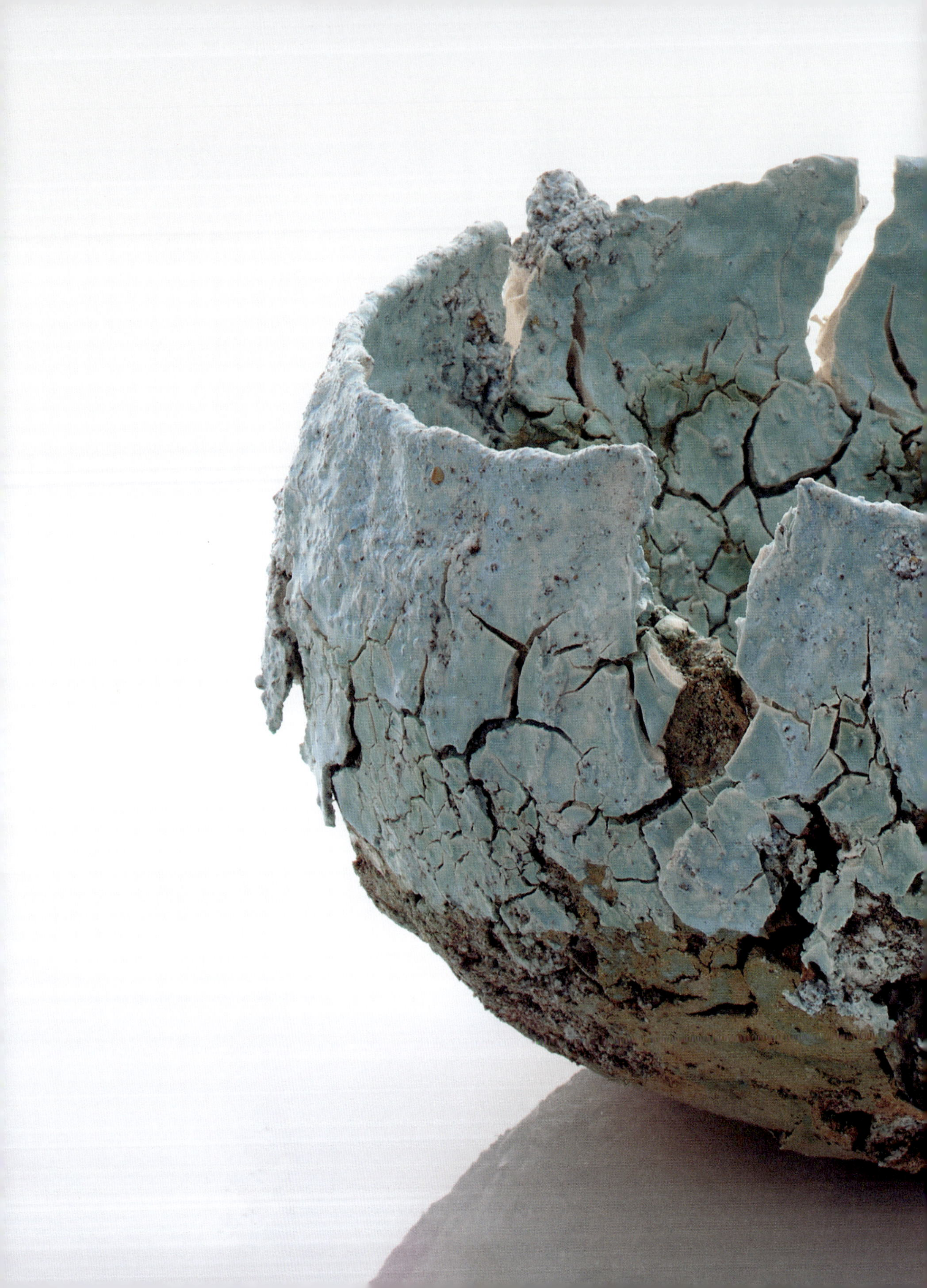

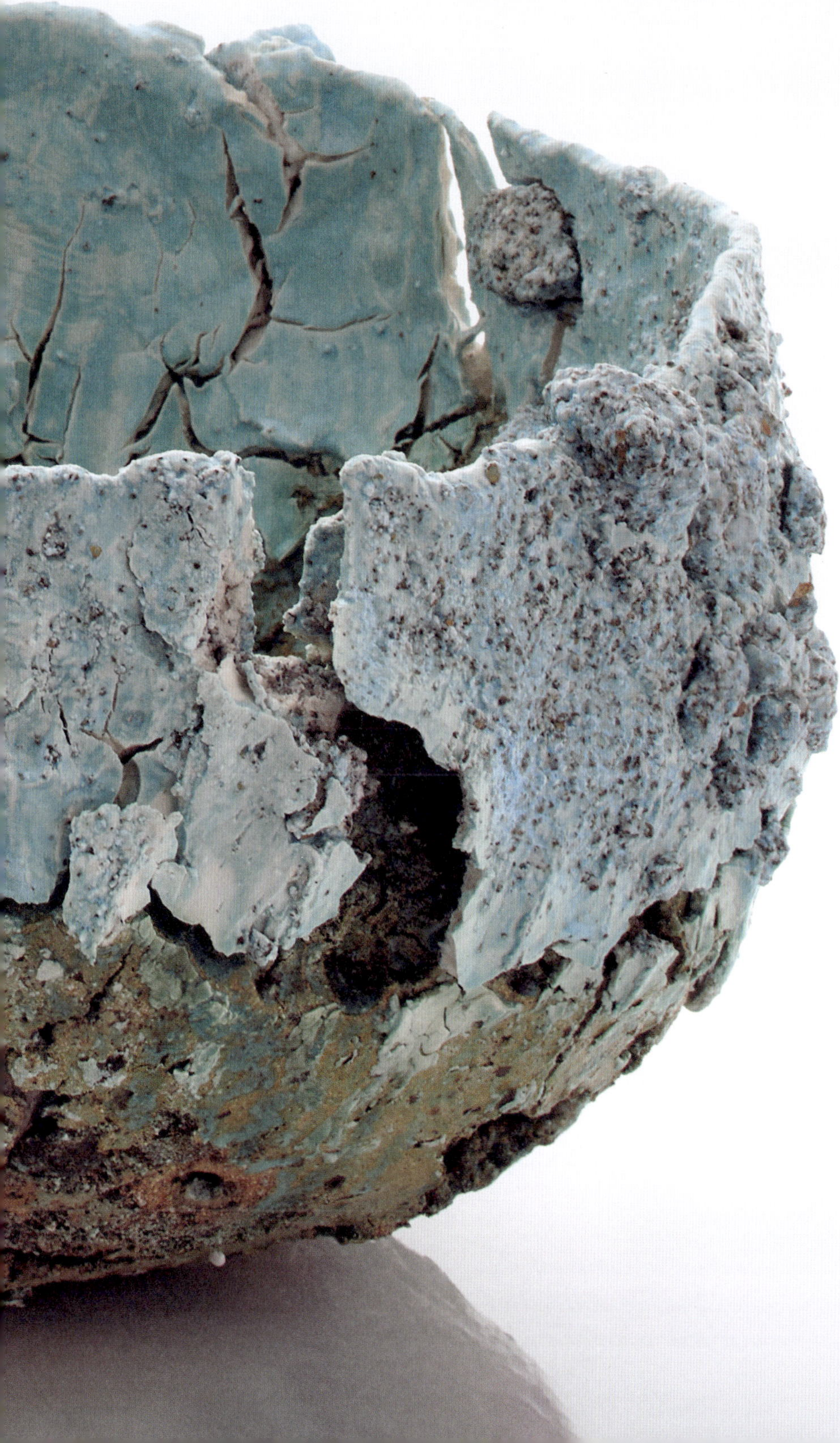

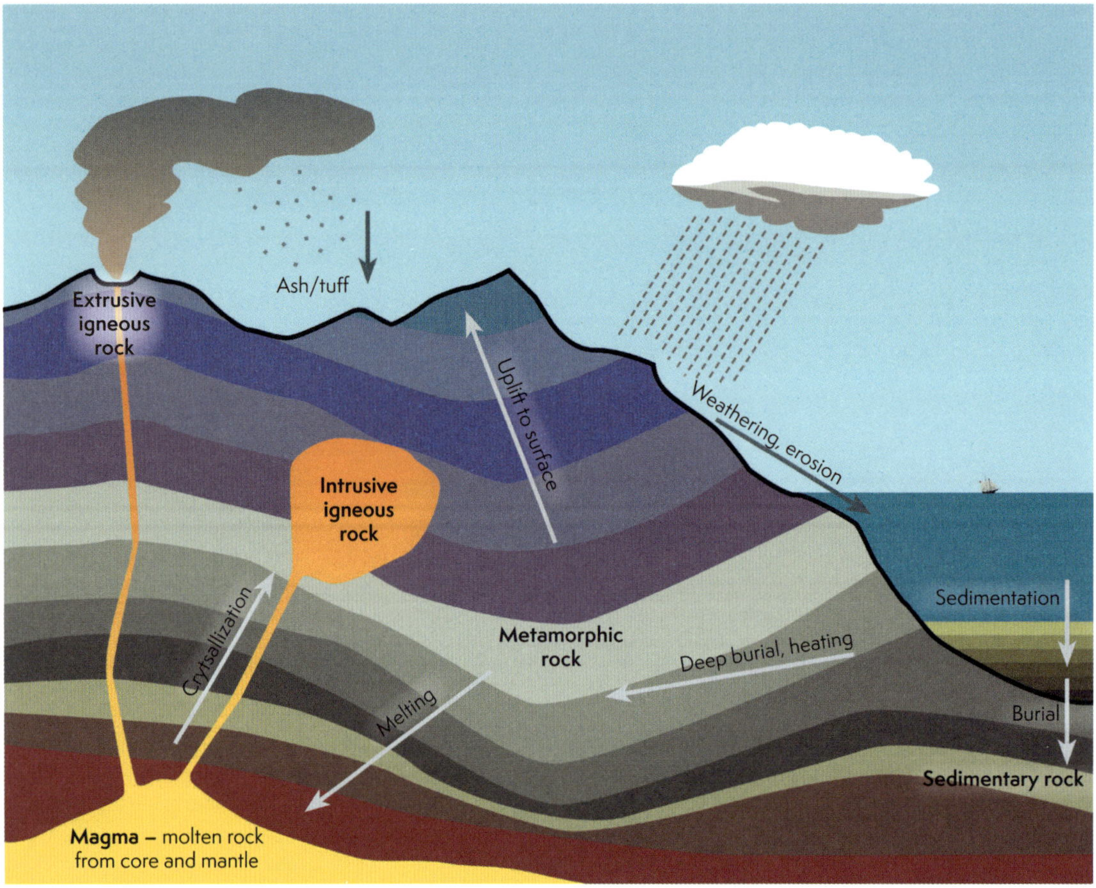


Igneous rocks play the most significant role in the formation of clay minerals. Igneous rocks are formed from magma that has risen from deep underground. Extrusive igneous rocks are formed from openings above ground from magma that has flowed to the surface and been spewed as lava or ash. Intrusive igneous rocks are formed from magma that remains underground, cooling more slowly and crystallizing. The type of rock known as feldspar is an intrusive igneous rock that precipitates out of magma as it cools and crystallizes. Feldspar can eventually morph into the clay mineral known as kaolinite. The kaolinite mineral plays a part in many different types of clay.

**Above:** Rock cycle
**Opposite left:** Grey granite rock
**Opposite right:** Pink granite rock

Two processes change rock into clay minerals: decomposition and disintegration. Disintegration is the mechanical breakdown of rocks and minerals and can be caused by multiple factors including weathering from freeze-thaw cycles, wind, lightning and fire, root action from plants, animal movements, glaciers and the crystallization of soluble salts. Decomposition is the chemical breakdown of rocks and minerals from acids including acid rain, underground hydrothermal acidic solutions and gases, or carbonic acid from decaying organic matter.

The rock granite contains feldspar, and it offers a good example of how clay is naturally formed out of rock. Granite is composed of three minerals: quartz, feldspar, and mica. These three components can be distinguished by their colour and finish; quartz is glassy clear or grey, feldspar is white, dull pink or yellow and mica is shiny black. Even though granite is quite hard, its coarse-grained texture can break apart over time, exposing and separating these three distinct minerals. Given the right conditions, the feldspar mineral of the granite can be altered through weathering and hydrothermal processes into the mineral kaolinite. Kaolinite is the primary component of kaolin and other clays.

Feldspar breaks down physically from granite initially by disintegration, breaking apart from the quartz and mica. The feldspar becomes further reduced through decomposition which happens on a molecular level. As feldspar decomposes, the soluble alkali compounds of sodium and potassium dissolve and wash out of the feldspar. After their removal, only two compounds remain: alumina and silica. Having lost the alkali components, the alumina and silica are primed to bond with water, also known as a hydrous aluminosilicate. This new substance is also called kaolinite which is the mineral basis of clay. The ideal formula for clay matches the chemical formula of kaolinite: one molecule of alumina bonded to two molecules of silica and two molecules of water.

| | | |
|---|---|---|
| $Al_2O_3$ | $2SiO_2$ | $2H_2O$ |
| Alumina | Silica | Water |
| (Alumino) | (Silicate) | (Hydrous) |

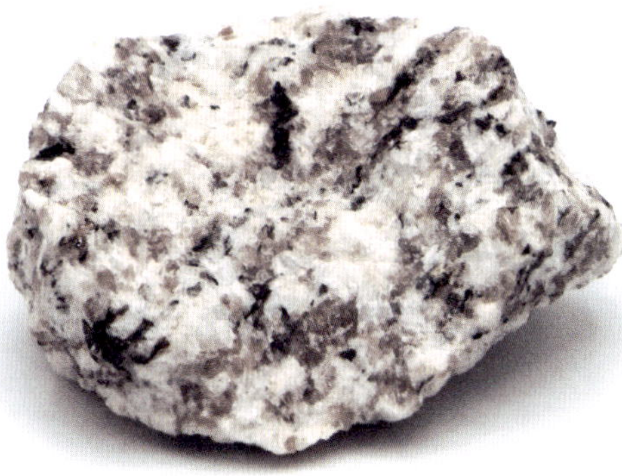

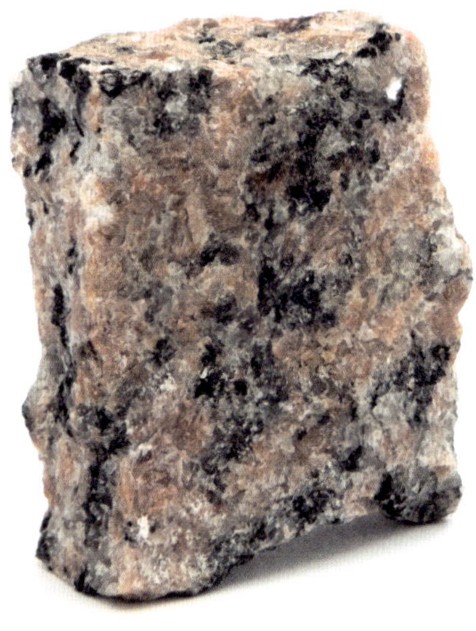

## CLAY MINERALS

Clays can be classified according to their mineral composition. In ceramics, the three commonly encountered clay minerals are kaolinite, illite and smectite (montmorillonite).

Kaolinite is the most common clay mineral in ceramics. Despite appearing as a powder, kaolinite is a crystalline mineral formed from feldspathic rocks. As it is transported from its initial deposit site by erosion, kaolinite combines with other minerals to form the basis of most clays.

Illite is formed from decomposed mica, one of the three minerals found in granite, though it is also found individually. Illite comes primarily from sedimentary clays that have hardened into shale.

Smectite, also known as montmorillonite, is a group of minerals derived from decomposed volcanic ash or tuff. Molecular bonds in montmorillonite are easily broken by water and the structure can absorb large amounts of water which causes high expansion when wet and high shrinkage when dried. Bentonite is the most common smectite or montmorillonite clay.

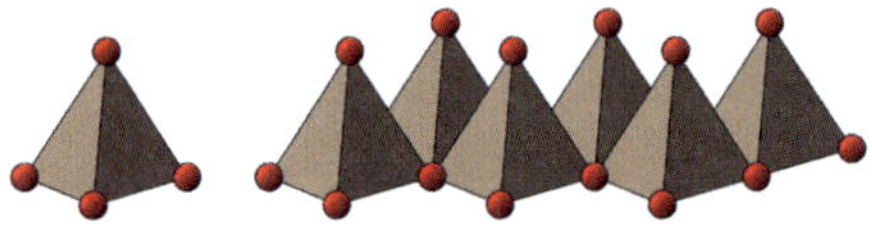

Silica tetrahedra

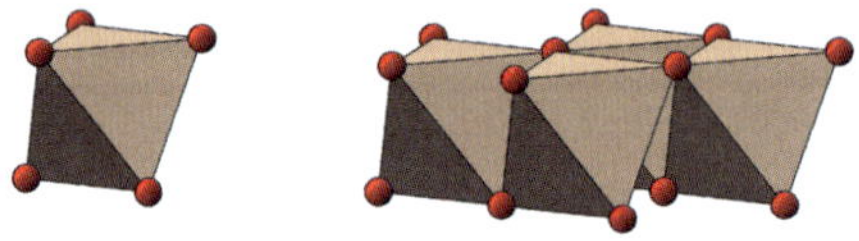

Alumina octahedra

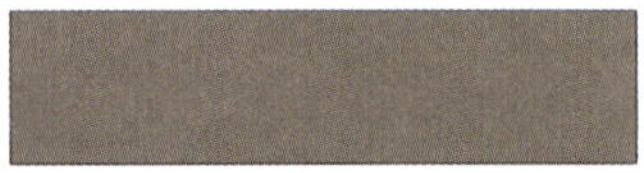

Schematic of alumina layer

Schematic of silica layer

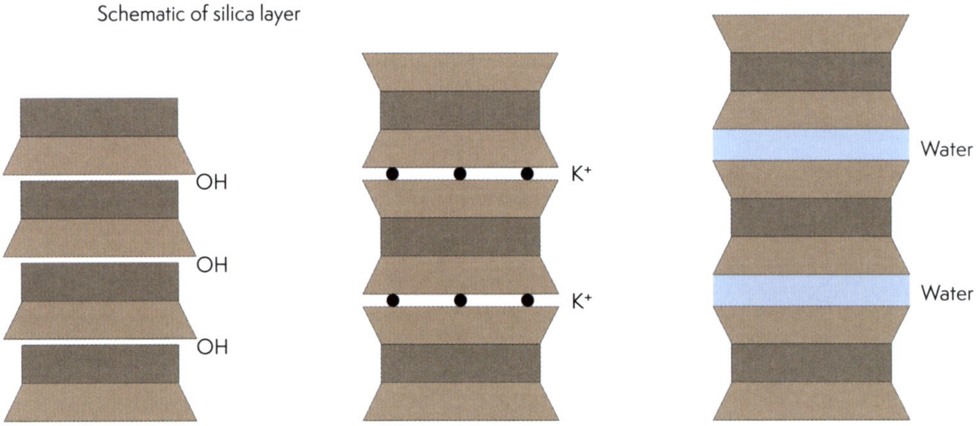


Kaolin Illite Bentonite

**Above:** The structure of clay. Silica tetrahedra and alumina octahedra. The red atoms at the corners are oxygen. The silicon atoms are at the centres of the tetrahedra and the aluminium atoms are at the centres of the octahedra

Layers of silica tetrahedra, alumina octahedra and water. Kaolin has repeating alumina and silica layers bonded with weak hydroxyl OH bonds. Illite has three layers bonded strongly with potassium (K+) ions, while bentonite is similar but the layers are less strongly bonded and can swell by taking in water between each layer

## RESIDUAL AND SEDIMENTARY CLAY

While clay deposits vary widely one from another depending on the geology of the area where they form, it is helpful to divide clays into two general categories: residual and sedimentary. Residual clay remains in its original deposit and sedimentary clay has been transported away from this site. The place where clay minerals occur is an important factor in determining the attributes of the clay such as particle size, firing temperature and colour.

Residual or primary clay results from the disintegration and decomposition of rocks that remain at or near the site of the parent rock. Kaolin is a primary clay. The word kaolin is derived from the Chinese word *gaoling* which means 'high ridge', presumably because kaolin usually occurs at high elevations in residual deposits. Kaolin forms through decomposition of feldspar at the site of the original granite rock. Kaolin particles are somewhat larger than those of other clays because they have not been reduced by the abrasive physical action of erosion. While granites and feldspars produce the whitest and most reliable kaolin, the rock diorite can also produce a pinkish red kaolin that is heavily stained with iron although much less common. When kaolin is transported by erosion from the primary site, its kaolinite minerals form the basis of other clays through sedimentary deposits.

**PRIMARY AND SECONDARY CLAYS**

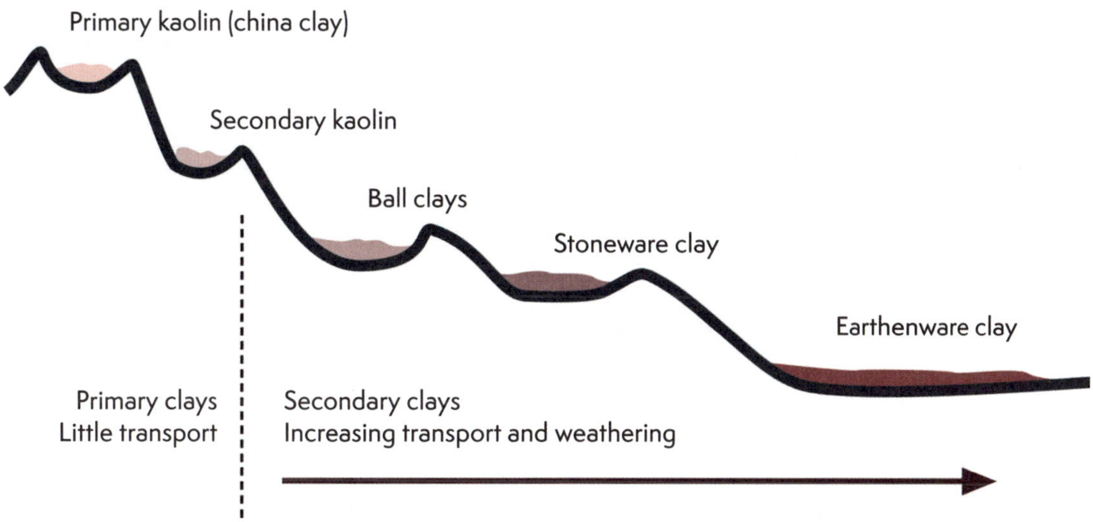

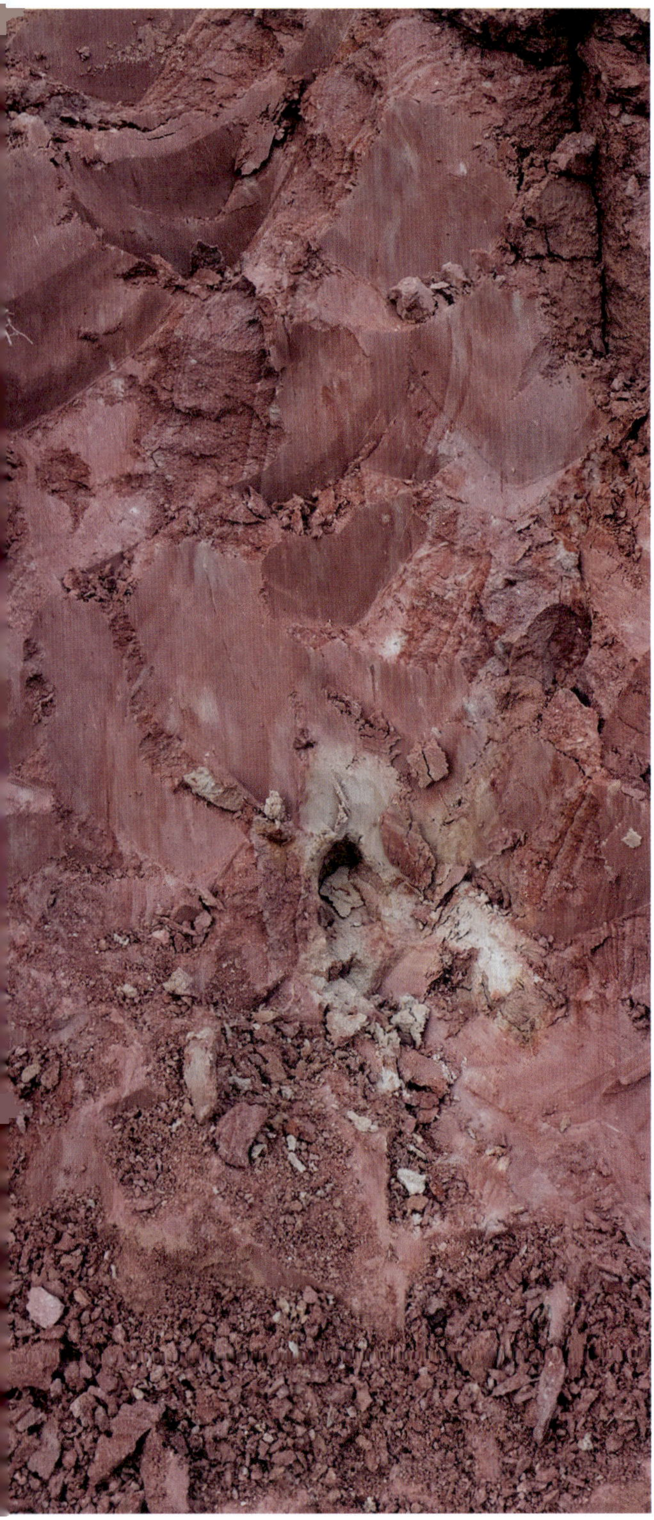

Sedimentary or secondary clay has moved from the original site of its formation at the parent rock. Secondary clays are generated when extremely small and lightweight clay particles are moved through weathering processes like wind and rain. While there are few deposits of primary kaolin remaining, secondary kaolins and other sedimentary deposits of clay are common. As clays are transported by these natural processes, their particle size becomes reduced which renders the clay more plastic. Secondary clays vary from low to very high plasticity, in colour and their refractoriness. Most clays are sedimentary, including ball clay, fire clay, common surface clay, and secondary kaolin.

Sedimentary clays are further described by the location of their deposit: alluvial clays are in streams, riverbanks or on flood plains, lacustrine clays are in lakes, estuarine clays are in bays or river mouths, and marine clays are near or offshore. These clays may also be referred to by the type of movement that they have undergone: glacial clays are moved by glaciers, aeolian clays are moved by wind, and colluvial clays are moved by gravity.

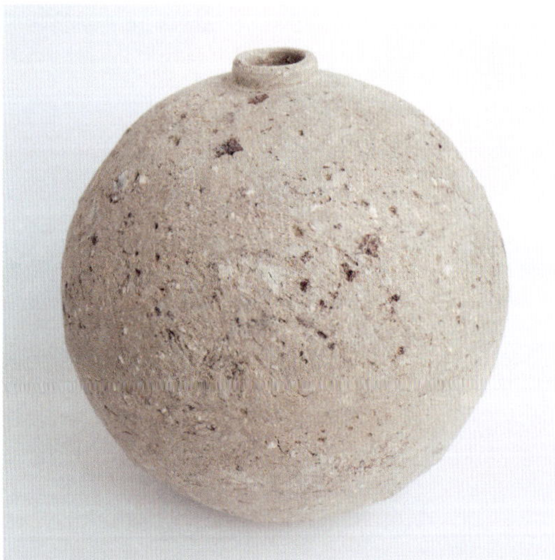

## SEDIMENTARY AND METAMORPHIC PROCESSES

Once a rock has morphed into clay, the clay may remain as clay or it may reform into sedimentary or metamorphic rocks. Clay that hardens into sedimentary rock is called shale. Shales can be ground back into clay and, while lacking plasticity, shales are used in tile and brick production that do not require plasticity. Shales are hydrous alumino silicates and contain water bonded at a molecular level. Once shale has been changed by heat and pressure it becomes the metamorphic rock known as slate. This process removes the chemically combined water such that ground slate can no longer exhibit the properties of clay.

| TIME | SEDIMENTS | CLAY/MUD | SAND | CALCIUM CARBONATE (WHITING) |
|---|---|---|---|---|
| ↓ | Sedimentary rocks | Shale | Sandstone | Limestone |
| | Metamorphic rocks | Slate | Quartzite | Marble |

*Chart adapted from Yvonne Hutchinson Cuff*

**Opposite:** Nina Salsetto, *Blatusa, Croatia*, 2023. Image of site and vessel made from pink kaolin collected by the artist in Croatia conveys a story of this particular place. *(Image credit: Unurgent Argilla)*
**Below:** Slate rock

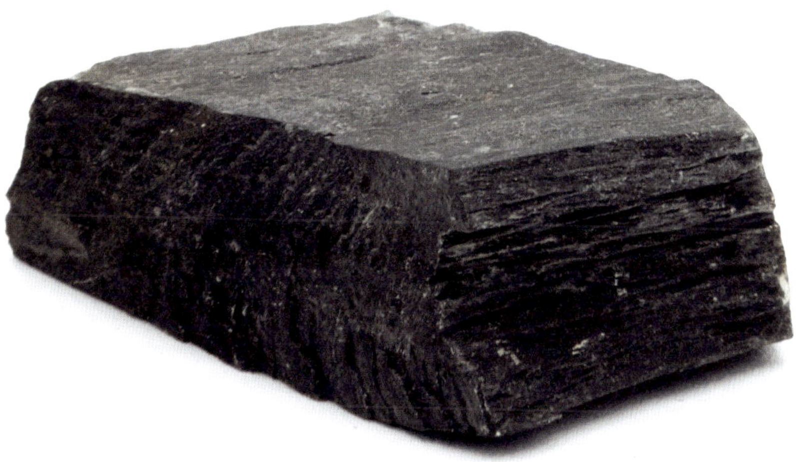

## ROCKS AND CHEMISTRY

To understand how the geology of rocks and minerals relates to the chemistry of clay and glaze materials, it is helpful to review a few terms that describe different forms of matter, going from rocks down to atoms and elements.

| | |
|---|---|
| **ROCKS** | Mixtures of one or more minerals. |
| **MINERALS** | Compounds with a definable chemical formula and molecular structure and distinct crystal formation. |
| **OXIDES** | Compounds in which oxygen atoms are combined with another element, for example, calcium oxide ($CaO$). |
| **COMPOUNDS** | Substances composed of two or more elements bonded together that create a new substance with different characteristics from its constituent elements. All ceramic materials are compounds, the most common being OXIDES. |
| **MOLECULES** | Units of combined atoms, either of the same or different elements. |
| **ATOMS** | The smallest units of an element. |
| **ELEMENTS** | Material of an element is made up of atoms of a single type and has properties distinct from material of other elements. |

*All elements are listed in the periodic table (Appendix 1).*

Natural sources of rocks and minerals are often complex with multiple elements that create the character of a material when it is used in ceramics. The following chart demonstrates how an individual element like aluminium relates to one of its many sources, granite.

| | | |
|---|---|---|
| **ROCK** | Granite | Granite is comprised of feldspar, mica, quartz |
| **MINERAL** | Feldspar | $K_2O\ Al_2O_3\ 6SiO_2$ |
| **CLAY MINERAL** | Kaolinite | $Al_2O_3\ 2SiO_2\ 2H_2O$<br><br>Feldspar can convert to the clay mineral kaolinite that contains alumina ($Al_2O_3$), silica ($SiO_2$) and water ($H_2O$) |
| **OXIDE/ COMPOUND** | Alumina | $Al_2O_3$<br><br>Aluminium + oxygen = aluminium oxide or alumina |
| **ELEMENT** | Aluminium | Al |

A challenge in ceramics arises from the fact that there are multiple sources for any single element. Elements combined with oxygen occur in ceramics as oxides and oxides often occur in combination. Alumina appears in many ceramic materials including clay, feldspars, kyanite, some frits that contain silica and other ceramic oxides. Combined oxides tend to melt better but accounting for how much of any single oxide occurs in a recipe requires some calculation as explained in Chapter 7.

| OXIDE | CERAMIC MATERIALS | NATURAL SOURCES |
|---|---|---|
| **Alumina ($Al_2O_3$)** | Clay | Clays |
| | Feldspar | Feldspars |
| | Some frits | Igneous rocks (granite) |
| | Alumina hydrate | Bauxite |
| | Kyanite | Kyanite |

Oxides are formed when elements combine with oxygen. Oxygen is the most abundant element in the crust of the Earth. Oxygen is extremely reactive and combines with many other elements. Ceramic materials most often occur in their oxide form. Clay bodies and base glazes are made up primarily of the 13 oxides listed in the chart below. Knowing these 13 oxides and how they interact and occur in materials establishes a foundation for understanding ceramics.

| | ELEMENT | SYMBOL | OXIDE | OXIDE NAME |
|---|---|---|---|---|
| 1 | Aluminium | Al | $Al_2O_3$ | Aluminium oxide or alumina |
| 2 | Silicon | Si | $SiO_2$ | Silicon dioxide or silica |
| 3 | Boron | B | $B_2O_3$ | Boron oxide or boric oxide |
| 4 | Phosphorus | P | $P_2O_5$ | Phosphorus or phosphoric oxide |
| 5 | Sodium | Na | $Na_2O$ | Sodium oxide or soda |
| 6 | Potassium | K | $K_2O$ | Potassium oxide or potassia |
| 7 | Lithium | Li | $Li_2O$ | Lithium oxide or lithia |
| 8 | Calcium | Ca | $CaO$ | Calcium oxide or calcia |
| 9 | Magnesium | Mg | $MgO$ | Magnesium oxide or magnesia |
| 10 | Barium | Ba | $BaO$ | Barium oxide or baria |
| 11 | Strontium | Sr | $SrO$ | Strontium oxide or strontia |
| 12 | Zinc | Zn | $ZnO$ | Zinc oxide |
| 13 | Lead | Pb | $PbO$ | Lead oxide |

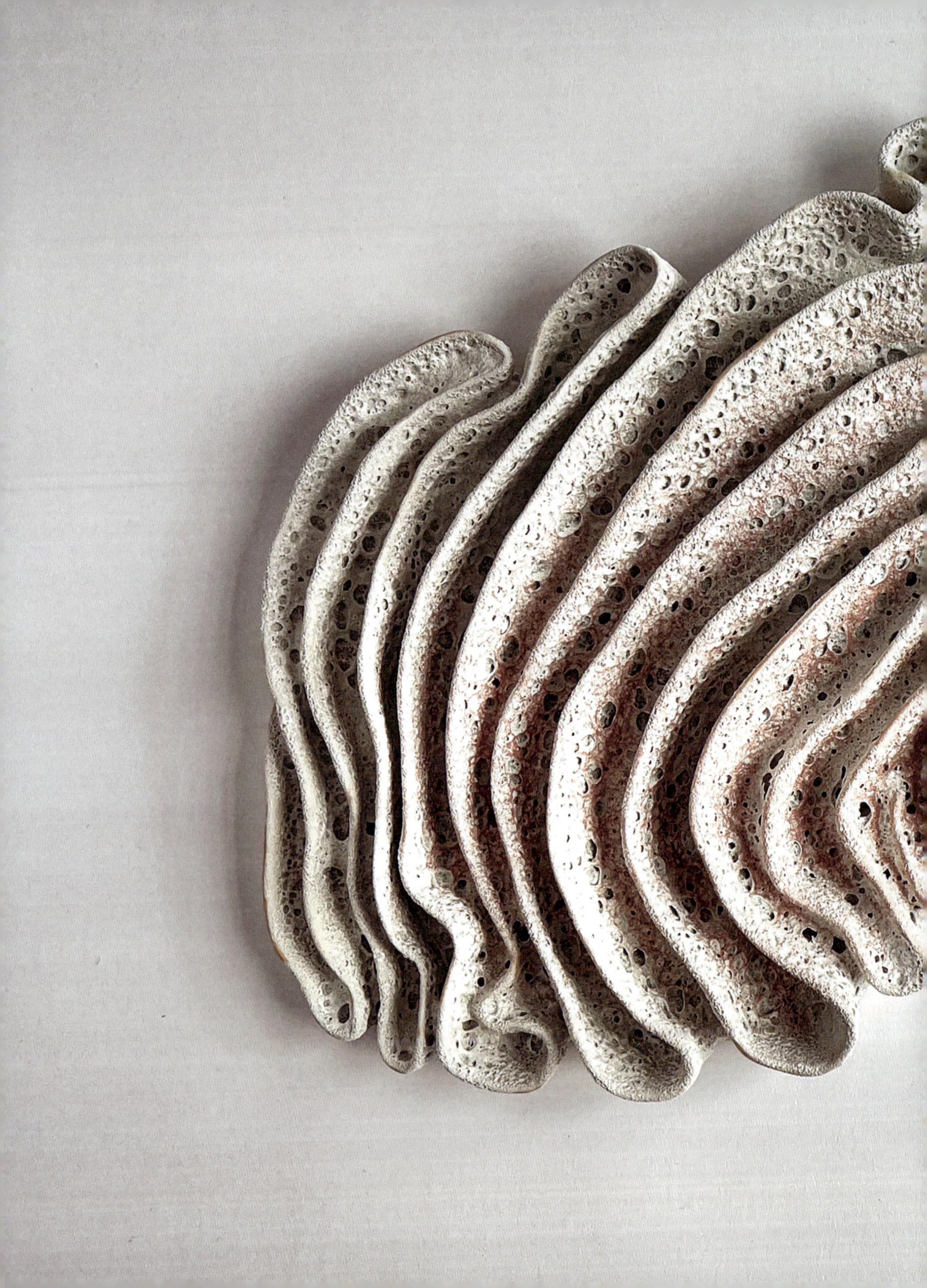

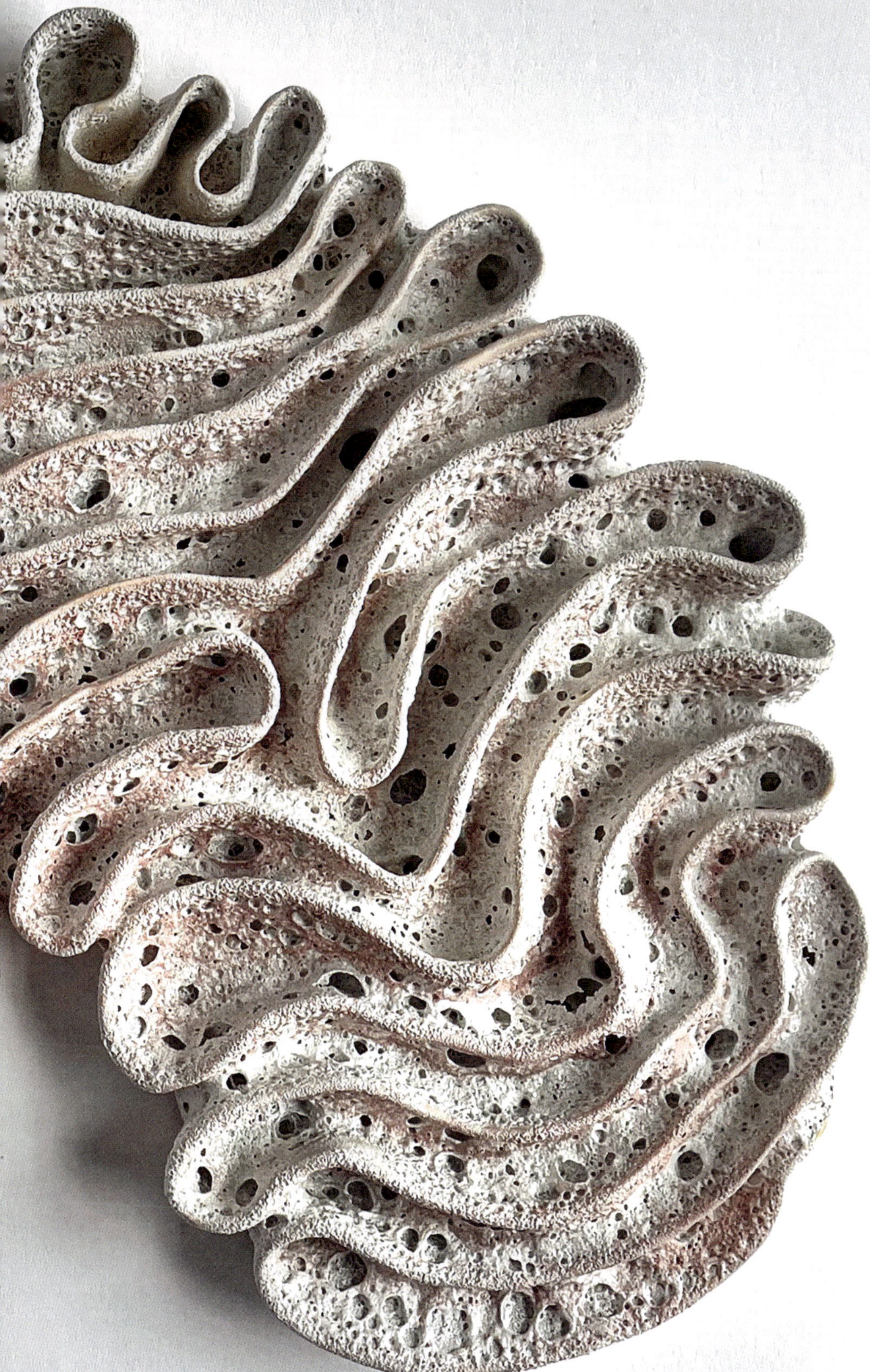

Aside from oxygen, elements may also combine with other elements such as carbon, sulphur, chlorine, nitrogen, and silicon. When elements occur with carbon, they are called carbonates or carbides. Carbides contain carbon, as in silicon carbide (SiC), and carbonates contain carbon as well as oxygen, as in cobalt carbonate ($CoCO_3$). Carbon does not usually remain in the ceramic matrix after firing but is released as a gas during firing in the form of carbon monoxide (CO) or carbon dioxide ($CO_2$). Elements combined with sulphur are sulphides or sulphates. Most sulphates are highly soluble and hazardous when absorbed through the skin or inhaled as fumes. Chlorides and chlorates have combined with chlorine, and nitrides or nitrates have combined with nitrogen, and these may release hazardous fumes during firing. Silicates have combined with silica.

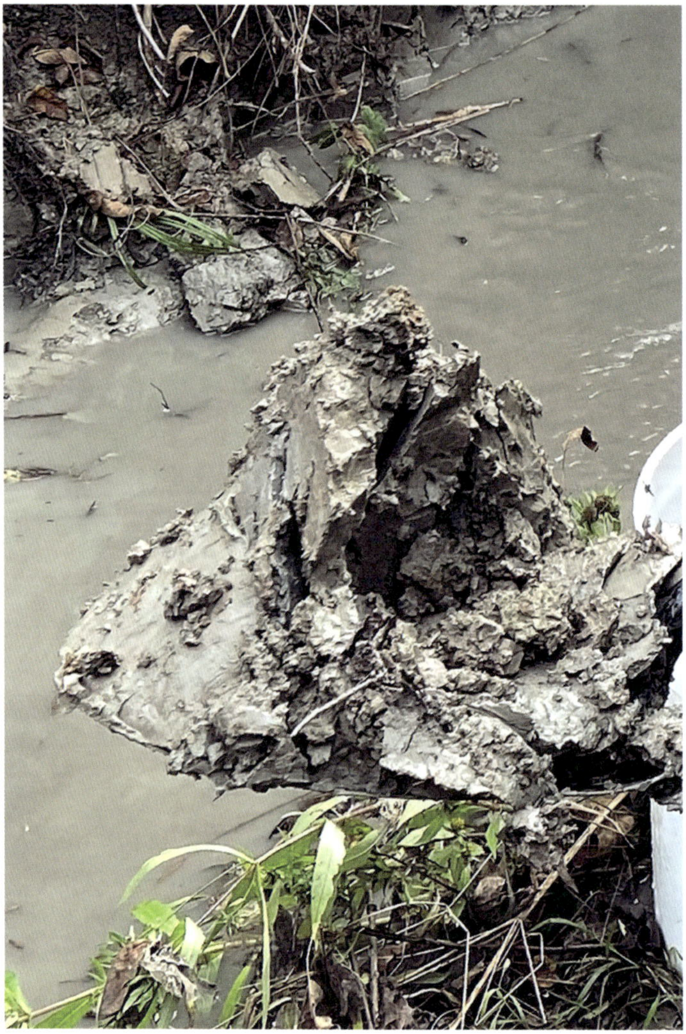

**Previous:** Thora Finnsdottir Søe, *Mimicry*, 2023. Lava glaze traps the bubbles created when silicon carbide releases carbon dioxide gas during firing. *(Image credit: Thora Finnsdottir Søe)*

**Above:** Local clay collected in Quebec, Canada

**Opposite:** The Search for Icelandic Porcelain Project, 2017. Porcelain sourced from found resources in Iceland with granite as a substitute for feldspar. *(Image credit: Axel Sigurdarson)*

## COLLECTING NATURAL MATERIALS

Finding and incorporating rocks and minerals that occur locally can be an effective strategy to lower the environmental impact of making ceramics. Clay is common but it is usually underground below layers of soil and vegetation. An efficient way to find clay is to look in locations where cuts have already been made into the land from waterways, roads, construction or farming, to reveal underlying layers of clay. When clay is exposed, it erodes more easily than rocks with a more crumbly and irregular appearance. In flat areas like former lakebeds, clay can be identified where cracks appear on the surface of the dried ground. Geological surveys, geologists, and those with local knowledge of the land such as farmers or rock collectors may aid in finding local clay and rock sources for ceramics.

Materials that are collected from local sources can be complex in composition, containing multiple minerals that produce variable results when fired. It is important to be able to identify these materials to avoid hazardous ones. Another important aspect to collecting materials is to respect and preserve the existing geological features of any site where materials are collected. Permission from landowners and local authorities should be requested and the laws governing the area respected. Common rock and mineral sources for glaze can be found in Chapter 6 and in Appendix 2.

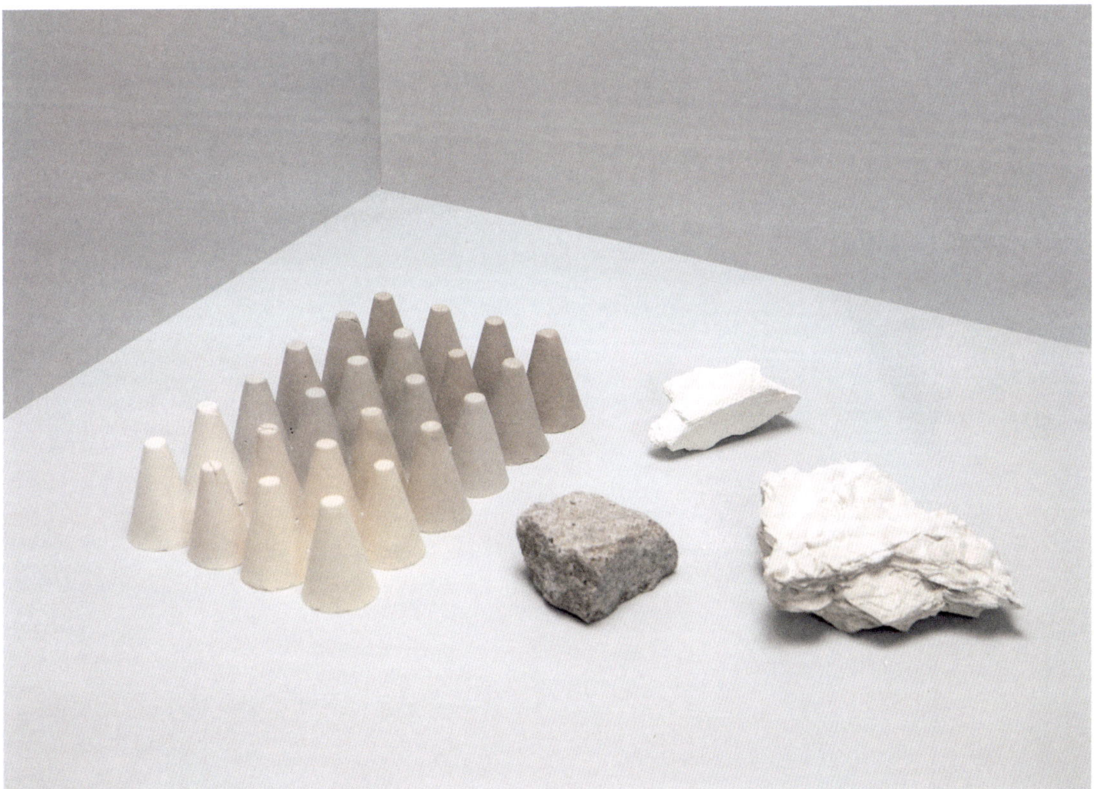

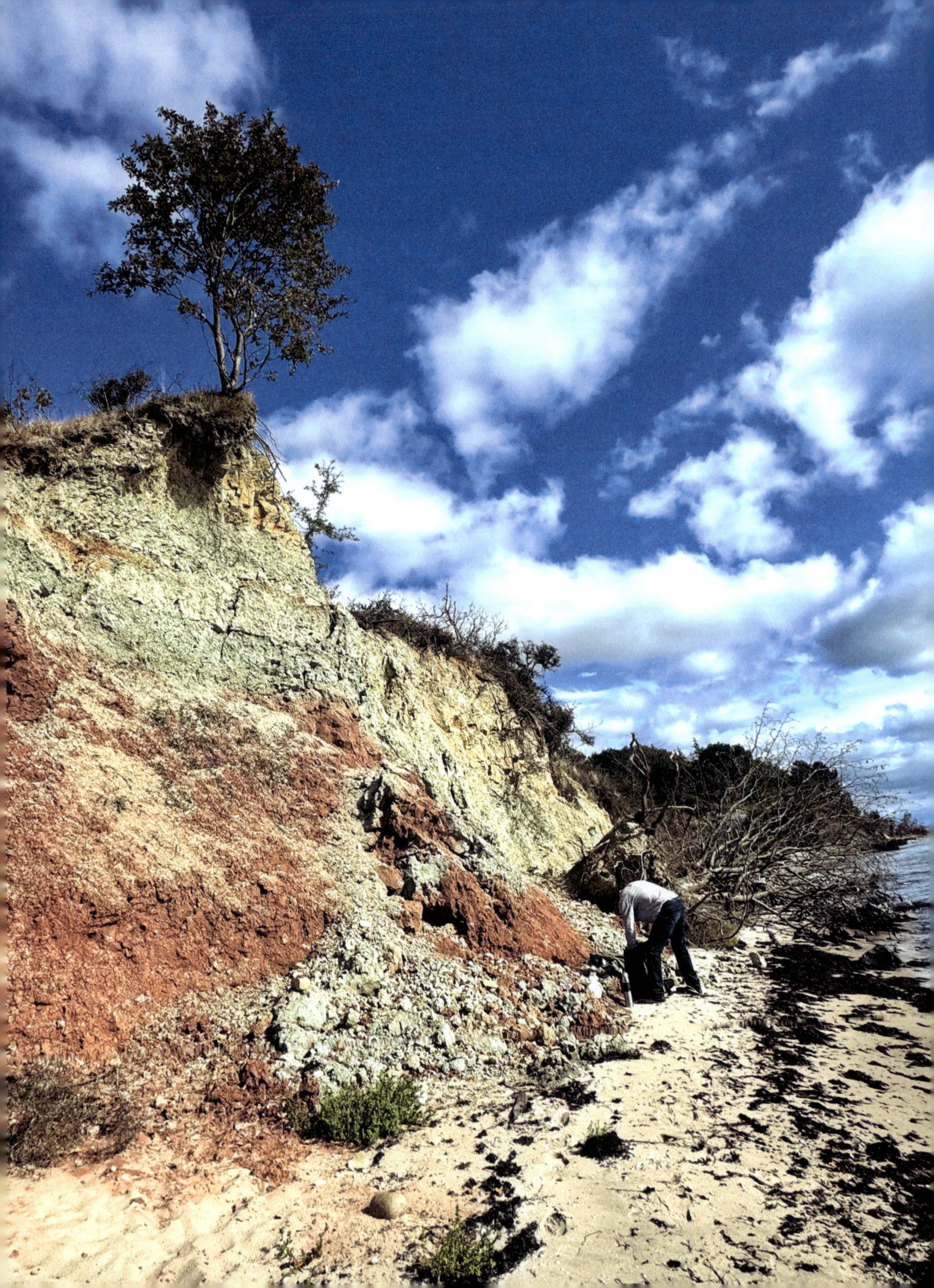

CHAPTER TWO

# CLAY

## UNDERSTANDING CLAY

What is clay? In geologic terms, clay is comprised of very fine particles of crystalline clay minerals. On a chemical level, clay is composed of silica, alumina, and water, or hydrous aluminosilicate. In practical terms, clay is endlessly malleable when wet but irreversibly hardens when fired. Each of these descriptions contribute to understanding how clay holds the potential to transform and endure.

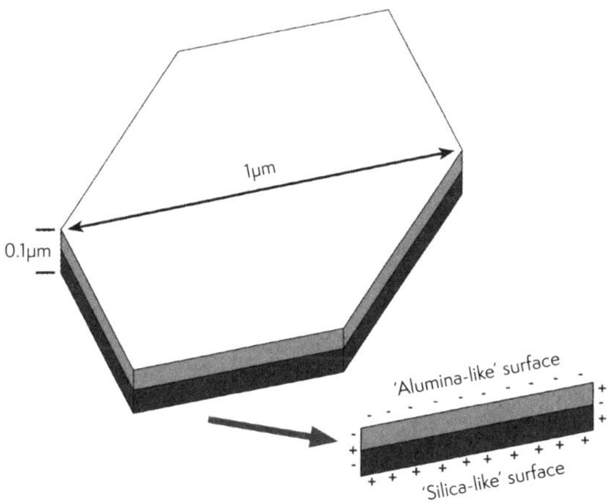


**Opposite:** Elysia Athanatos, *Bornholm Wild Clay*, 2024. *(Image credit: Elysia Athanatos)*
**Above:** Clay particles

Clay particles are extremely small. A particle of clay may be up to a thousand times smaller than a grain of sand. While sand is produced from rocks that are physically broken down into tiny granules, clay is derived from processes that decompose feldspathic rock or volcanic ash on a molecular level. Through decomposition, alumina and silica bond with water. This chemically combined water becomes part of the crystalline structure of clay, and appears in the chemical formula of clay: $Al_2O_3 + 2SiO_2 + 2H_2O$.

Clay particles are thin, flat, and shaped like platelets. For example, kaolinite particles are ten times wider than they are thick. Positive and negative electrical charges on the opposite faces of clay particles encourage them to stick together in stacks. As clay becomes wet, water enters between the particle layers, enabling the particles to slide without breaking apart. This added water is known as the water of plasticity, and it evaporates out of clay as it dries.

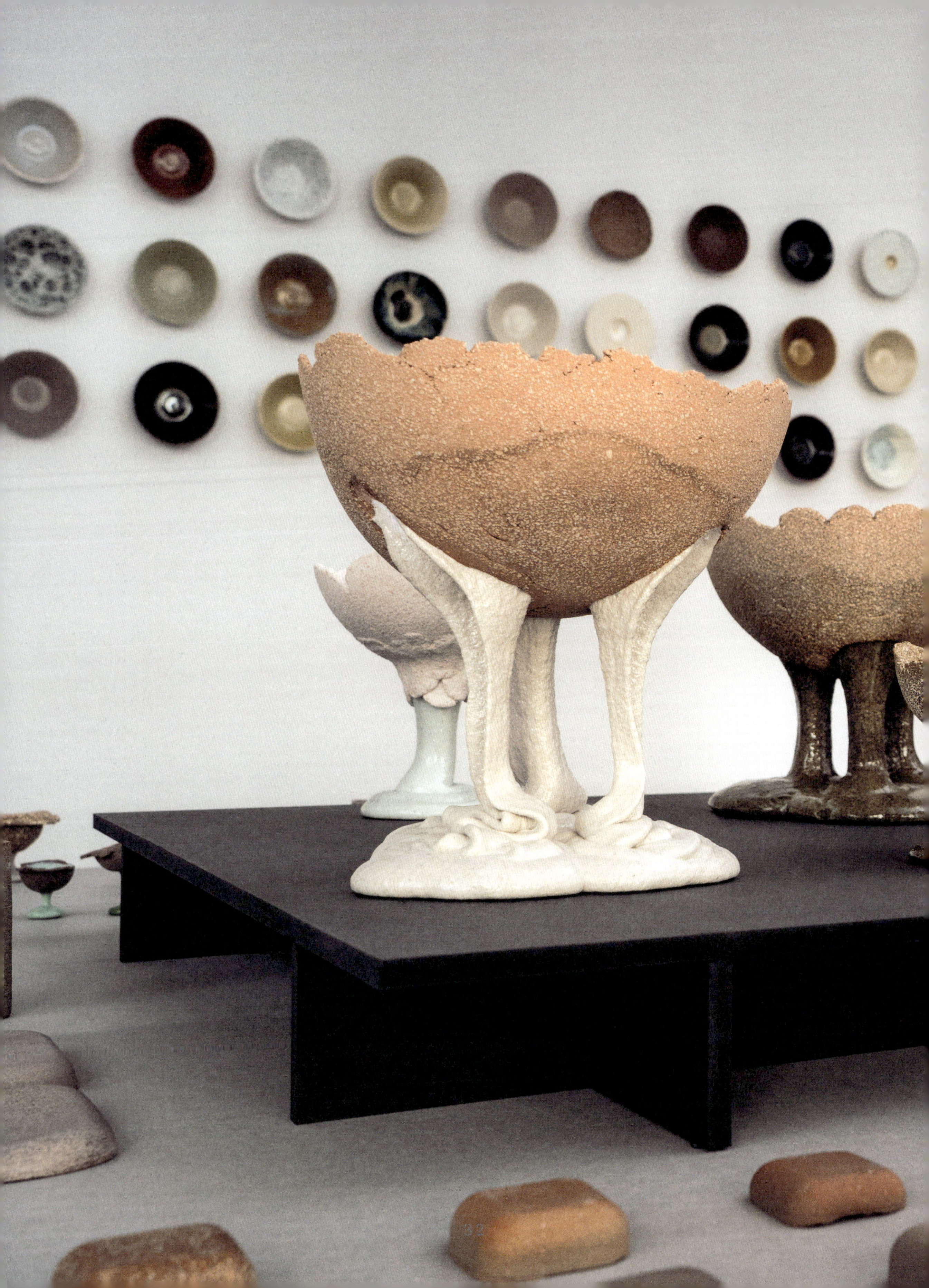

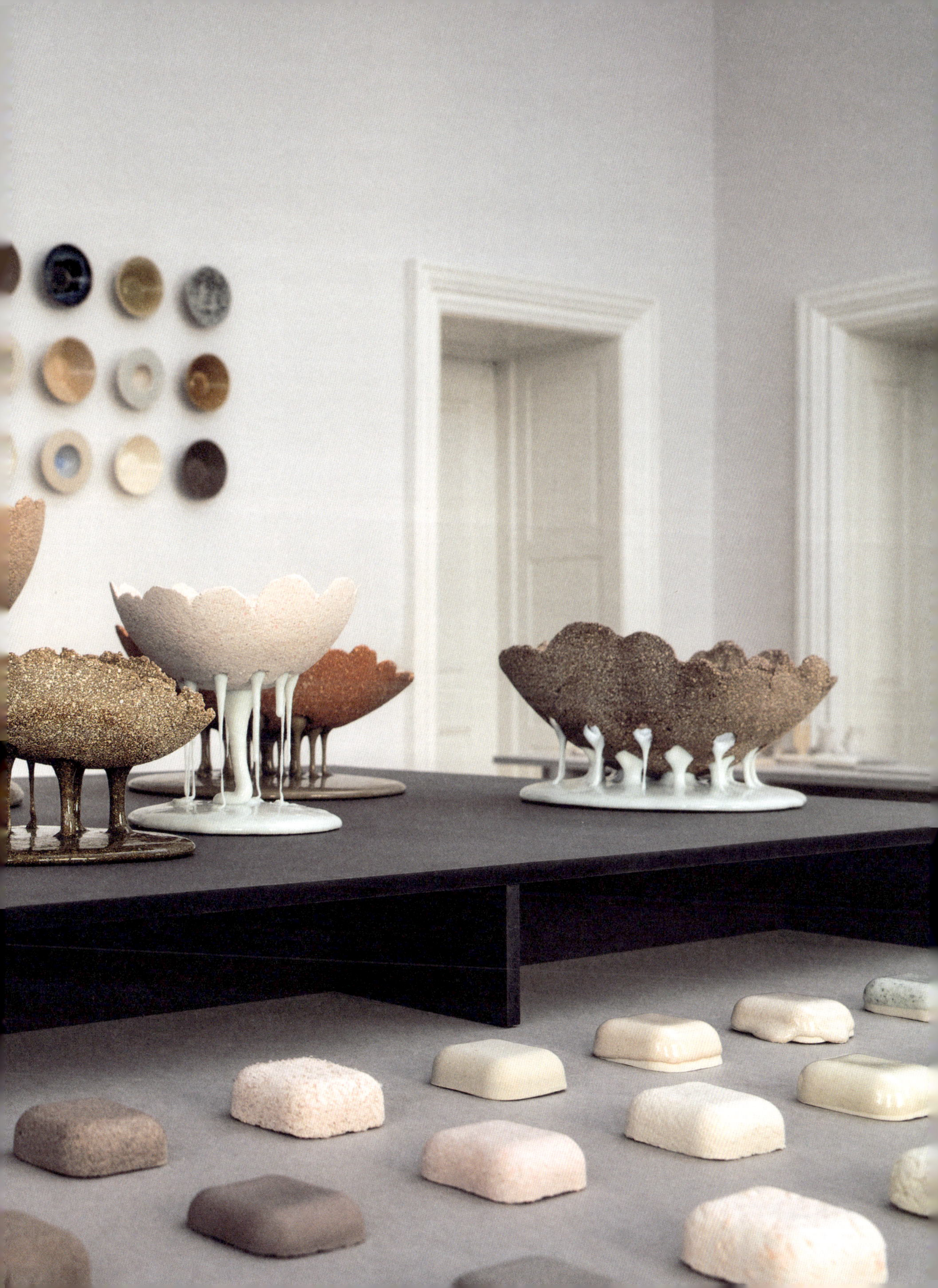

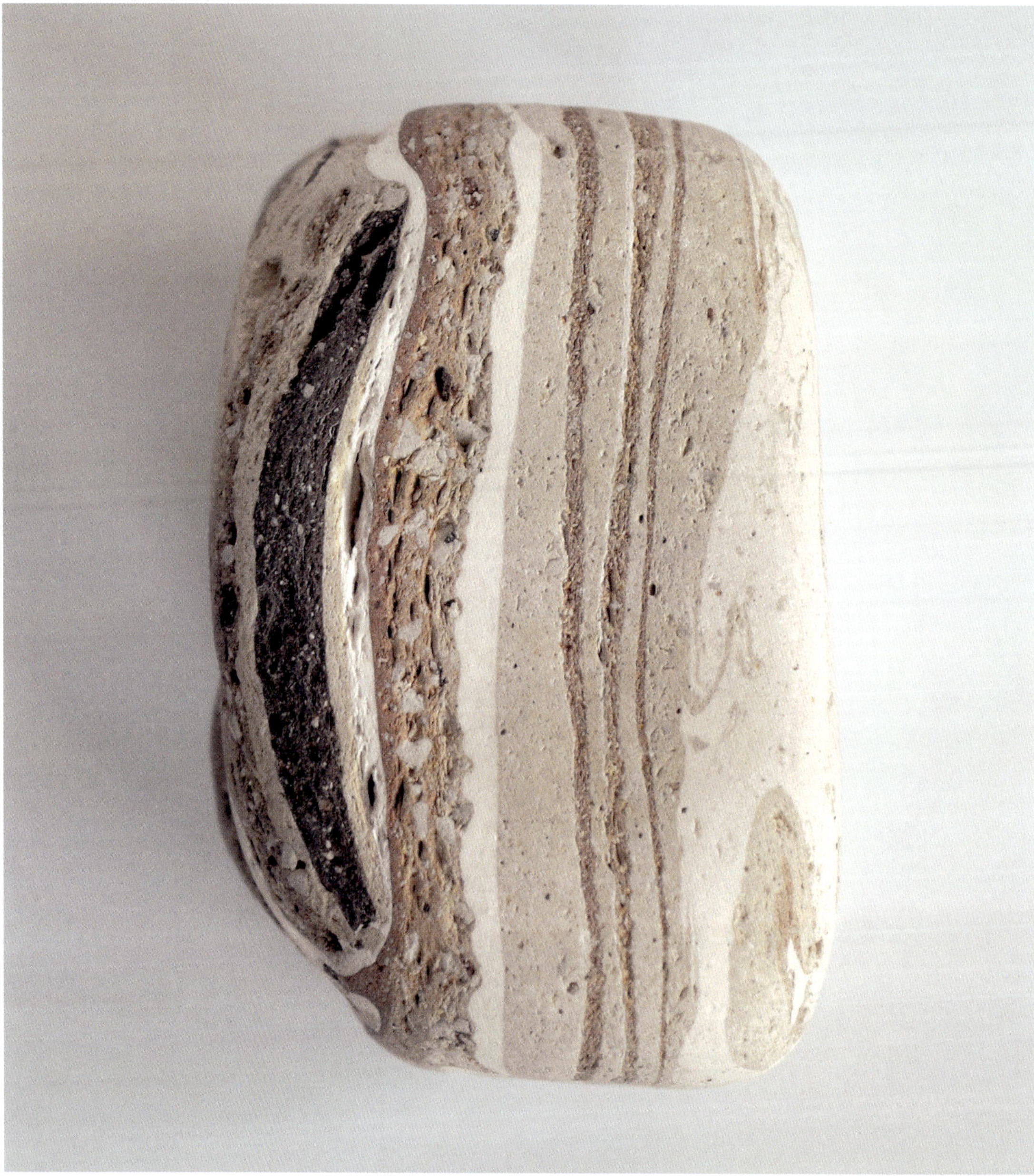

**Previous:** *ØDELAND*, 2025. A collaborative project of Christina Schou Christensen, Signe Fensholt and Thora Finnsdottir Søe to experiment reusing materials from the waste stream, including crushed sanitary ware, medical glass, studio waste and local granite powder with minimal stoneware or porcelain clay. *(Image credit: Christoffer Askman)*

**Above, opposite and overleaf:** Katrine Køster Holst, *Fragment*, 2016. The deposition, sedimentation and stratification of discarded clay materials are choreographed to parallel the way clay forms and appears in the landscape. *(Image credit: Katrine Køster Holst)*

Clay loses flexibility or plasticity as it dries and shrinks. As clay dries, it becomes increasingly resistant to deformation until it hardens completely. Moist clay is soft and malleable and when dried slightly it becomes suede leather hard or stiffened but still bendable. After more drying, clay becomes leather hard, a state when it breaks when bent yet still contains moisture. Finally bone dry is the stage when all moisture has evaporated out of the clay at which point it becomes fragile. Plasticity can be restored by rewetting the clay. When clay has dried completely to the bone-dry stage, adding water will dissolve the clay into a liquid slip which can then be reconstituted by slightly drying it again to a moist, workable consistency.

As the clay dries, evaporation removes the water of plasticity in the clay. The chemically combined water can only be eliminated through firing clay at high temperatures. Once the chemically combined water is removed from clay, fired clay no longer exhibits the same properties as unfired clay because the water has been eliminated from its chemical structure. This transition is known as the ceramic change. Even if it is finely ground, fired clay particles no longer stick together and their plasticity cannot be recovered.

All clays are made up of one or more of the clay minerals – kaolinite, montmorillonite, or illite – combined with additional mineral and metallic elements. Different combinations of these components create unique qualities in any single clay deposit relative to the local geology. While clay is abundant, it is not limitless. Important kaolin clay reserves in the UK will be depleted in only 40 years at the present rate of extraction. Beyond ceramics, clay is used in numerous other industries such as paper and textiles manufacturing, petroleum products and food.

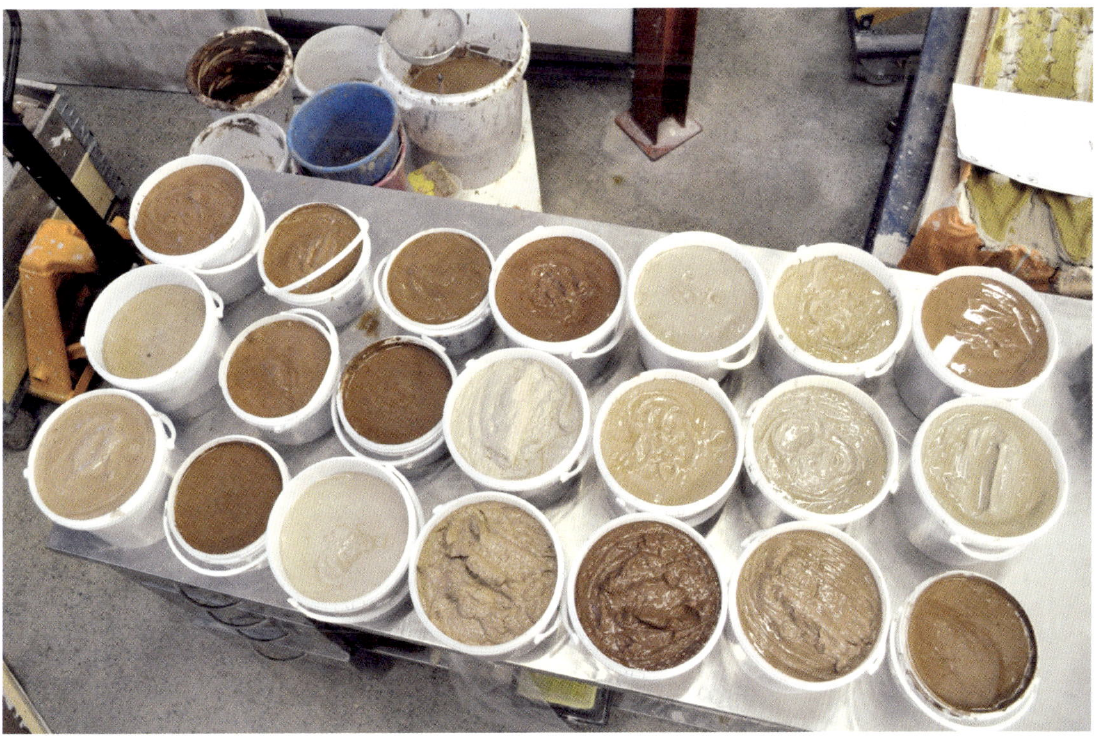

Clay usually lies below a layer of soil called overburden. Once the overburden is removed, clay can be extracted by blasting, shovelling, hydraulic jets or tunnelling. The least expensive method of clay production takes place in open pits above ground. Underground mining increases the cost of materials as does processing the clay which involves refining, purifying and homogenization. Few clays work well for ceramics just as they are dug out of the ground. Clays are modified by adding other ceramic materials with them to improve their workability, visual characteristics or firing properties. This is looked at in more detail in Chapter 3.

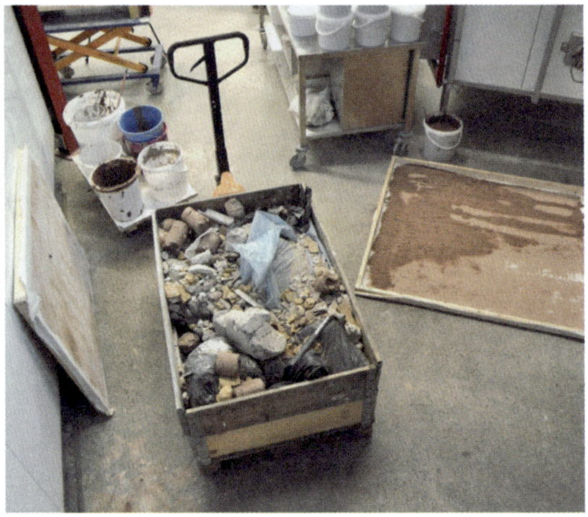

Different types of clay exhibit different characteristics and working properties. As noted in Chapter 1, clay can be categorized in relation to its site of formation as either residual/primary or sedimentary/secondary. Clay suppliers and ceramic technical literature more commonly define clays based on mining nomenclature into six clay types: kaolin, ball clay, fire clay, stoneware clay, surface clay, and bentonite clay. Only kaolin and bentonite form in situ as residual clay. Other clays have been transported and deposited as sedimentary clay.

| RESIDUAL/ PRIMARY CLAY | SEDIMENTARY SECONDARY CLAY |
|---|---|
| Kaolin | Secondary kaolin |
| Bentonite | Ball clay |
| | Stoneware clay |
| | Fire clay |
| | Earthenware or surface clay |

Clays are distinguished and analysed according to their plasticity, particle size, colour, refractoriness, vitrification and strength. The distinctive characteristics of various types of clay are described using these terms.

| | |
|---|---|
| **PLASTICITY** | plastic, fat, semiplastic, or short |
| **PARTICLE SIZE** | fine or coarse-grained |
| **COLOUR** | raw and fired |
| **REFRACTORINESS** | heat range, softening or melting point |
| **VITRIFICATION** | porous or vitrified |
| **STRENGTH** | low, medium or high |

## KAOLIN

Kaolin was first used to make vitrified ceramics in China around 600 CE. Also known as china clay, kaolin is used to make porcelain and is typically expensive. Because kaolin is a residual or primary clay that has formed at the site of the weathered feldspathic parent rock, it is relatively free of impurities. The composition of kaolin clay closely resembles the pure mineral kaolinite. It is typically whitish in colour, though sometimes contaminated by very small amounts of iron, titanium, quartz or mica. Primary kaolin is somewhat uniform and relatively coarse grained (under 10 microns) because it has not been physically eroded. Kaolin matures at 1480°C (2696°F) and exhibits high shrinkage.

Secondary kaolin forms when kaolin clay has been transported from its original site while maintaining relative purity. This movement tends to increase its plasticity relative to primary kaolin and to reduce is particle size (under 5 microns). Secondary kaolin may contain additional contamination from iron and titanium to become off-white when fired.

## BALL CLAY

The name ball clay refers to chunks of ball clay that rounded as they were handled and moved during mining. With a high degree of plasticity and high shrinkage, the particles of ball clay are extremely fine (under 1 micron). Ball clays are sedimentary clays comprised primarily of kaolinite with illite and low amounts of montmorillonite. Ball clays are higher in silica, calcium and iron than kaolin. They contain about 1–3% organic matter which adds good dry strength and colour (grey, blue or blackish) that burns away leaving a light-coloured clay. Ball clays are primarily used to add plasticity to other clays and to make casting clays that are easily deflocculated. Ball clays complement kaolin in light workable clay bodies, but even in low amounts, they will inhibit the translucency of porcelains and create a grey colour due to the presence of iron and titanium. The illite content in ball clay contributes to a lower maturing temperature than kaolin at 1300°C (2400°F).

**PARTICLE SIZE**

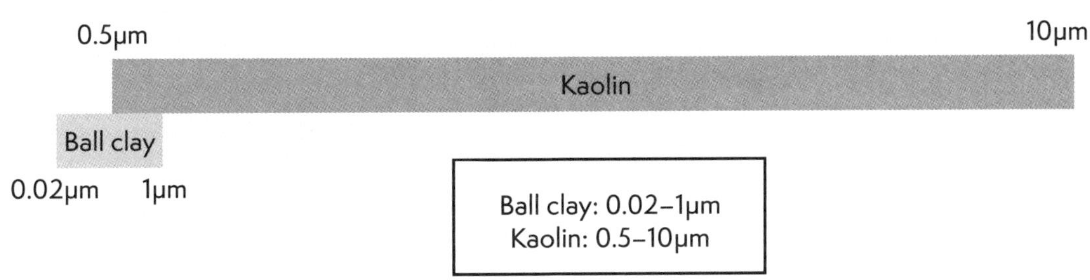

## STONEWARE CLAY

Naturally occurring stoneware clays are workable as found, yet rare. Stoneware clays have a mixed particle size with good plasticity like ball clay and more 'tooth' or texture like fire clay. Stoneware clays vary widely in colour from buff or tan to brown with higher impurity levels than ball clay. With low alkali and calcium levels, stoneware clays vitrify at temperatures over 1200°C (2192°F).

## FIRE CLAY

Fire clays are distinguished by their refractory quality and can withstand temperatures up to 1400°C (2552°F) due to their high alumina content coupled with low alkali levels. Fire clays resist deformation and are used to make refractory products like firebricks, saggars, kiln furniture, wadding and kiln mudding. The open, coarse texture of fire clay particles leads to good drying properties, yet plasticity can range widely from low to high. Colouration in fire clays varies from white, grey and pink to red brown. While generally containing low amounts of iron, fire clays may contain iron deposits that cause spots or speckling. Found underneath present-day coal seams, fire clay can contain volatile impurities such as sulphur and carbon that lead to black coring or bloating if improperly fired.

**Right:** Elysia Athanatos, *Bornholm Wild Clay*, 2024. *(Image credit: Elysia Athanatos)*

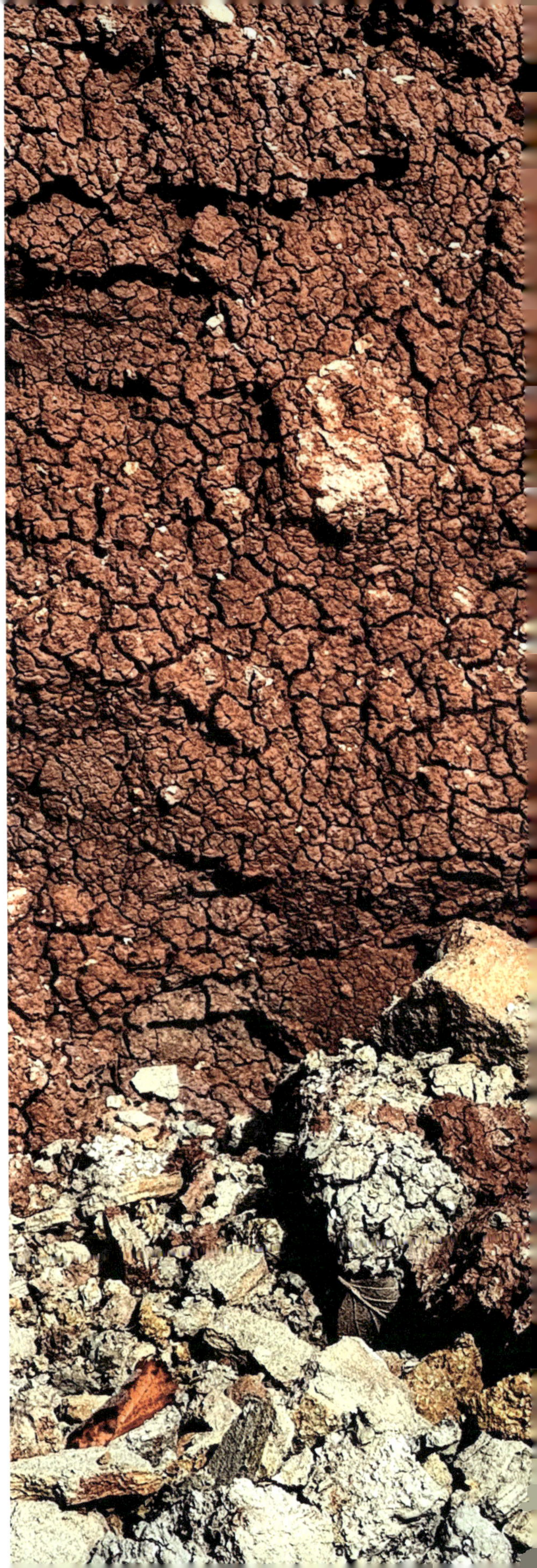

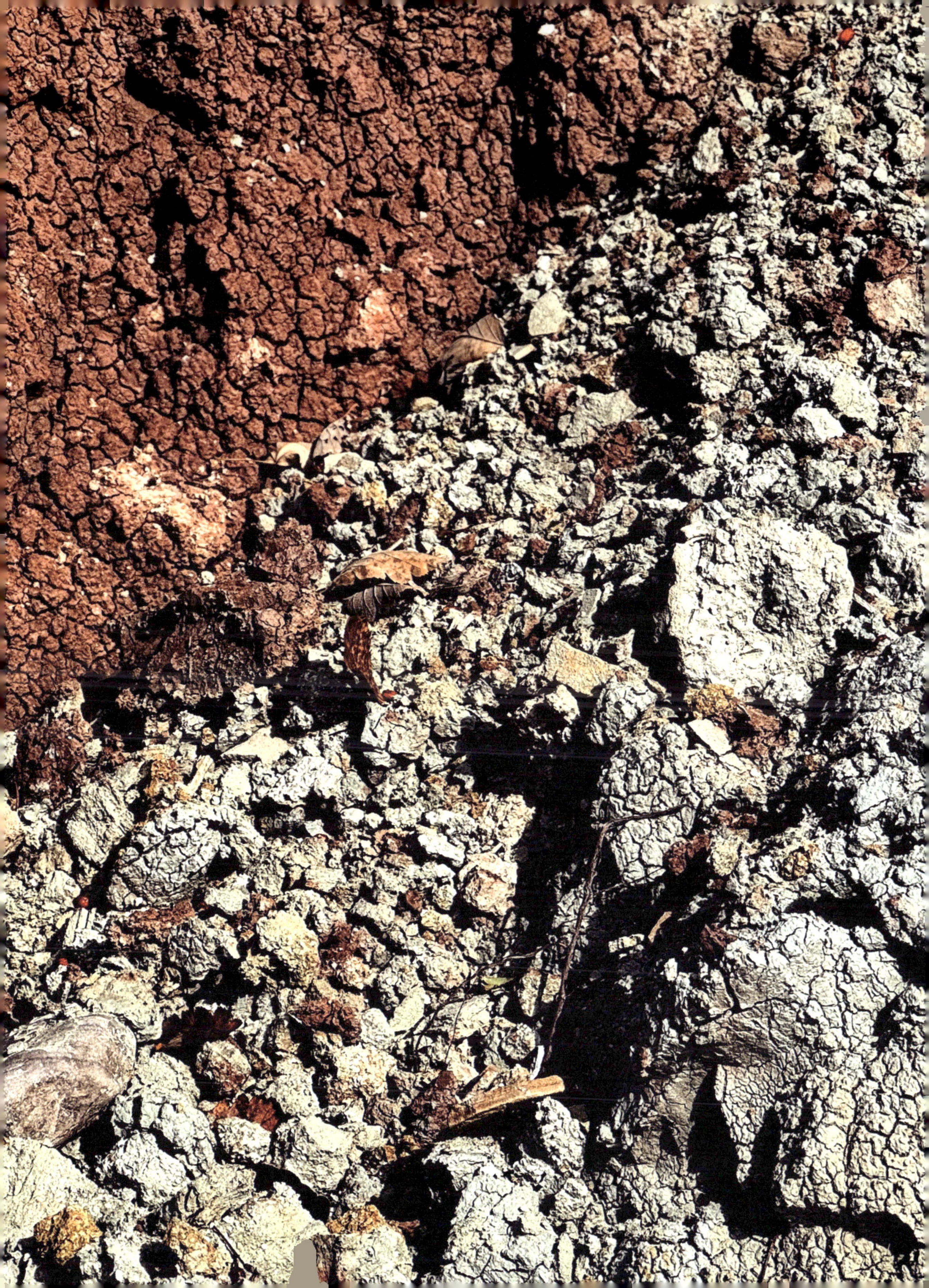

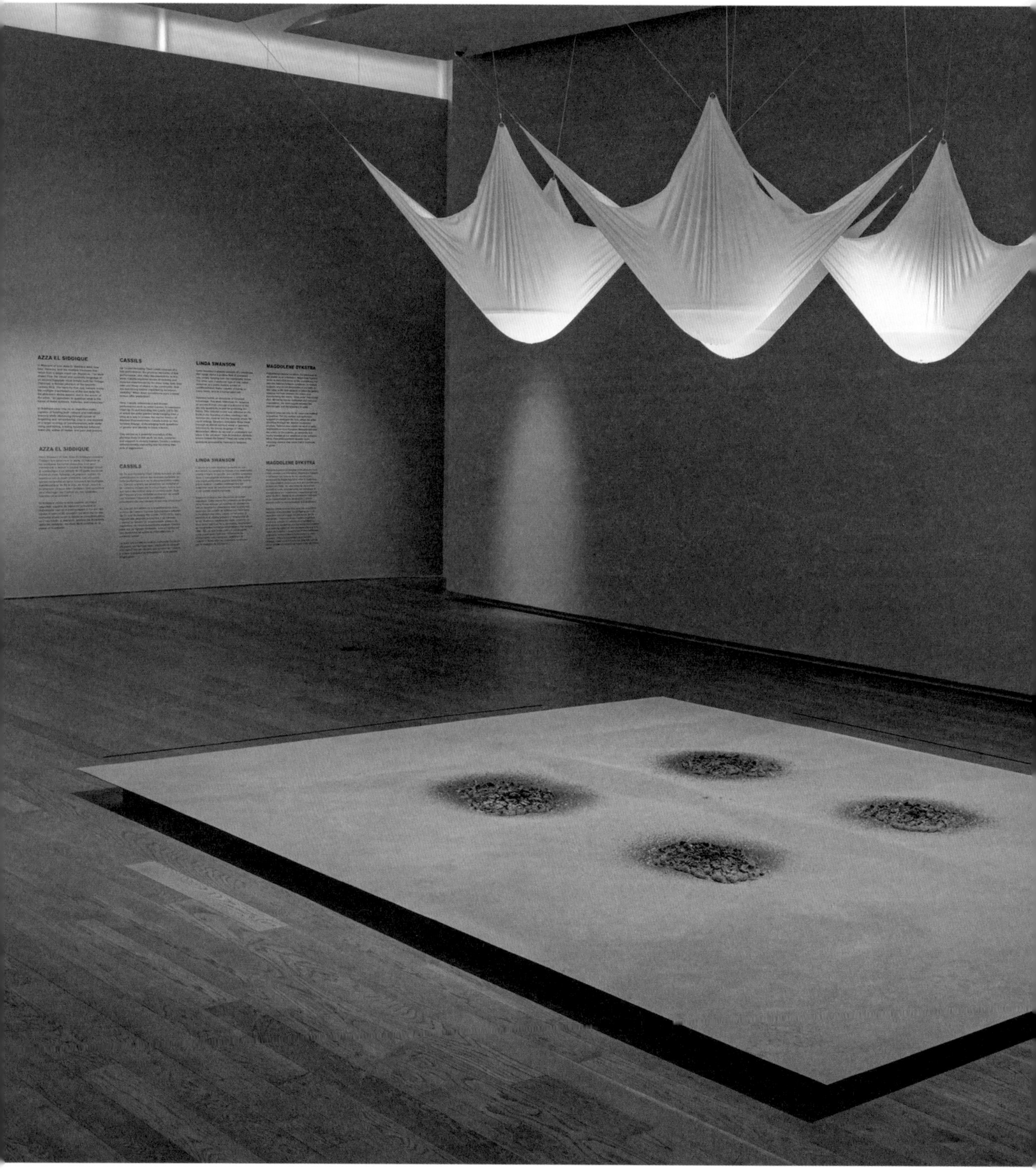
AZZA EL SIDDIQUE
CASSILS
LINDA SWANSON
MAGDOLENE DYKSTRA

## EARTHENWARE CLAY

Earthenware clay, also known as surface clay, makes up most of the usable clay in the world. Earthenware clay is formed through extensive weathering and transportation, resulting in a mixed particle size (up to 5 microns), a mixture of clay minerals and significant impurities that lower the melting temperature to about 1100–1200°C (2000–2200°F). Earthenware clays contain 5–10% iron, some calcium and titanium. Earthenware clay is typically highly plastic, but its plasticity may be reduced by the presence of sand or stones. Raw earthenware clay is found in many colours including greens, blues, reds, browns, greys, and black. Fired earthenware clays vary from pink to red, buff to tan, and brown to black. Earthenware clays are commonly used to make pottery, bricks, tiles, pipes and architectural ornamentation. Some earthenware clays are unusable due to high amounts of soluble alkaline salts that cause extensive scumming or calcium contamination from calcite causing lime pop-outs.

Shale clays are metamorphic rocks that have formed out of earthenware clays. Sub-surface clays and shales are much harder than typical surface clay and require mechanical pulverization. After crushing, shales can be used as a non-plastic raw material for bricks or addition to clay bodies.

**Left:** Linda Swanson, *Templum of a Precious Thing of No Value, a Shapeless Thing of Many Shapes*, 2020. A field of dry bentonite clay transforms as it is moistened by water dripping from a suspended membrane. *(Image credit: Toni Hafkenscheid)*

## BENTONITE

Originating from decomposed volcanic ash, bentonite is comprised of the clay mineral montmorillonite, also known as smectite. Bentonite is the finest-grained clay (less than a micron) and is valued for its extreme plasticity. Bentonite can become gummy and thixotropic when water is added in high quantities. Due to its dramatic shrinkage, bentonite is generally incorporated in low amounts (2–3%) to add plasticity into clay bodies or as a suspension agent for glazes.

## ADDITIONAL TERMS FOR CLAY

Clays may also be designated by an element that occurs in particularly high amounts. High alumina clays contains at least 38% alumina. Siliceous clays are high in silica. Calcareous clays contain calcium in significant amounts such as 2% or more lime (calcium oxide) or 4% or more calcium carbonate. Ferruginous clays contain 8% or more iron. Carbonaceous clays are high in carbon content like shales.

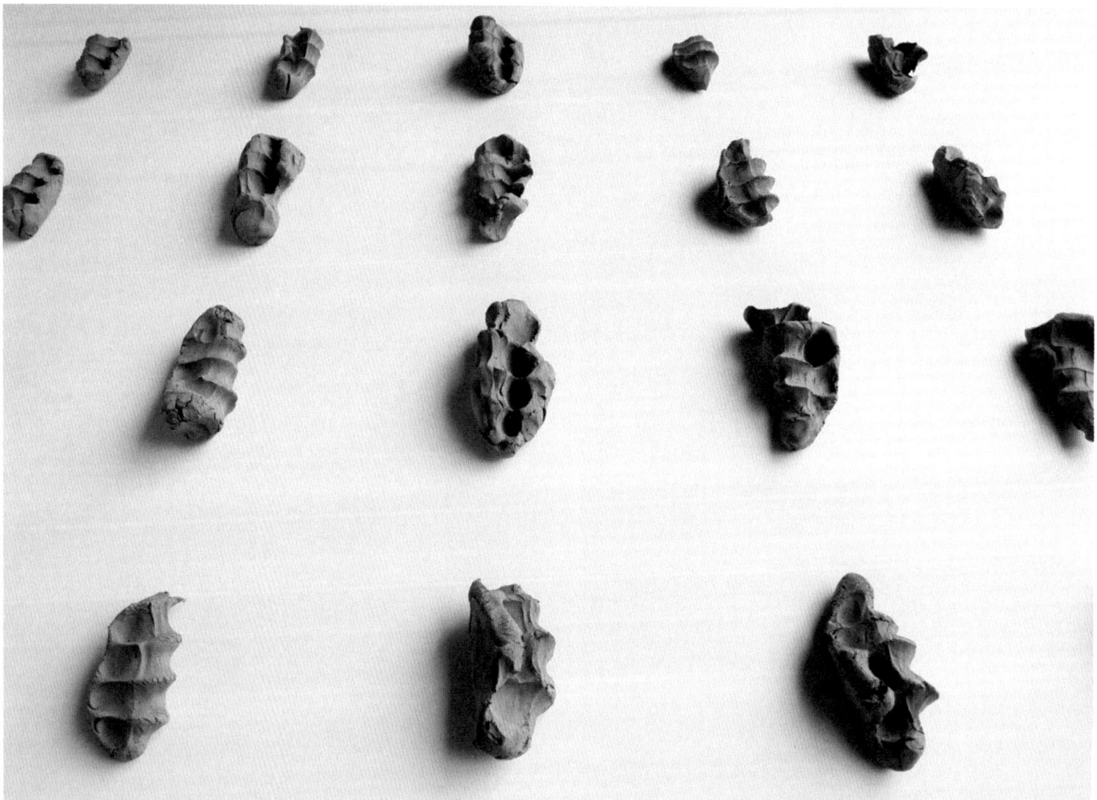

**Above:** Mat Chivers, *Migration*, 2018. Detail of 1480 unfired clay hand prints and impactite. *(Image credit: Corine Lemieux)*

**Opposite and overleaf:** Studio ThusThat, *Red Mud Ceramics*, 2019. Red mud is the residue that remains when aluminium is extracted from the high-alumina clay rock bauxite. This reclaimed waste material lends its distinctive colouration to this series of vessels. *(Image credit: Studio ThusThat)*

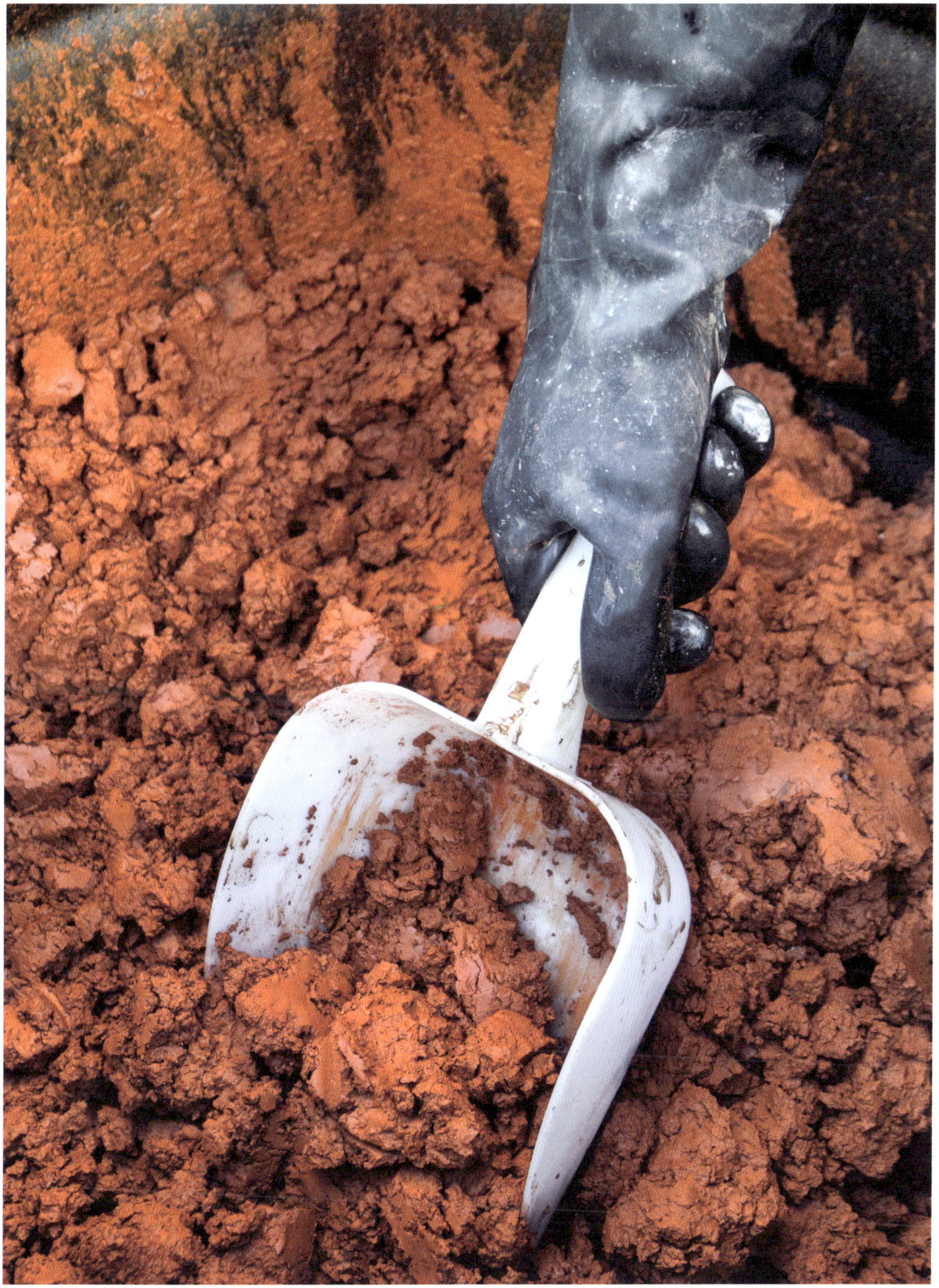

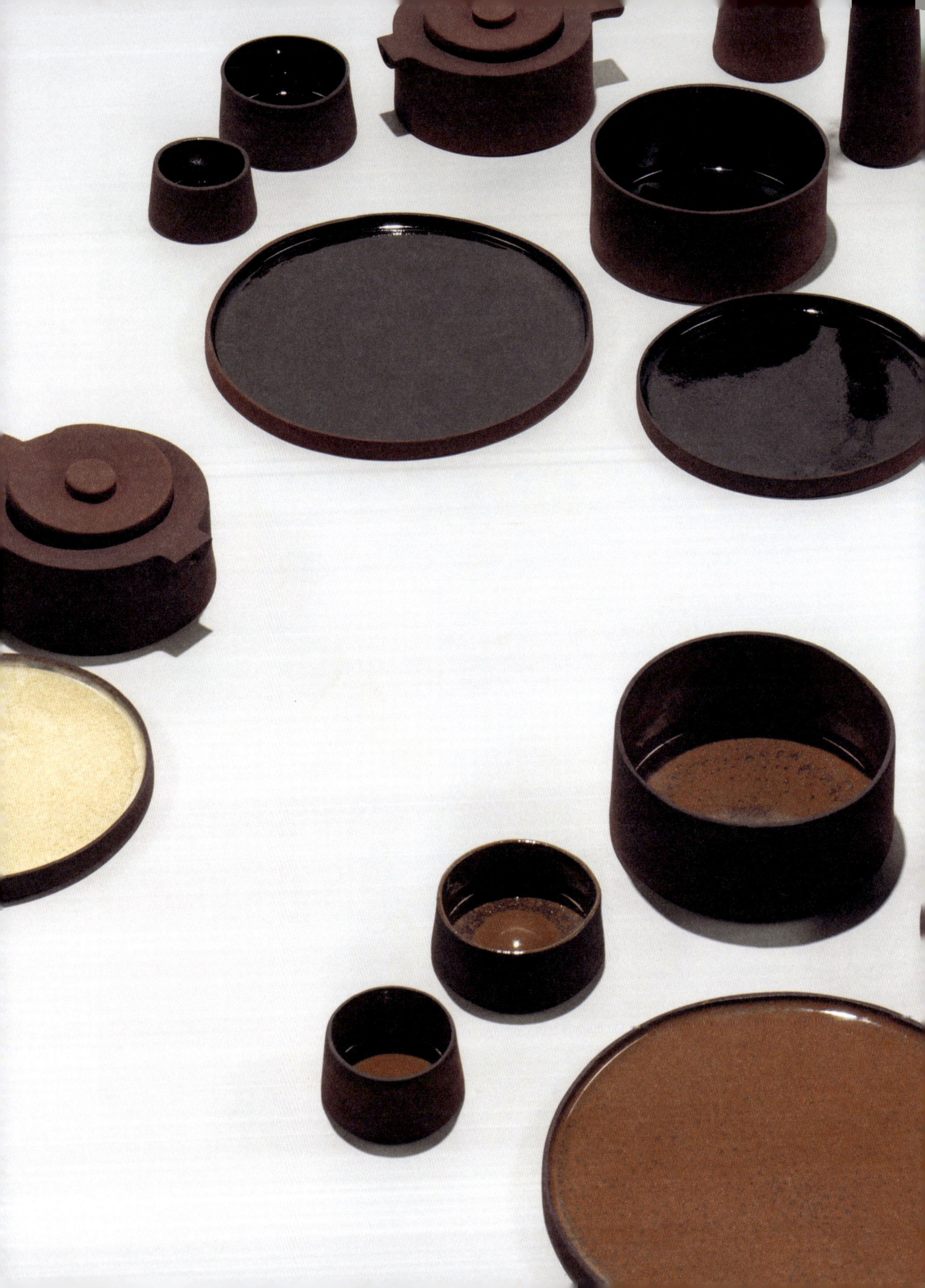

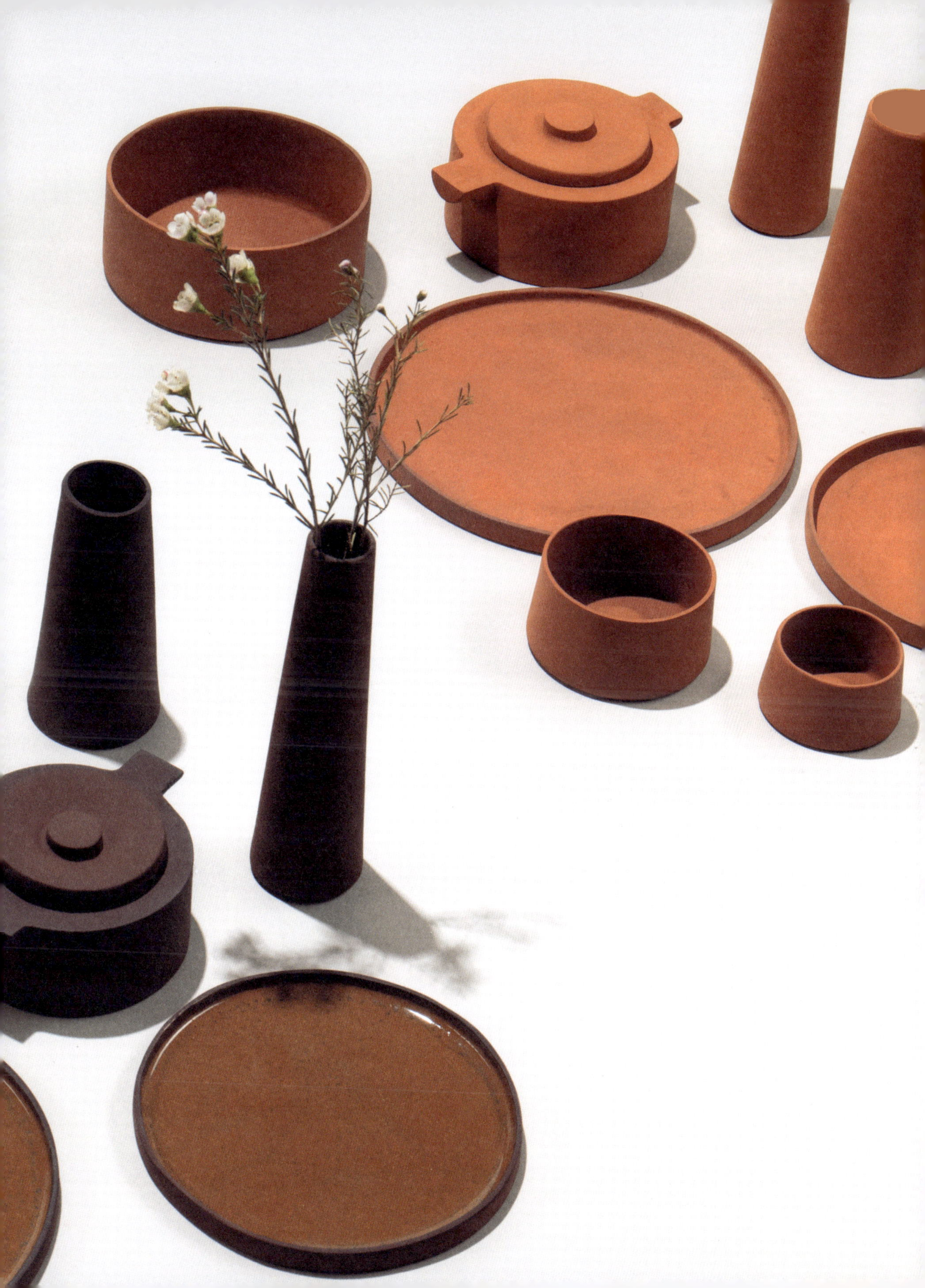

CHAPTER THREE

# CLAY BODIES

Derived from the German word *kleben* meaning 'to stick', clay can be too sticky to be workable as it comes from the ground. Mixing clay with other ceramic materials to create clay bodies helps to improve upon the inherent qualities of clay.

Clay bodies can be designed for specific purposes – including throwing, hand building, or slip casting – and formulated for different temperatures including low, mid or high fire. There are three basic types of clay bodies: porcelain, stoneware, and earthenware, as well as speciality types such as flameware. Any single clay body will usually be described with a combination of these designators, an example of which would be a mid-range, porcelain throwing body.

**Opposite:** Sara Howard, *Circular Ceramics Collection*, 2023. This clay body is made from 100% waste reclaimed from a ceramics factory wastewater treatment system. The glaze is derived from post-consumer glass retrieved from Bali's rivers and industrial stone slurries diverted from landfill. *(Image credit: Sara Howard)*

Clay bodies contain five ingredients: clay, flux, glass former, filler and colourant. Clay bodies are primarily formulated around the clay in a recipe. Naturally occurring clay can be modified to alter its plasticity, firing temperature and vitrification, texture, shrinkage and colour. Some naturally occurring earthenware and stoneware clays are usable as found, since they contain mixtures of clay minerals and other elements consolidated through erosion and transportation. Most clays need additional materials to work well through all stages of the ceramic process.

| **CLAY BODY =**<br>**CLAY + FLUX + GLASS FORMER + FILLER (+ COLOURANT)** | |
|---|---|
| **CLAY** | brings plasticity |
| **FLUX** | lowers the melting point of the other ingredients |
| **GLASS FORMER** | creates more glass to increase density and vitrification |
| **FILLER** | alters texture, aids in drying, and reduces shrinkage and cracking |
| **COLOURANT** | shifts the colour |

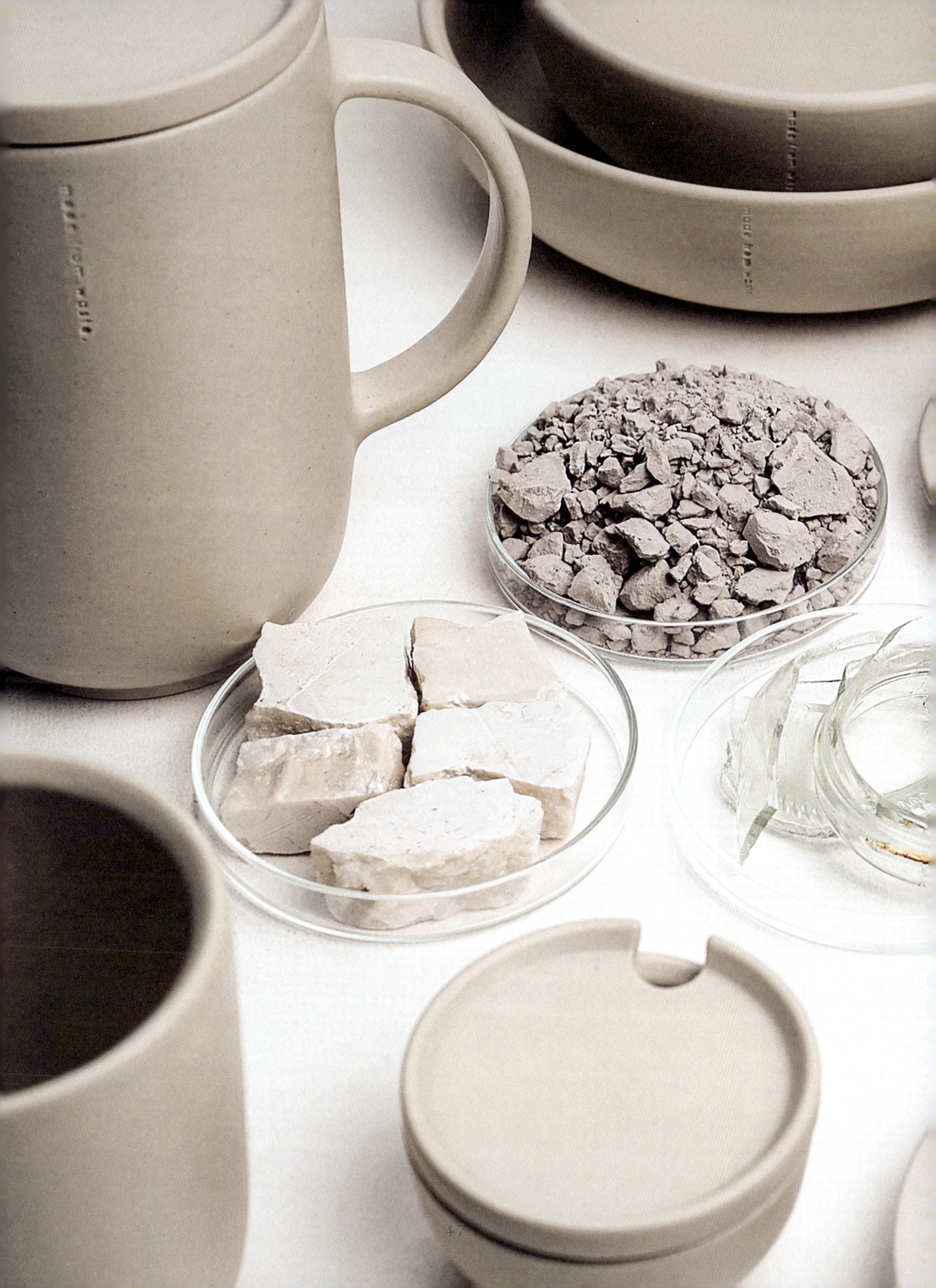

## CLAY IN CLAY BODIES

Clay is the primary ingredient in clay bodies and the type of clay used sets a baseline for the plasticity, the firing temperature, and colour. A clay body requires enough clay, in relation to the other ingredients, to hold shape through the firing. Each type of clay (kaolin, ball clay, fire clay, stoneware clay, earthenware clay, and bentonite) offers distinctive properties to a clay body.

Clay is the only ceramic material with plasticity that renders it malleable. As clay particles become wet, they cling and slide which enables clay to stretch instead of break. A clay has good plasticity when it is capable of absorbing large amounts of water and creating forms without cracking. Throwing bodies require the most plasticity, hand building bodies moderate plasticity, and casting slips do not require plasticity.

Clay bodies need to balance the plasticity from clay with the desirable qualities of other ingredients in the recipe. The plasticity of a clay body is inherently reduced by adding non-plastic clays, fluxes, glass formers, and fillers such as sand or grog. If a clay body needs more plasticity, it needs more clay relative to other ingredients. Plasticity can also be enhanced through wet mixing by blunging, wedging and ageing.

The refractory nature of clay in a clay body enables it to hold shape when fired. Clay solidifies when heated to temperatures approaching red heat (900°C (1652°F)) and maintains its shape to higher temperatures depending on the type of clay, in combination with the other ingredients in the clay body. Low fire clay bodies are more absorbent or porous while higher fired clay bodies become more vitreous.

**CLAY PROPERTIES**

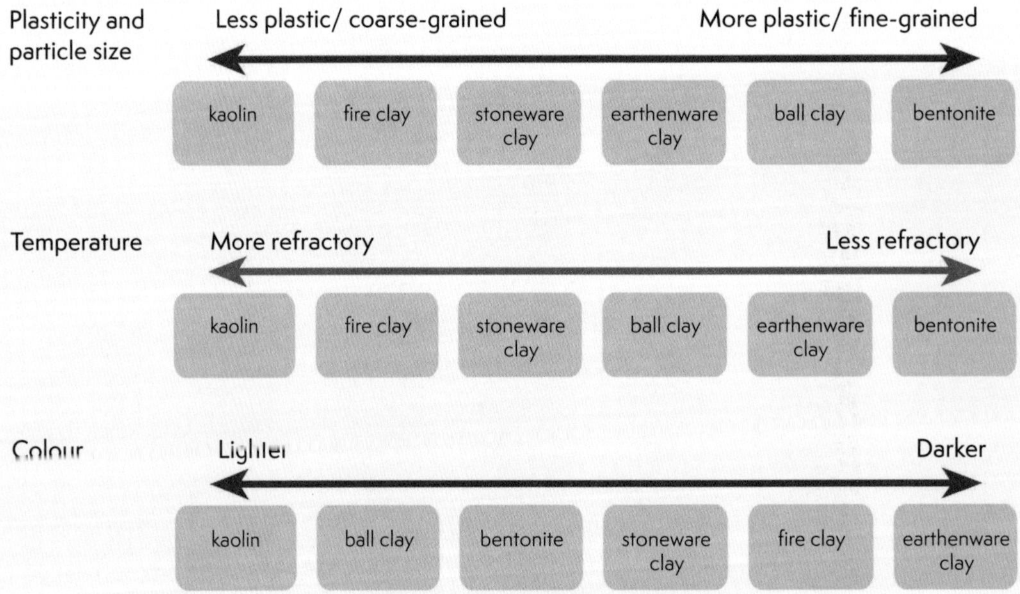

| | | |
|---|---|---|
| **HIGH FIRE CLAY BODIES** | cones 8–10 | 1260–1305°C (2300–2381°F) |
| **MID FIRE CLAY BODIES** | cones 5–6 | 1150–1260°C (2100–2300°F) |
| **LOW FIRE CLAY BODIES** | cones 06–02 | 900–1150°C (1700–2100°F) |

**Below:** Nicole Seisler, *Holding Patterns*, 2019. The plasticity of clay is simultaneously increased and recorded on the wall as the artist methodically wedges it across the surface. *(Image credit: Charlie Schneider)*

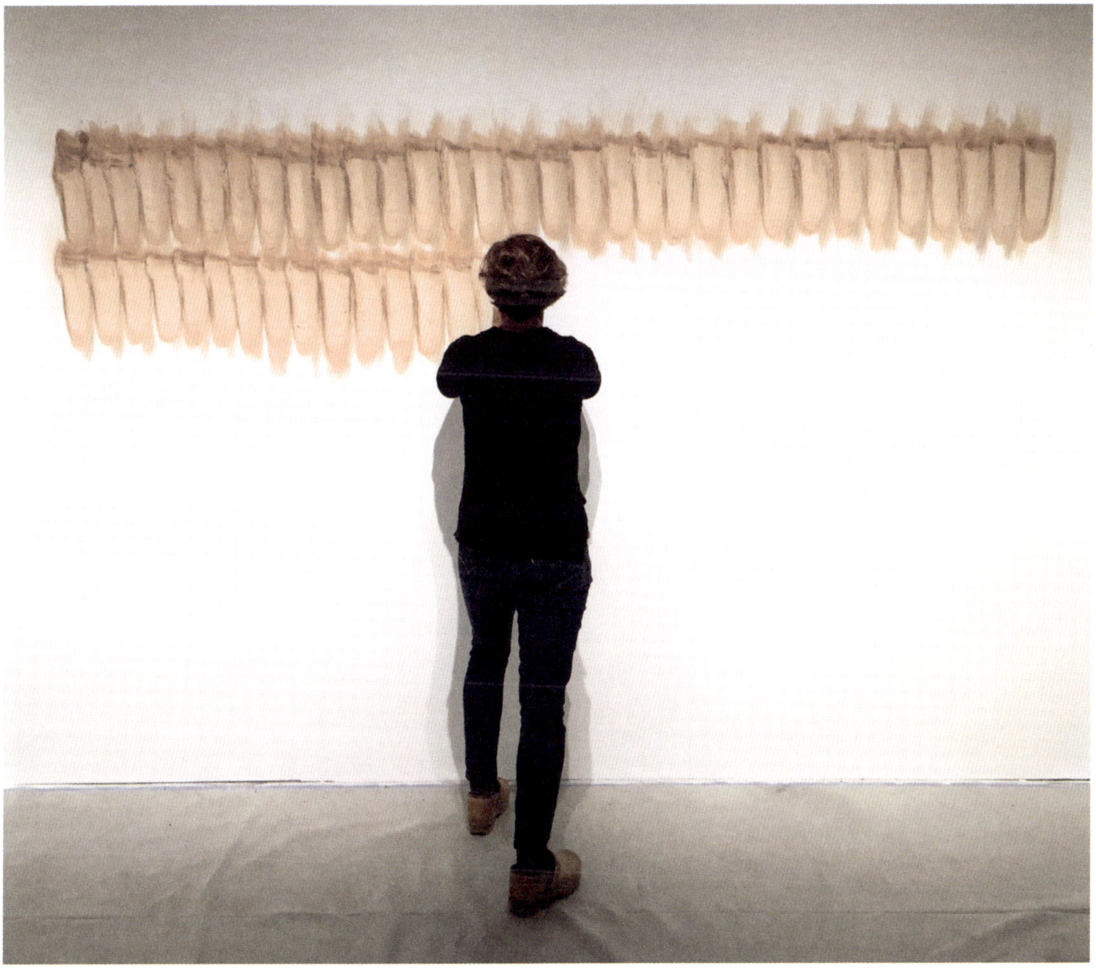

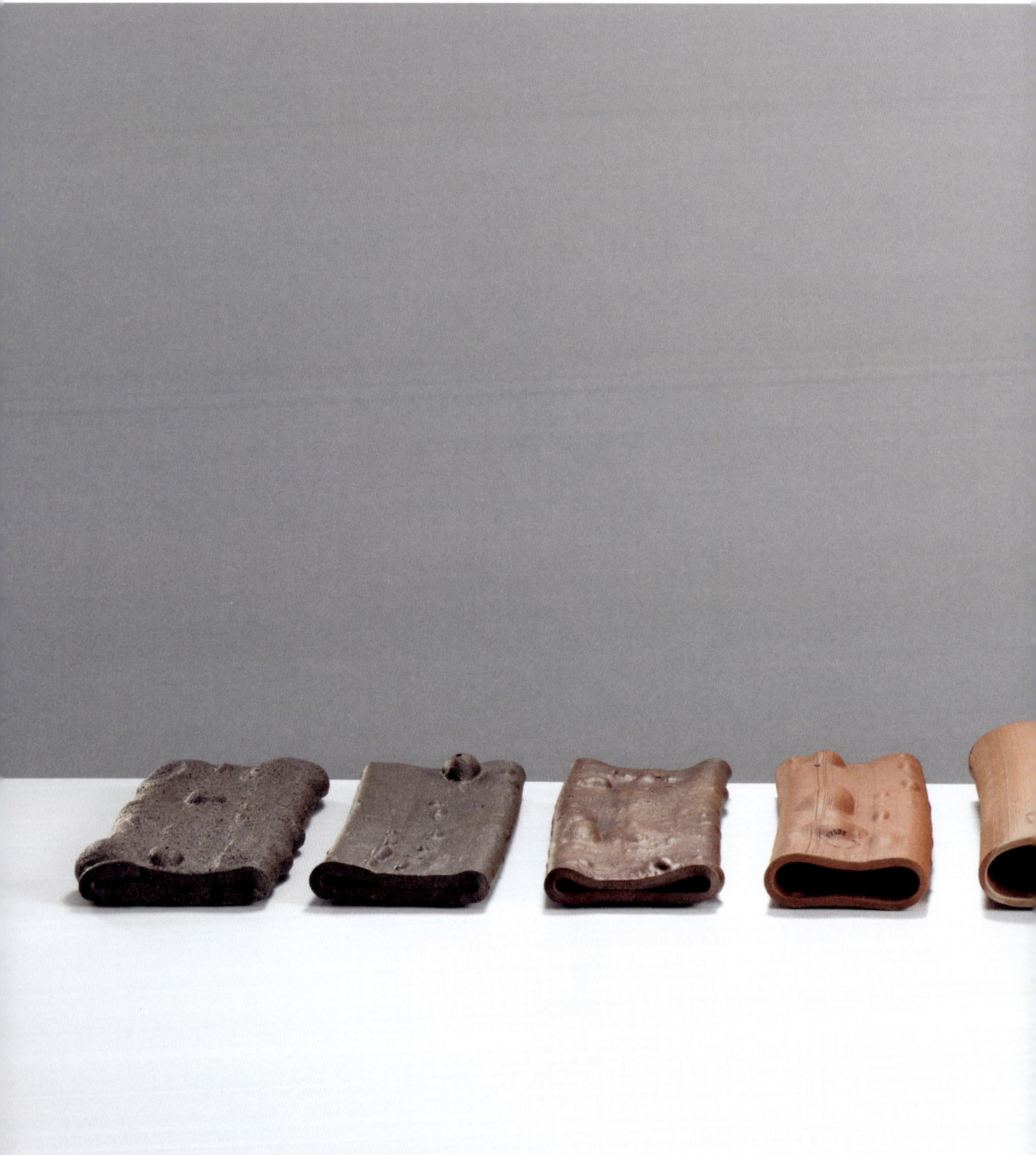

The firing temperature of a clay body is primarily determined by a combination of the clay and flux components in the recipe. Increasing the flux content will effectively decrease the overall melting point of the mixture. Fluxes sometimes occur naturally in combination with clays, like in earthenware clay. The fluxes in clay bodies include the alkaline oxides of sodium, potassium, lithium, magnesium, and calcium, or high quantities of metallic oxides such as iron oxide. Clay can also be added to adjust the firing temperature of a clay body by adding a clay high in alkalis to lower the maturing temperature or a more refractory clay to raise the maturing temperature.

**Left:** Allison Cooke, *Pipeline*, 2018. London surface clay fired at various temperatures reveals the impact of heat on clay. *(Image credit: Michael Harvey)*

## FLUX IN CLAY BODIES

Fluxes are ceramic oxides that promote the fusing of a clay body and increase density or hardness at a desired temperature. Alumina and silica in clay form a refractory structural network within which glass is produced during the firing. With enough added heat, the silica content in a clay body combines with flux to fuse and densify. Fluxes induce glass formation with silica at lower temperatures thus effectively lowering the melting temperature of a clay body. Clay body fluxes include the alkaline oxides of sodium ($Na_2O$), potassium ($K_2O$), lithium ($Li_2O$), magnesium ($MgO$), and calcium ($CaO$).

As a general guideline, high fire clay bodies use feldspars or nepheline syenite as fluxes. Mid-range clay bodies use feldspars, nepheline syenite, talc or frit. Low temperature clay bodies use talc or frit.

### Potassium and sodium feldspars

Feldspars are added as a flux to clay bodies to help melt silica into glass. Feldspars are the most common source of flux oxides in clay bodies due to their natural abundance, low cost and wide melting range. Typical feldspars contain both potassium and sodium in varying amounts as well as lesser amounts of calcium. Potassium

**COMPOSITION OF FLUXES**

| | | Flux oxide | | | | | Stabilizer | Glass former | | |
|---|---|---|---|---|---|---|---|---|---|---|
| | | Li | Na | K | Mg | Ca | Alumina | Silica | Boron | Phosphorus |
| Stronger fluxes used alone or in combination with weaker fluxes | Feldspar – Potassium | | L | H | | T | H | H | | |
| | Feldspar – Sodium | | H | L | | L | H | H | | |
| | Feldspar – Lithium | H | | | | | H | H | | |
| | Nepheline syenite | | H | H | | | H | H | | |
| | Talc | | | | H | | | H | | |
| | Frits | H | H | H | | H | L | H | H | |
| Weaker fluxes used in combination with stronger fluxes | Whiting (calcium carbonate) | | | | | H | | | | |
| | Bone ash | | | | | H | | | | H |
| | Wollastonite | | | | | H | | H | | |

Flux oxides often occur in combination with silica or both silica and alumina as in feldspars and frits.

Key: H=high amount, L= low amount, T=trace

feldspars, or potash feldspars, contain a higher proportion of potassium than sodium and are more commonly used in clay bodies. Sodium or soda feldspars contain a higher proportion of sodium than potassium and are sometimes used in clay bodies but more commonly in glazes. Along with alkaline oxides, feldspars also contain silica and alumina in a proportion of 6:1. Feldspar provides an additional source of silica that forms glass in the clay body.

| | $K_2O$ | $Na_2O$ | $Al_2O_3$ | $SiO_2$ |
|---|---|---|---|---|
| Potassium feldspar | 0.75 | 0.25 | 1 | 6.35 |
| Nepheline syenite | 0.25 | 0.75 | 1 | 4.65 |

### Lithium feldspar

Lithium is another strong alkaline flux that can work across all temperature ranges and is valued for its low thermal expansion. Spodumene and petalite are referred to as lithium feldspars since they contain lithia, silica and alumina. Lithium feldspars are an important ingredient in flameware and ovenware, specialized clay bodies that withstand high thermal shock. Two disadvantages of lithium feldspars are the relatively high cost and difficulties with glaze fit due to their low expansion.

### Nepheline syenite

Nepheline syenite is considered a feldspathoid, similar in composition to feldspar with less silica ($SiO_2$) and more sodium ($Na_2O$). The high sodium content more effectively lowers the melting

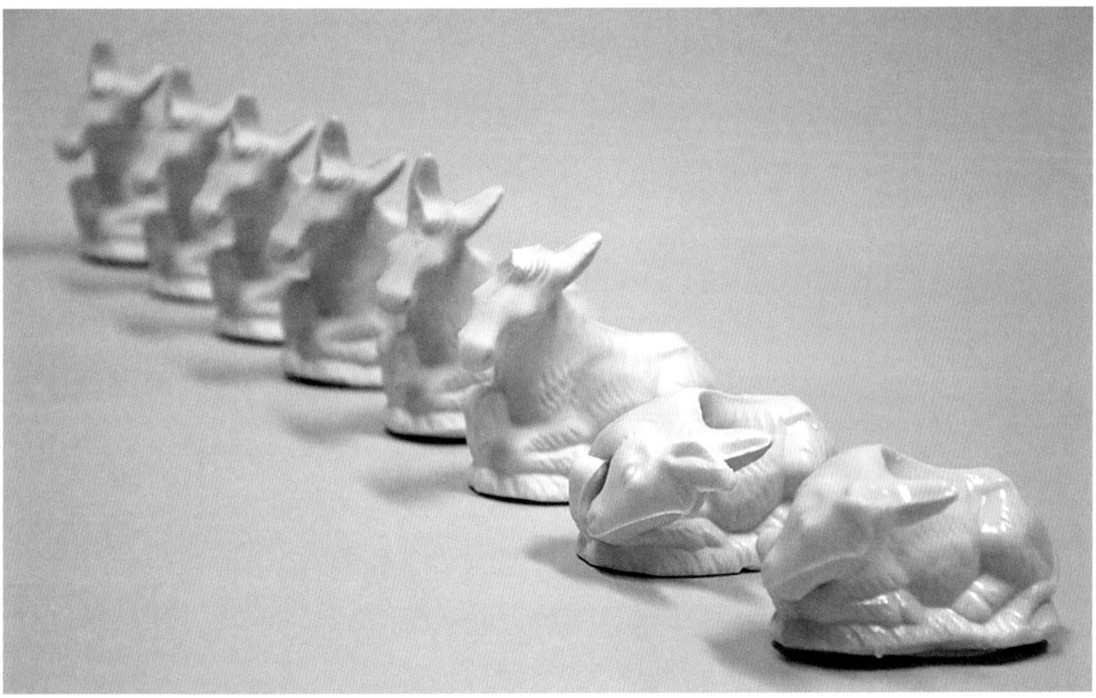

**Above:** *Glaze Bodies*, 2019. The recipes for unusual clay bodies that soften during firing due to high amounts of flux are featured on the website HotAlchemy.ca

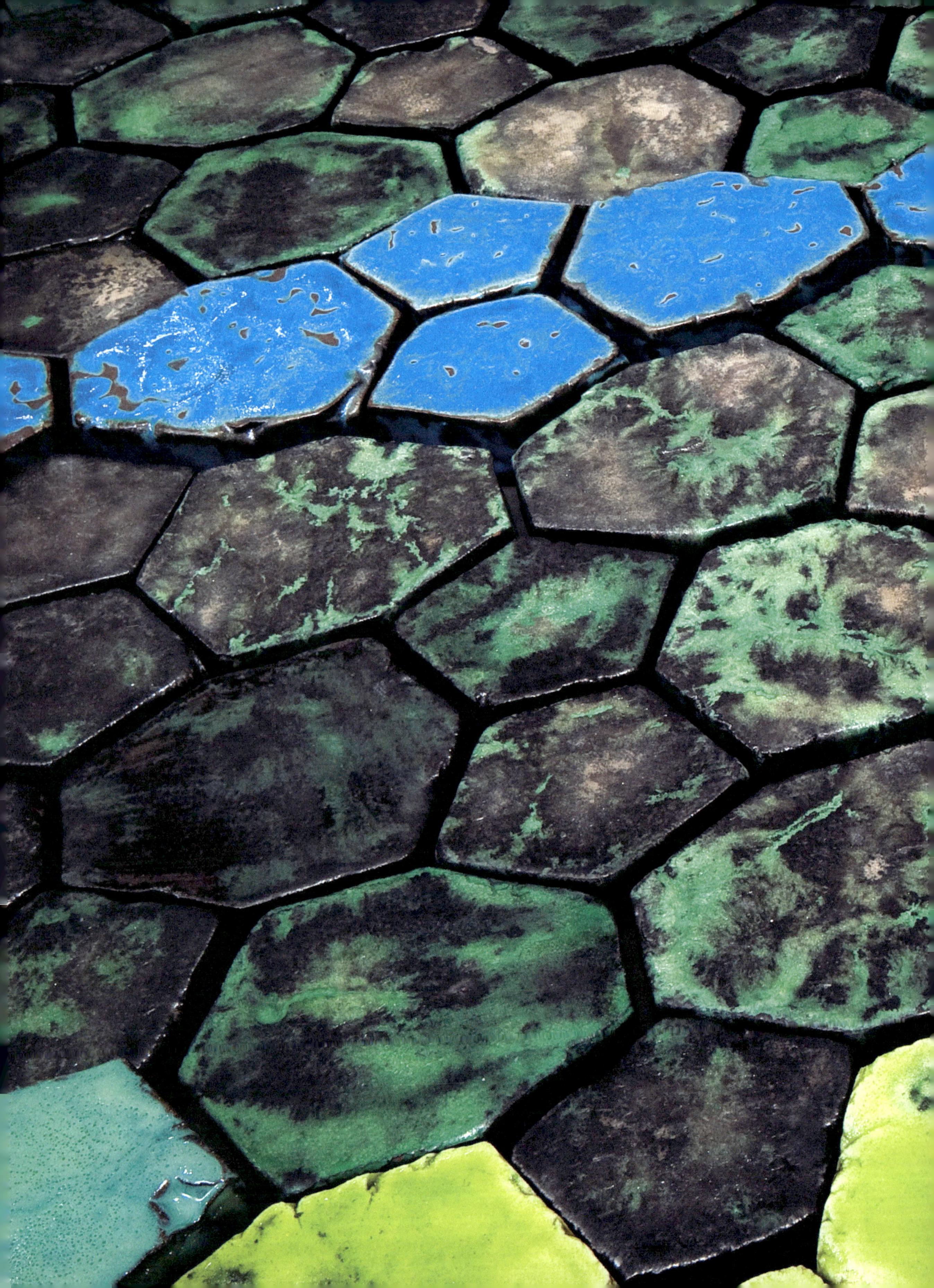

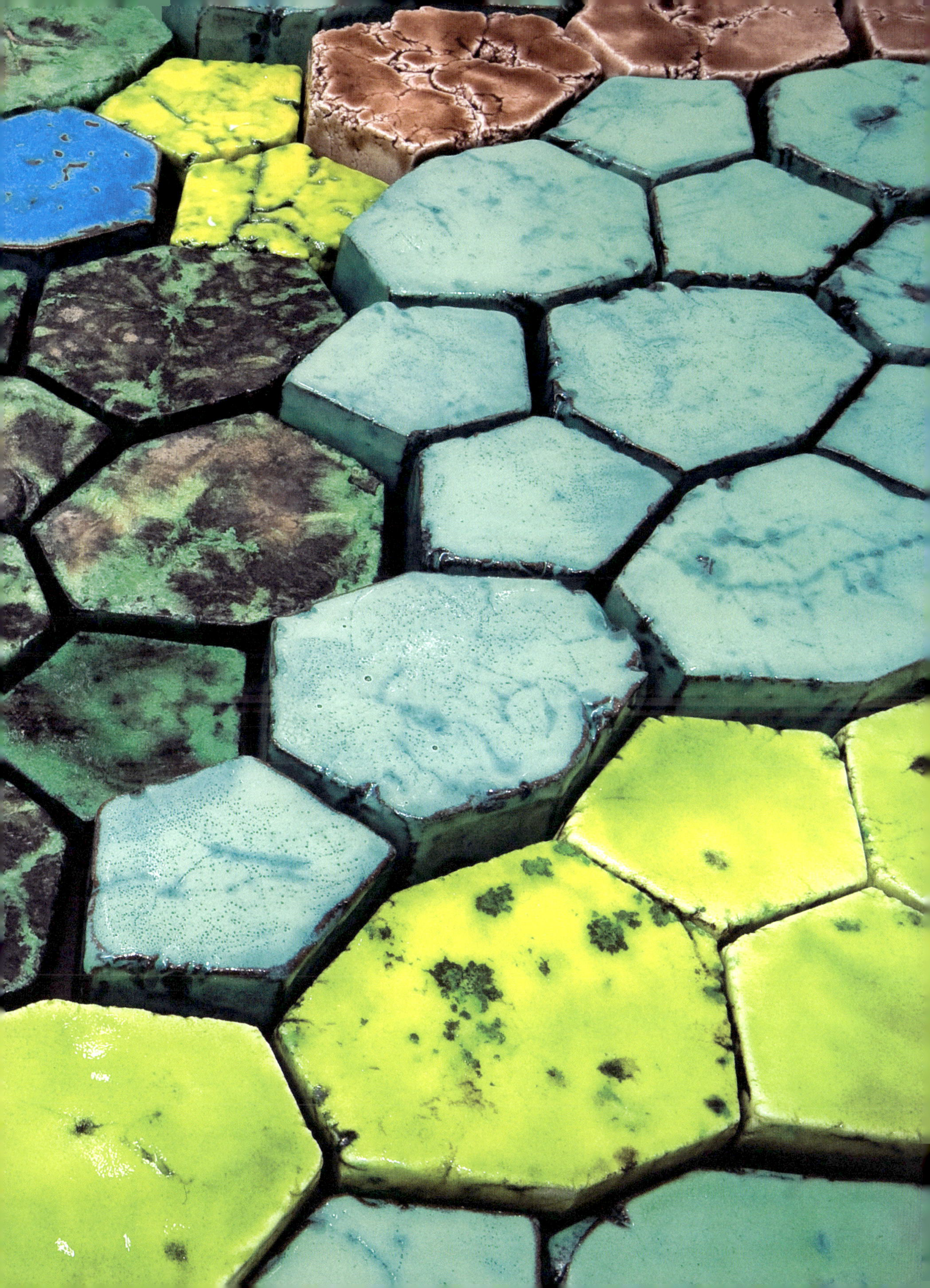

temperature of a clay body when compared with feldspar. Since nepheline syenite can be added to help lower the overall firing temperature of a clay body, a simple way to lower a cone 9 clay body to cone 6 would be to partially replace the feldspar in the recipe with nepheline syenite. However, the slight solubility of the sodium in nepheline syenite can cause the clay to become deflocculated and less plastic. To correct this, add a small amount of a saturated solution of Epsom salts and wedge into the clay.

### Talc

Talc is a naturally occurring form of magnesium combined with silica and is commonly used as a flux in low fire casting slips.

### Frits

Frits are not naturally occurring but are industrially formulated mixes of flux oxides combined with alumina and silica. Frits are made by combining these elements into a glass that melts at a specific temperature. The mixture is then cooled, solidified and crushed into a powder.

While frits are powerful melters that work at lower temperatures, their melting range is much shorter than feldspars. Frits are most useful in combination with other fluxes to decrease the maturing temperature of porcelain and stoneware to mid range. Frits are also useful in making low-temperature clay bodies although the sudden and short range of vitrification is more difficult to control in a clay body than in a glaze. Frits are a more expensive source of flux than feldspar.

### Calcium

Calcium is used as a flux in clay bodies as bone ash, wollastonite or whiting. Bone ash, or calcium phosphate, is a combination of calcium and phosphorus that acts a powerful flux at high temperatures yet is ineffective as a flux at mid-range and low fire temperatures.

Bone ash is primarily used in the production of bone china. Wollastonite provides calcium in a silicate form that is stable and insoluble. This combined form of calcium and silica is considered to be both a flux and a filler due to its fibrous, needle-like crystalline shape which is useful at high fire to improve fired strength. Whiting, or calcium carbonate, is added to clay bodies only as a secondary flux because it is very strong and difficult to control. Small additions (up to 2%) of whiting can increase strength and even translucency in high fire porcelain recipes.

Other forms of calcium, such as lime (calcium oxide) are not used as clay body fluxes because their range of vitrification is short, producing a sudden melt. Calcium can naturally occur in clays from shell fragments which cause lime popping or calcium oxide hydration, a serious problem that appears in clay bodies after firing above 700°C (1292°F) when calcium carbonate ($CaCO_3$) converts into quicklime ($CaO + CO_2$). Lime may absorb atmospheric moisture to expand and crack a clay body over time.

**Previous:** Cary Esser, *Landscape I* (detail), 2009. Fritware, red earthenware, low temperature glazes, press moulded. *(Image credit: E. G. Schempf)*

## GLASS FORMER IN CLAY BODIES

Glass formation within a clay body increases strength and vitrification or density. The most common glass former in clay bodies is silica, also known as quartz or flint, which is naturally present in all clays, fluxes and many fillers or can be added as an additional ingredient in the form of ground quartz or flint. Silica on its own forms a glass at 1710°C (3110°F), beyond the temperatures of ordinary ceramic kilns. Silica will melt into glass at lower temperatures in combination with sufficient flux. The presence of alkaline fluxes in a clay body help to lower the melting temperature of silica to the standard firing range of a kiln.

Glass formation happens to a lesser extent in clay bodies than glazes because of the high clay content and lower flux levels in clay bodies. Clay bodies are more refractory than glazes due to their higher alumina content from higher amounts of clay. While clay bodies contain fewer alkali flux oxides than glazes, the fluxes that are present will combine with silica to form some glass content in a clay body. High temperatures lead to increased glass formation that densifies the clay and lowers the overall level of thermal expansion.

Silica makes up a large percentage of clay bodies as well as glazes. Excess silica in clay bodies, beyond that contained in clay structure and glass formation, may recrystallize into cristobalite. Cristobalite develops from very fine particles of silica at high temperatures (1250–1350°C (2282–2462°F)). Above 1000°C (1832°F) kaolin decomposes into cristobalite as well as mullite. The formation of mullite is positive because the fine needle-like form creates a network that provides structural support in clay bodies. Cristobalite formation is less desirable due to its thermal expansion. At 226°C (439°F) cristobalite expands dramatically. Very high or low thermal expansion levels in clay bodies can impact the ability of glazes to 'fit' or adhere to a clay body.

Boron and phosphorus can also be utilized as glass formers for more specialized clay bodies with specific applications. Boron is added in frits to clay bodies to dramatically lower the firing temperature of a clay body. Frits are relatively expensive and lower plasticity in clay bodies when used in large quantities. Phosphorus is present in bone ash, as an ingredient in bone china, augmenting silica as a glass former to create translucency. A typical bone china recipe uses 50% bone ash, 25% feldspar and 25% kaolin.

Glass formation increases vitrification and density while reducing porosity. During firing, the process of vitrification involves the development of mullite and glass in the clay body. A clay body can be tested for its rate of absorption by making a test bar and a simple calculation (see Appendix 4). The percentage of absorption indicates the level of porosity or the degree of vitrification, the strength and durability. The average absorption for a mature, well-vitrified clay body, at any temperature, will be below 5%. Low fire earthenware clay bodies with more than 5% porosity remain porous and must be glazed to become waterproof.

## FILLER IN CLAY BODIES

Fillers perform several desirable roles in a clay body but they are not a required component in a recipe. Except for silica, which is included in a clay body recipe that adds up to 100, the filler amount is usually added as an additional percentage on the recipe. Fillers impact the physical texture of clay bodies, workability and visual appearance. Fillers can reduce shrinkage, decrease cracking and warping, adjust expansion, provide structural support and generate texture. Common fillers are listed in Appendix 5.

Fillers can improve the functionality of clay bodies by reducing certain faults. One of the difficulties with fine plastic clays is their ability to retain water, which results in the exterior drying more quickly than the interior. The moisture differential caused by uneven drying can lead to warping and cracking in the clay body.

Fillers in a clay body create a ventilation network, generating pathways that allow moisture to escape during drying and gases to exit during firing. Within a matrix of tiny clay particles, the larger filler particles open small spaces within the dense clay structure that allow water vapour to more easily evaporate into the atmosphere. Ingredients with relatively large particle sizes like some fluxes may act like fillers that help in the drying stage. While fillers may add strength during the drying process, they also create tension within the clay body as the clay shrinks, and microcracks may form around non-plastic particles.

Ceramic-based fillers including silica, silica sand, and grog among others, are high in either silica or alumina. These fillers remain embedded in the clay body matrix after firing and stiffen the clay body to reduce slumping at peak firing temperatures.

While fine silica may function as a glass former in a clay body, larger mesh silica and silica sand do not melt but function as a filler in a clay body. Silica sand reduces shrinkage, supports a clay body and facilitates the escape of gases during the early phase of a firing. Sand particles may cause cracks in porcelain bodies due to its high expansion. For low fire clay bodies, sand reduces shrinkage and is regularly used in the manufacture of bricks.

Grog is fired clay that has been ground down into small particles that are graded into fine, medium and coarse sizes. Combining different sizes of grog can be an effective way to create structure within a clay body, providing support for large work or for reducing shrinkage. Grog can be abrasive in throwing bodies. The colour of grog varies according to the clay from which it is made. Grog made from porcelain is called molochite.

Organic fillers serve their primary purpose during forming and drying and then burn out during firing leaving small voids. In proper proportion, these voids do not jeopardize the integrity of the clay form.

Cotton, flax and other fibres can be added to clay to create a temporary structural network within the clay body that provides support during the forming and drying stages. The fibrous network binds the

**Opposite:** Matt Wedel, *Flower Tree*, 2017. The addition of grog in sculptural clay bodies helps structurally when building large forms, aids in drying and reduces cracking. *(Image credit: Matt Wedel Studio)*

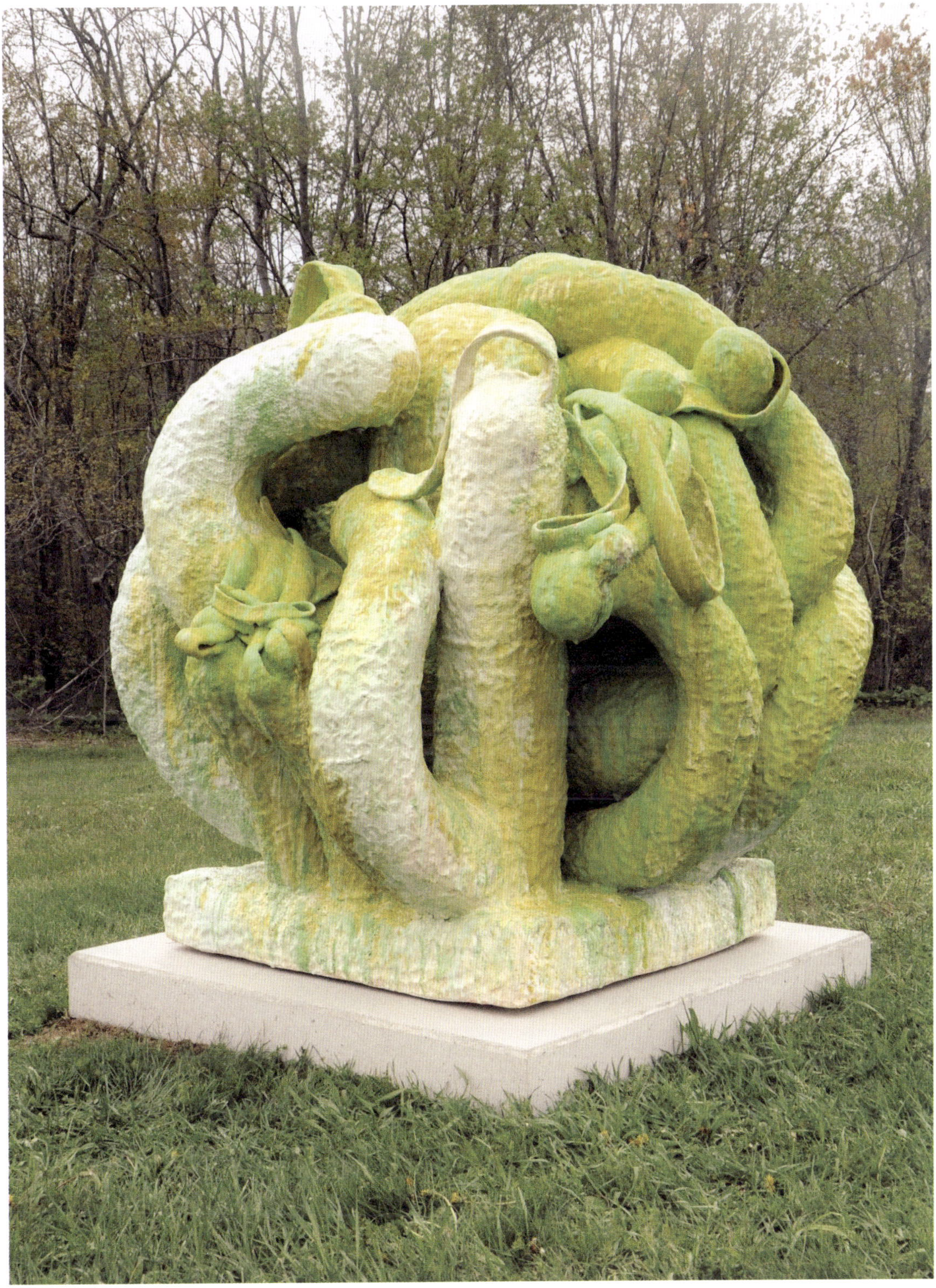

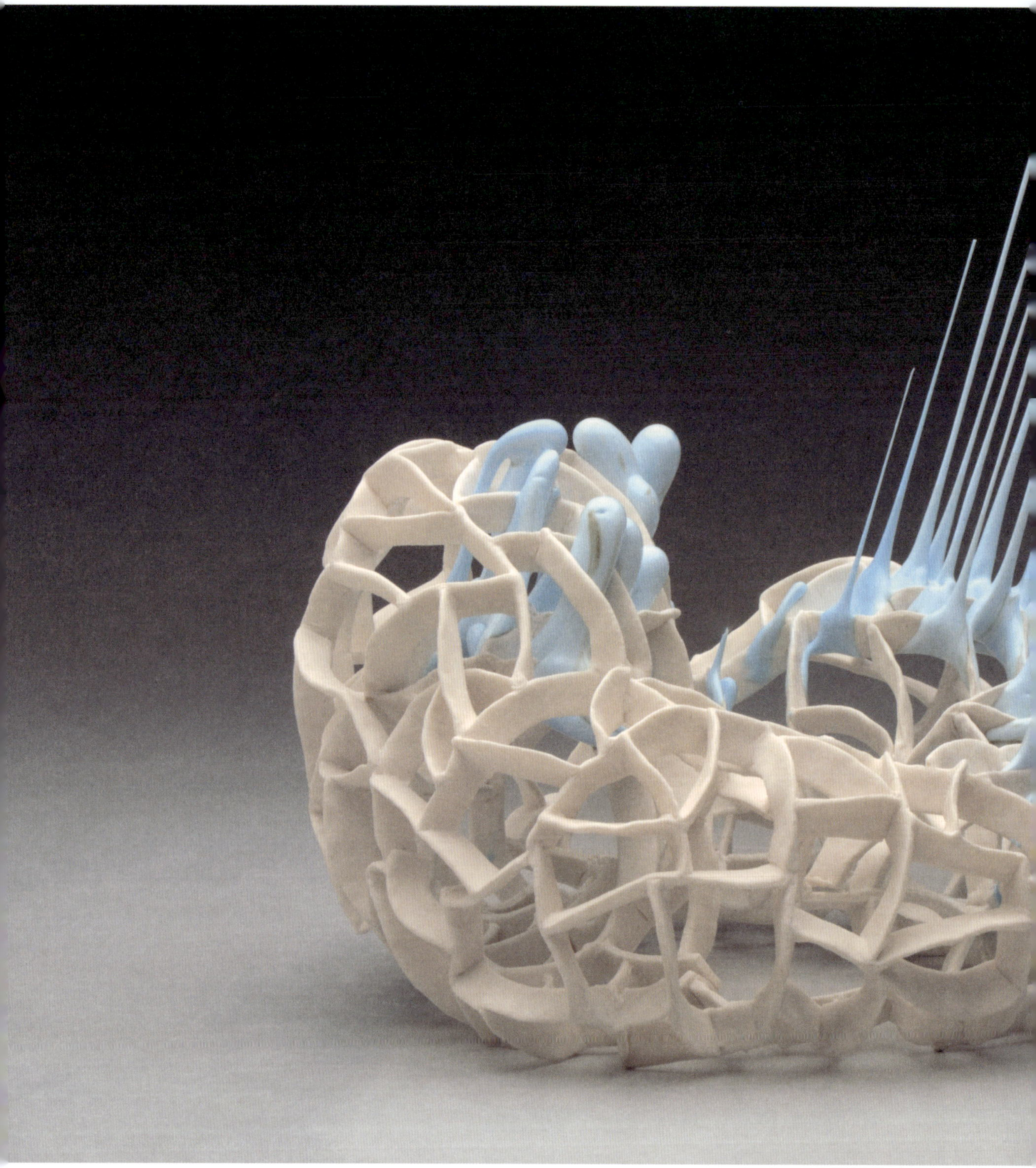

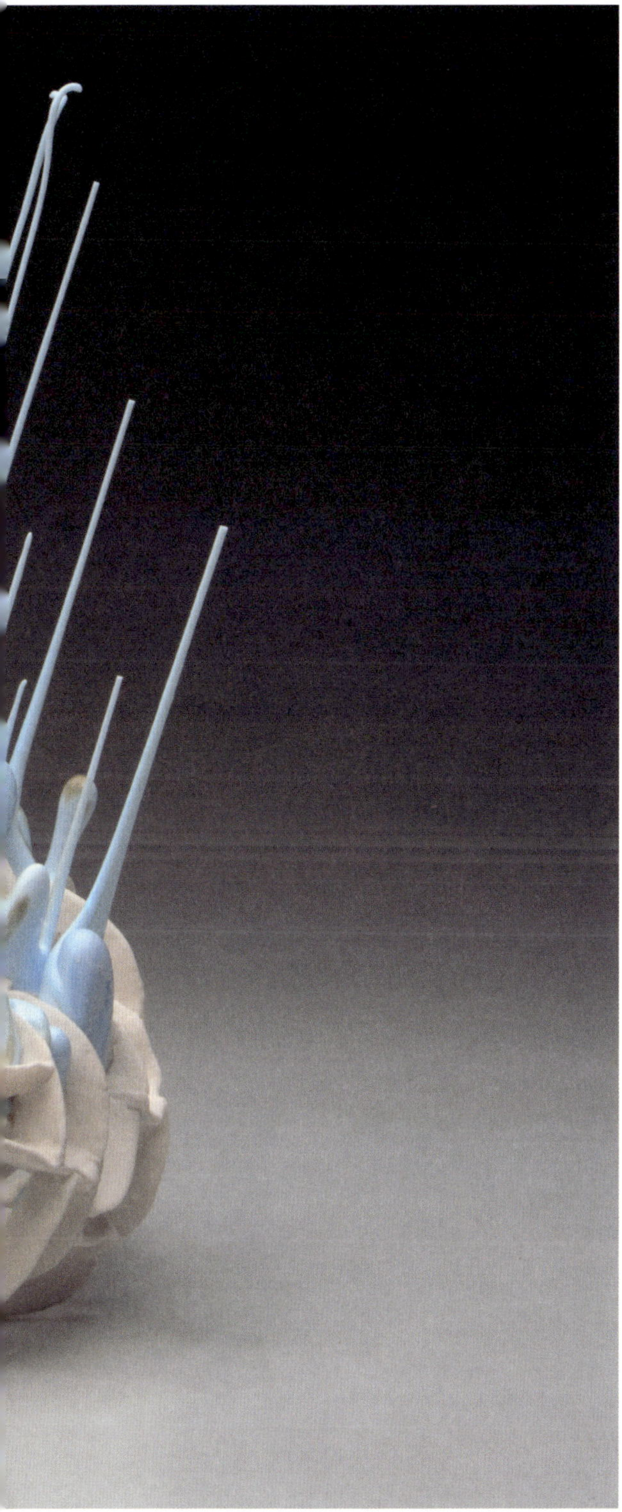

clay together as it dries to effectively prevent small cracks from developing into large ones. Very fine fibres in small amounts are invisible on the surface of the clay body. During firing the fibres burn away leaving imperceptible voids in the clay body. When high amounts of fibres burn out of a clay body, it may become weak and fragile. Adobe, also known as cob, is a mixture of clay, sand and straw that is formed raw, dries, and remains unfired for architectural applications.

Paperclay is any clay body that has added paper fibre. To make paperclay, soak paper in water, blunge it and add the paper slurry to your clay body while accounting for the water in the paper mixture. Toilet paper breaks down quickly but has short fibres while other types of paper require more time to dissolve. Longer fibres in higher quality paper create stronger paperclay prior to firing.

**PAPERCLAY**

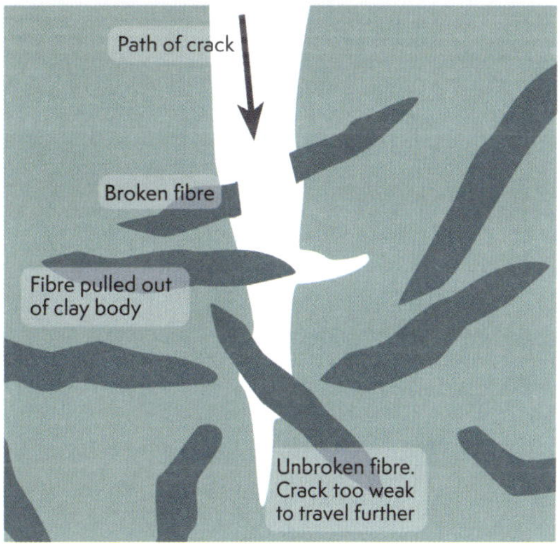


**Left:** Xu Shiyuan, *Growing*, 2016. Paperclay is useful when building delicate structures out of any clay body as the fibres help in preventing cracking during the drying and greenware stages. *(Image credit: Shiyuan Xu)*

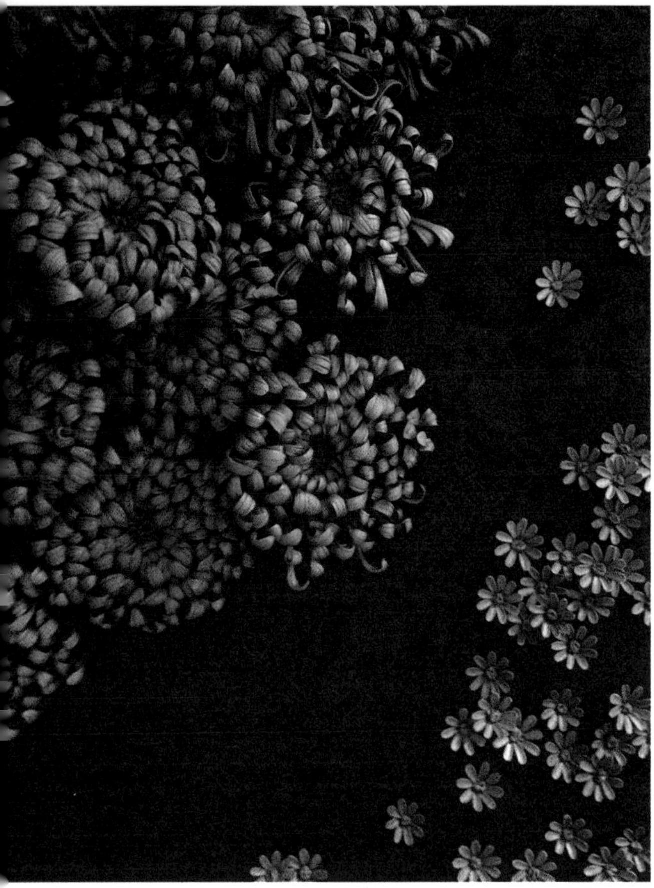

## COLOUR IN CLAY BODIES

The fired colour of naturally occurring clay is a range of earthy tones from light white to dark. The colour of a clay body can be changed by blending different colours of clay or by adding colouring oxides and stains. To achieve colours beyond the natural clay colour, ceramic oxides and stains can be mixed into a clay body. Suggested amounts of colourants for clay bodies are listed in Appendix 8. Clay bodies may require significant quantities of colourant to alter their original colour and very large additions may reduce the plasticity and overall workability of a clay body.

The colour of clay usually intensifies as a clay body fuses and becomes more vitrified with increasing temperature. With sufficient heat and/or reduction firing, colourants may behave like a flux to lower the overall melting temperature of a clay body. Colourants tend to be expensive and may be more efficiently utilized in surface treatments that coat a clay body such as terra sigillata or engobe.

**Above:** Jae Won Lee, *Seize the Flower in the Mirror* (detail), 2014. Porcelain coloured with black stain. *(Image credit: Jae Won Lee)*

**Opposite:** Erum Aamir, *Arcanus*, 2023. Handbuilt porcelain mixed with various percentages of stain and single fired to 1280°C (2336°F). *(Image credit: Erum Aamir)*

**Overleaf:** Amanda Salov, *Cyanometer*, 2023. Subtle colour gradation was achieved with small additions of blue stain mixed into porcelain. *(Image credit: Kyle L. Asfahl)*

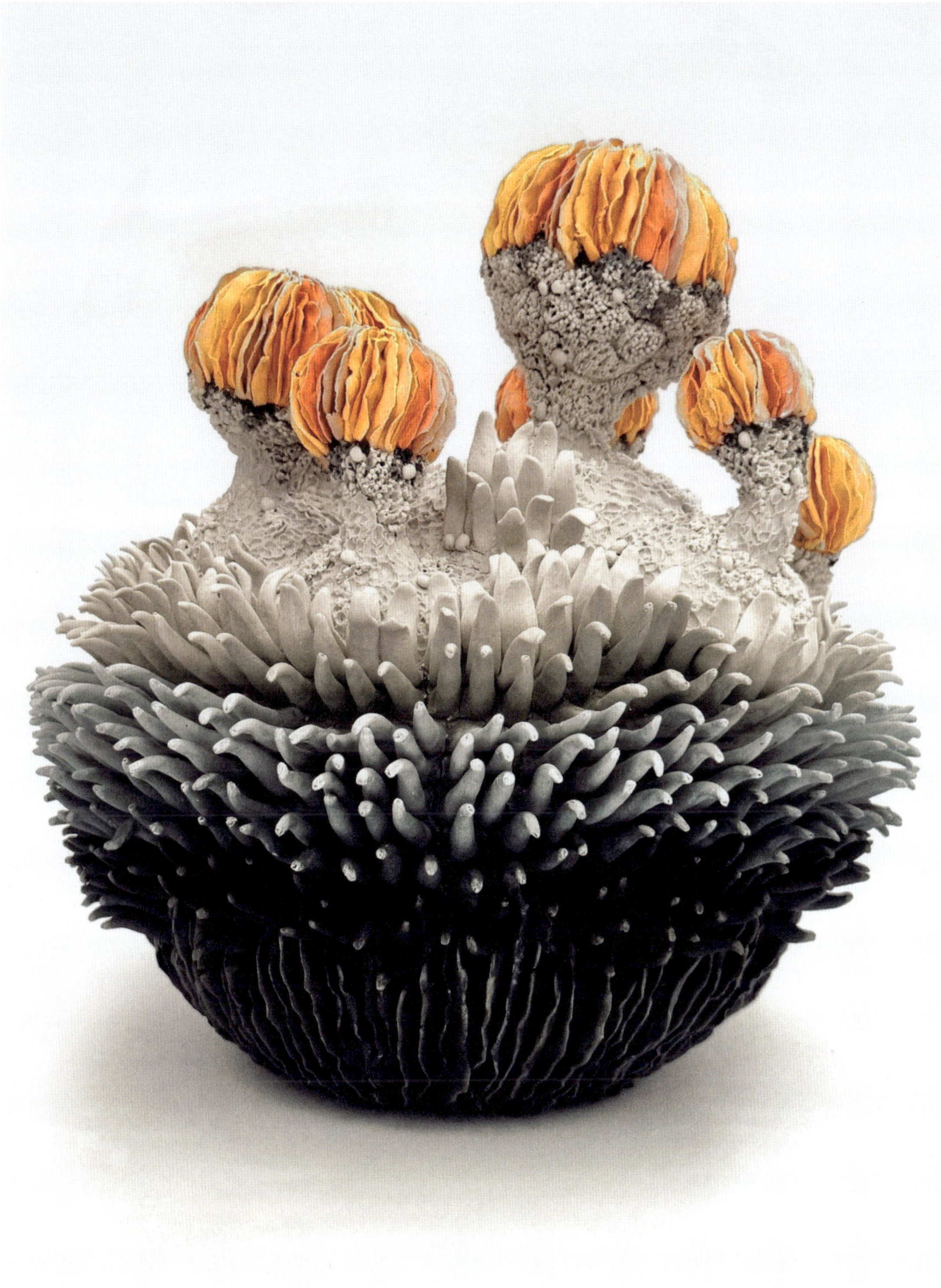

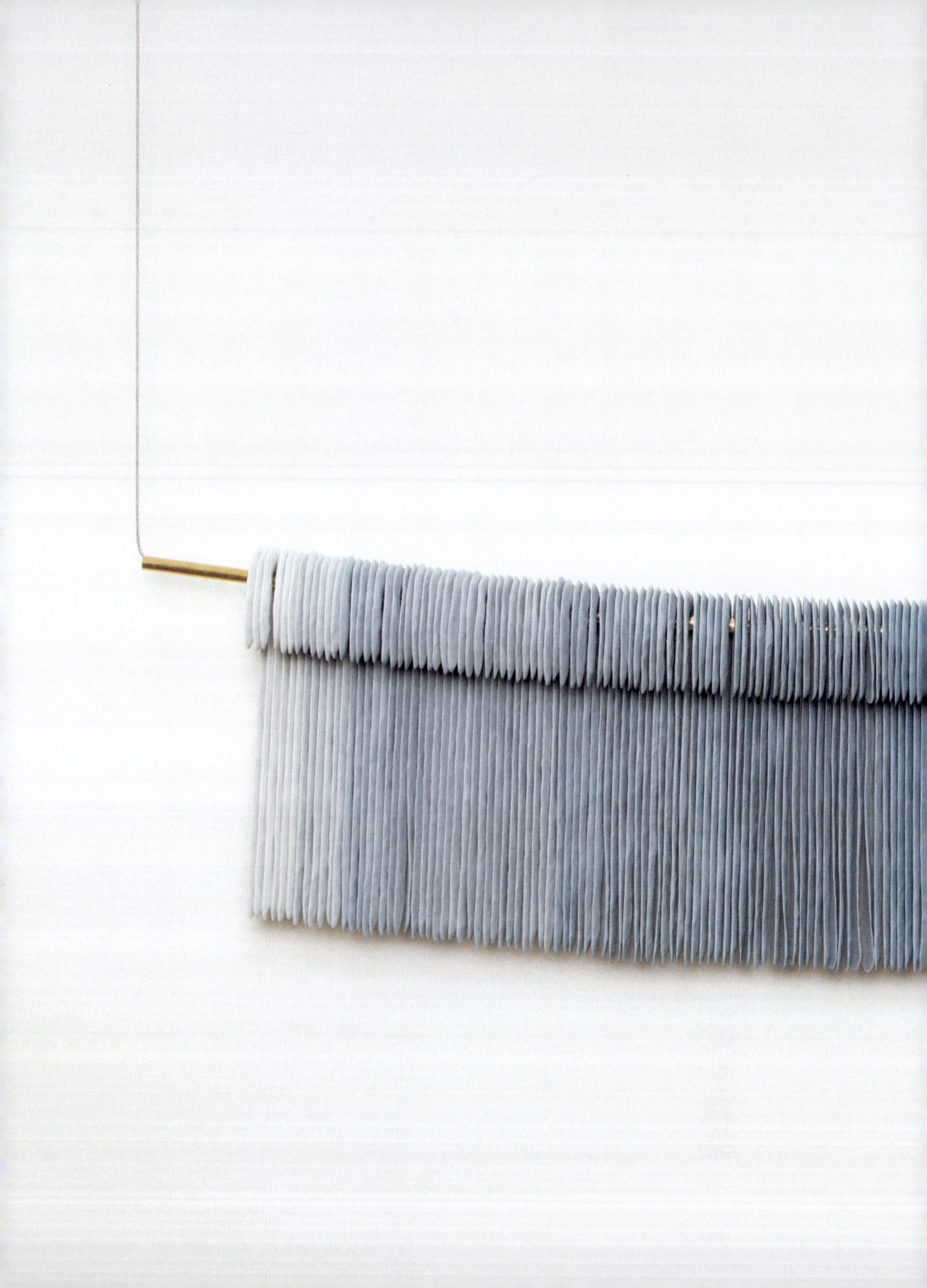

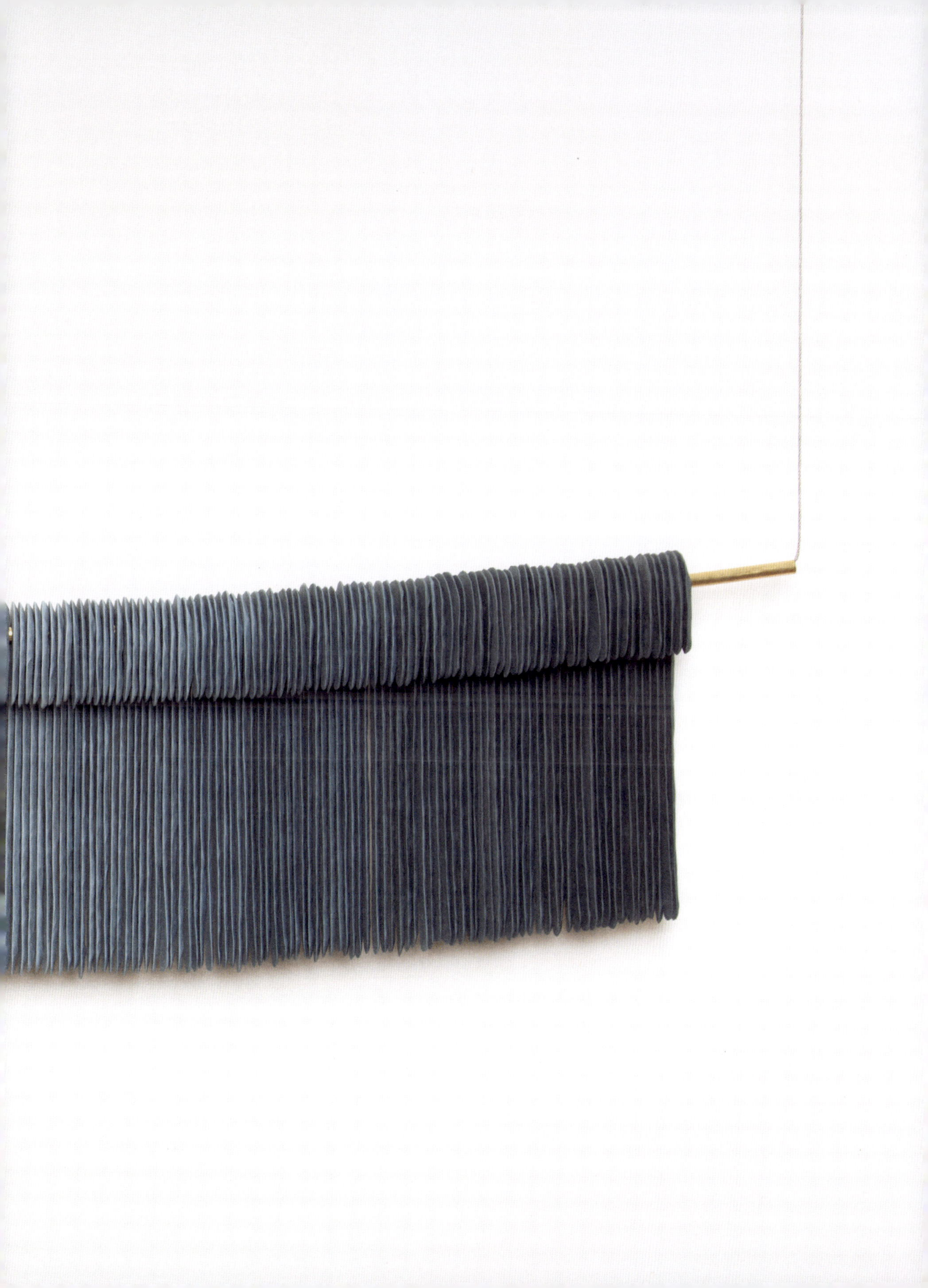

## CLAY BODY TYPES

There are three main types of clay body recipes: porcelain, stoneware and earthenware. Recipes for clay bodies usually contain between three and nine ingredients made up of clays, fluxes and glass formers in proportion to a total 100. Fillers and colourants that are variable are added on to the recipe as an additional percentage.

Porcelain clay bodies are a combination of white kaolin clay, feldspar, and silica. Porcelains are dense due to high glass formation and highly vitrified with very low absorption at 0–1% and high shrinkage. Because kaolin lacks plasticity, porcelains can work well as casting bodies. Hand building and throwing porcelains can be challenging due to warpage and cracking. Light-coloured ball clays or bentonite may be added into porcelain for added plasticity. Ball clays develop opacity in porcelain bodies and are not recommended for translucent porcelain. Pyrophyllite is an alumina-based filler that can be added to increase strength in porcelain bodies.

Stoneware clay bodies can contain various combinations of fire clays, ball clays and/or stoneware clays with feldspars as fluxes. Stoneware clay bodies are highly vitrified with low absorption at 2–3% and are useful as throwing, hand building and casting bodies.

**Notes on porcelain clay body composition**

| | |
|---|---|
| **Clay** | Total clay up to 50%<br>Kaolin clay 25–50%<br>Ball clay up to 25%<br>Bentonite up to 3% for plasticity |
| **Flux** | Total flux 25-40%<br>High fire: feldspar up to 25%<br>Mid-range: feldspar and frit up to 35%<br>Low fire: frit or recycled glass 40%<br>Bone China up to 40% bone ash<br>Whiting up to 2% to increase translucency |
| **Glass former** | Silica/flint up to 20% |

**Notes on stoneware clay body composition**

| | |
|---|---|
| **Clay** | Total clay up to 80%<br>Stoneware clay up to 50%<br>Fire clay up to 60%<br>Ball clay up to 50%<br>Bentonite up to 3% for plasticity |
| **Flux** | Total flux 10-35%<br>Highfire: feldspar 10%<br>Mid-range: feldspar and frit up to 35% |
| **Glass former** | Silica/flint up to 10% |

**Opposite:** Amy Snider, *Saskatchewan Glacier*, 2020. This fragile vessel disintegrates over time into tiny porcelain snowflakes, metaphorically referring to the precarious status of glaciers in the North. *(Image credit: Amy Snider)*

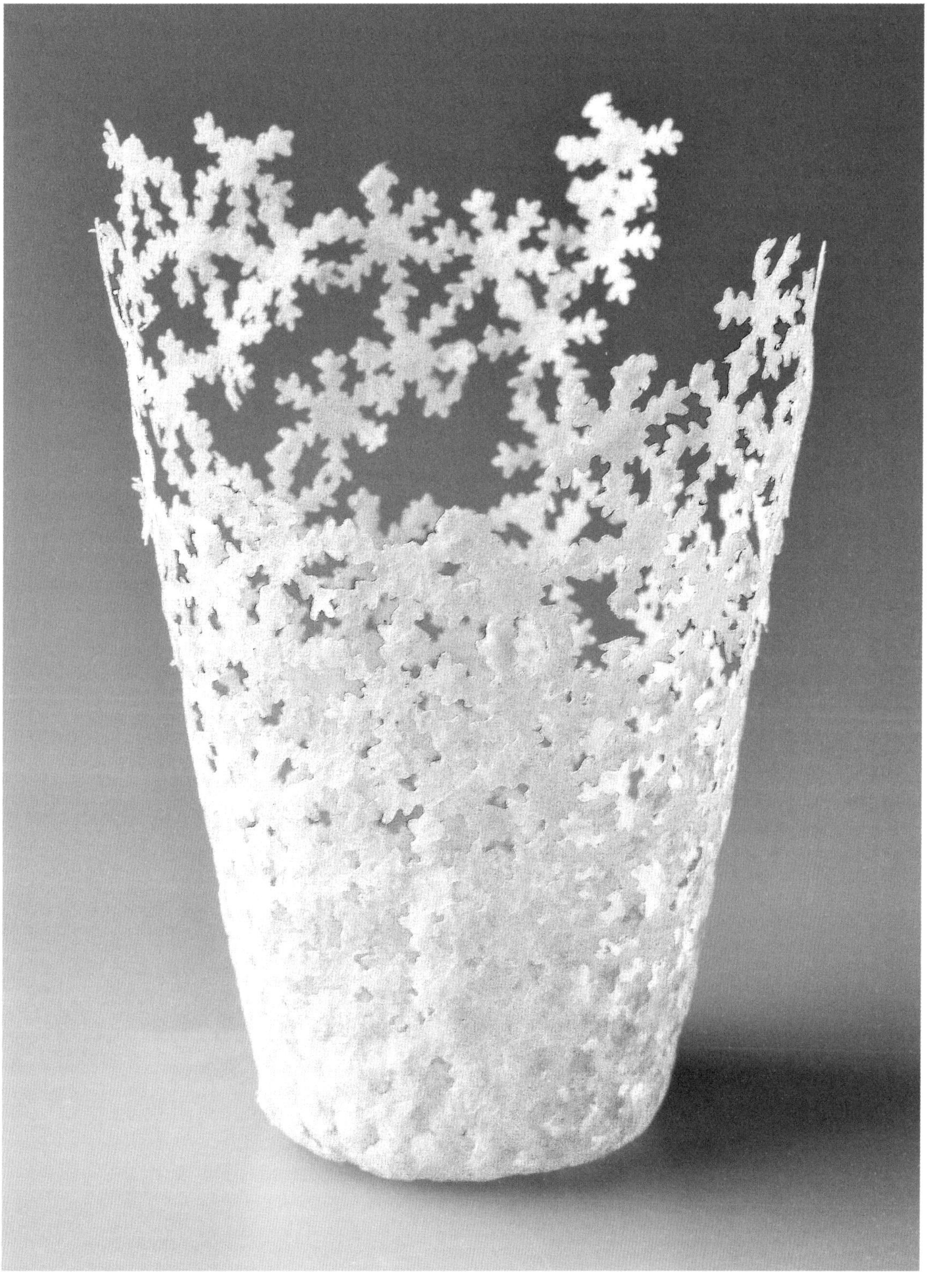

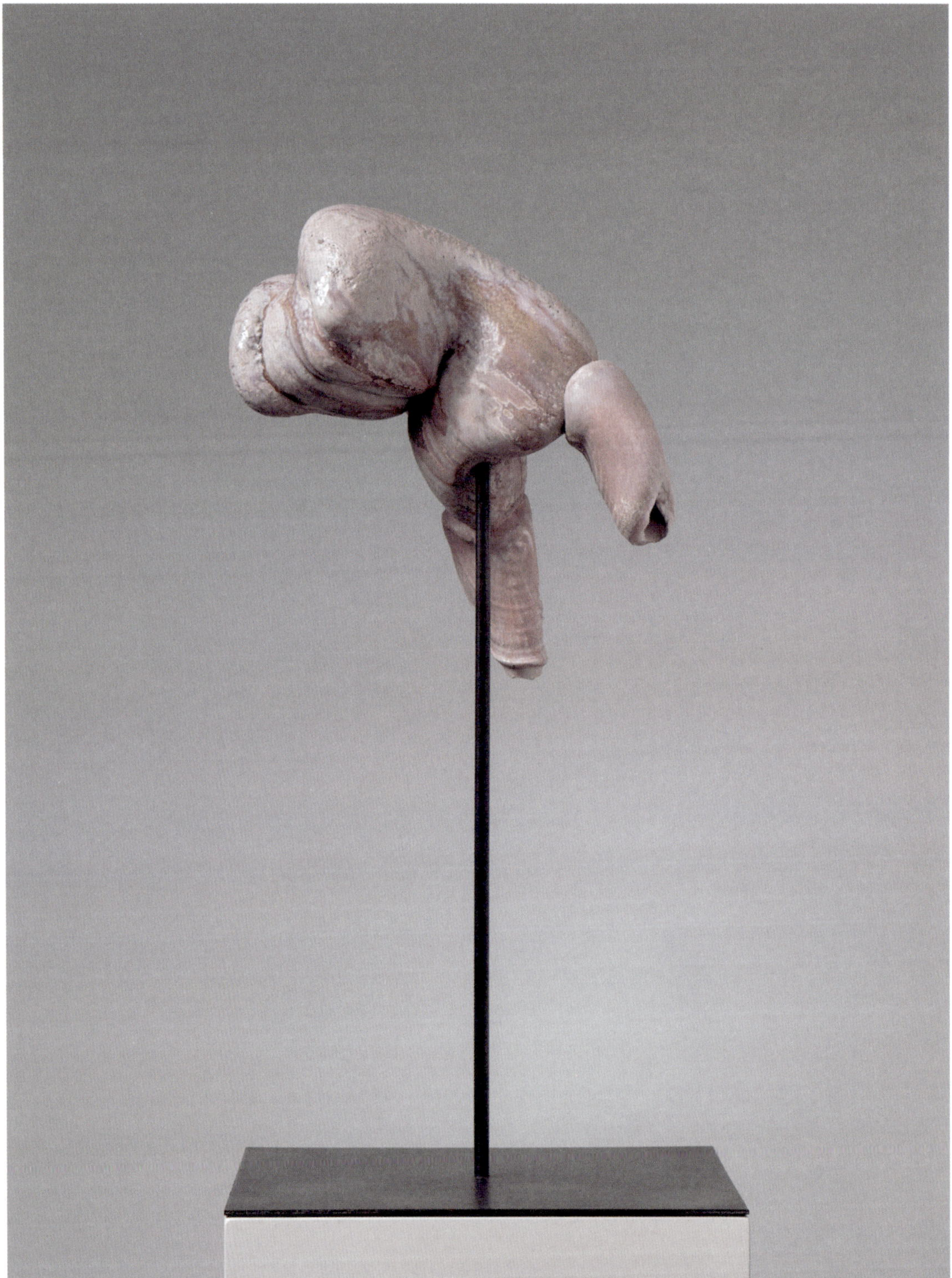

Earthenware clay bodies remain porous and unvitrified with high absorption at 8–12%. Naturally occurring earthenware clays and clay bodies made from them fire orange to red brown or darker. White earthenware clay bodies can be made from light-coloured clays with low fire fluxes such as talc and frit.

**Notes on earthenware clay body composition**

| | |
|---|---|
| **Clay** | Total clay up to 90%<br>Earthenware clay up to 50%<br>Ball clay up to 50%<br>Fire clay up to 30%<br>Stoneware clay up to 30% |
| **Flux** | Total flux up to 50%<br>Feldspar up to 30%<br>Talc up to 50% for white bodies |
| **Glass former** | Silica/flint up to 30% |

**Opposite:** Elsa Sahal, *Molly Bloom*, 2015. Glazed stoneware. *(Image credit: Hans Georg Gaul for Setereh Gallery)*
**Below:** Tony Marsh, *Perforated Totem Cluster*, 2008–9. White earthenware. *(Image credit: Tony Cunha, in the Permanent Collection MAD Museum NY)*

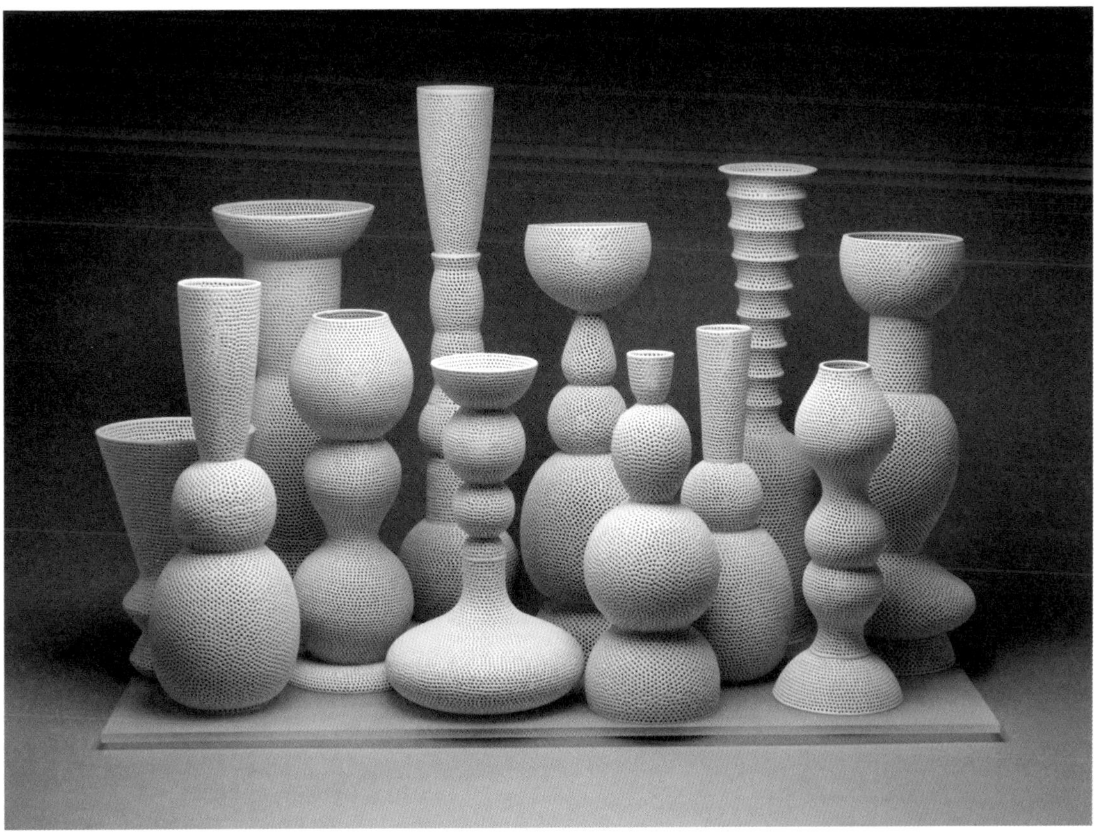

## CLAY BODY APPLICATIONS

Porcelain, stoneware and earthenware are formulated slightly differently when used for specific applications including hand building, throwing or casting.

Hand building clay bodies need to be open with good drying properties. The inclusion of fire clay and fillers like grog help support forms during building and aid in the drying of thick or large forms. Grading the size of fillers by including small, medium and large-sized particles helps to build a strong hand building body with lower shrinkage and less risk of cracking.

Throwing clay bodies require plasticity so the clay can stretch on the wheel without breaking. Between 20–25% ball clay may be used in a throwing body but higher amounts lead to shrinkage and cracking problems. Coarse fire clay or smooth fillers like sand are good additions for strength or support.

Casting clay bodies require little plasticity. Clays of different particle size may be combined to optimize colour, translucency and casting speed. Ball clays are often used in casting bodies but the fine particle size of ball clays in high amounts will decrease the speed of drying and casting.

## TESTING, FAULTS AND REMEDIES

It is advisable to test a clay body for plasticity, shrinkage, fired colour, firing temperature, vitrification, and glaze fit prior to using it. Plasticity can be tested by rolling out a coil of clay about the thickness of a pencil and wrapping it around a finger. If no cracks or small cracks appear as the clay is stretched, it has good plasticity. Non-plastic clays may be suitable for casting. Instructions for how to test a clay sample for shrinkage, glaze fit, firing temperature, vitrification and fired colour are in Appendix 4.

Clay body faults related to strength and durability can appear during forming or firing and can be remedied by substituting certain ingredients or changing their amounts. Common issues include low dry strength, brittleness, slumping, warping, bloating and cracking.

Bloating occurs from overfiring or fast firing. When organic, carbonate or sulphate materials decompose and burn out above 1200°C (2192°F) just as a clay body has become vitreous, off-gassing bubbles become trapped within the clay body.

Brittleness occurs when a clay body lacks vitrification from underfiring. To help strengthen a brittle clay body, increase flux content using feldspar for high fire and frit for low fire. Overfiring may also cause brittleness due to the over-development of glass in the clay body or lack of mullite which normally provides strength and support during firing.

Cracking can be caused by uneven drying and shrinkage. Clays need to be mixed with non-plastic ingredients such as fillers to open the body structure for better drying. Replace some plastic clay content with less plastic clay or increase non-plastic ingredients like silica or grog to reduce cracking.

Low dry strength occurs when a dry clay body breaks apart easily from lack of plasticity. To increase plasticity, substitute more plastic clays for non-plastic clays or non-plastic ingredients, replace fine, non-plastic ingredients (above 200 mesh) with coarser non-plastics. Plasticity will also be increased by wet mixing the clay body.

Slumping usually results from overfiring. To prevent slumping, decrease the flux quantity, eliminate soluble fluxes, substitute feldspar for other fluxes, or substitute kaolin for other clays. Slumping in raw clay can be improved by substituting coarser grade non-plastics for finer ones, adding sand or grog, or increasing plasticity by wet mixing clay.

Warping may occur from uneven drying and can be remedied by substituting kaolin for other clays, decreasing clay content, increasing non-plastics, using coarser grades of non-plastic ingredients, slower drying and avoiding movement during drying. To lessen warping during firing, decrease the amount of flux and substitute feldspar for other fluxes.

## GLAZE FIT AND THERMAL EXPANSION

When ceramics are fired, the clay body and glaze undergo thermal expansion. The individual ingredients in a clay body and glaze recipe determine the rate that they expand on heating and contract when cooled (Appendix 6). As they cool, the clay body and glaze need to contract at similar rates for the glaze to adhere well or 'fit' the clay body.

When the clay body and glaze contract at different rates, faults will appear in the glaze. If a fault is minor, it may be possible to fix it by adjusting the glaze recipe rather than the clay body recipe, since the volume is less. Other glaze fit problems may be remedied with modifications to a clay body.

If the glaze shrinks more than the clay body, small cracks called crazing will appear in the glaze as it stretches to fit the body. To remedy crazing, the thermal expansion of the clay body needs to be increased by one of the following: increase silica content in the clay body by adding very fine quartz particles; replace coarse silica with a finer mesh (200+); decrease the feldspar content; or reduce any flux containing sodium.

If the clay body shrinks more than the glaze, the glaze forms as a shell that can flake from the clay body surface, known as shivering. To remedy shivering, the thermal expansion in a clay body needs to be decreased by one of the following: decrease the silica content; eliminate silica as an ingredient in the recipe; replace fine silica with coarse silica; incorporate more than 20% feldspar in the clay body; increase overall flux content; or eliminate flux containing calcium or magnesium.

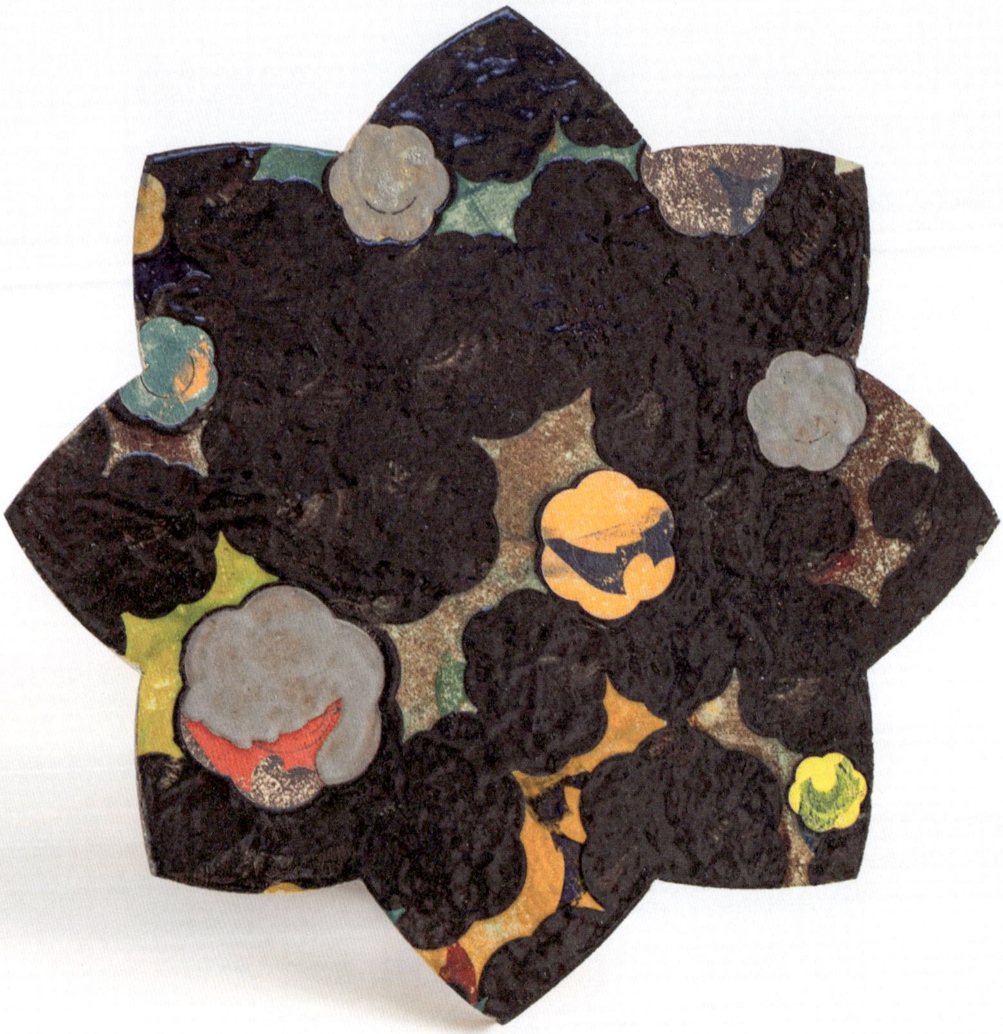

CHAPTER FOUR

# CLAY AS LIQUID

## SLIP

Clay made into a liquid state is called slip. Slip is a broad term that encompasses different formulations of liquid clay useful for a range of applications, from the most basic function of joining pieces of clay to creating surface design and casting clay forms. Clay slips are commonly applied to the surface of a clay body to alter its colour or texture, masking or changing the identity of a clay body. Slips are formulated to adhere to a clay body, or in the case of casting slip, to act as a clay body. The composition of slip varies according to its purpose, firing temperature, and surface quality.

Slips are more similar in composition to clay bodies than glazes because they contain a higher proportion of clay. The alumina content in clay creates higher viscosity and stability in slips than glazes. The refractory quality of alumina stabilizes the slip, preventing it from melting and moving on the surface of a clay body throughout the firing process. Even when formulated with increased fluxes and glass formers to make vitreous slips, such slips melt less than glaze. The stability of slip is advantageous for layering one over the other and for carving through layers. Casting slips are simply clay bodies that can be poured into plaster moulds to create clay forms.

Slips can be made from a single natural clay or from the same components as a clay body: clays, fluxes, glass formers, fillers and colourants. As a surface layer, slips need to fit onto clay forms that may already have undergone some shrinkage from drying. Therefore, the recipe for a clay body and the slip applied to it are rarely the same. Slips are usually formulated to have lower shrinkage than the clay body on which they are applied. Applying slips in thin layers or multiple thin layers rather than a single thick layer will reduce the likelihood of cracking or peeling. Casting slips require moderate rather than high shrinkage for the slip cast form to effectively replicate the mould shape then release without cracking.

**Opposite:** Sanam Emami, *Patterned Night Tile*, 2021. Coloured slips applied in layers with stencil designs. (*Image credit: Del Harrow*)

Slips are formulated for various purposes, effects and modes of application. Making a basic joining slip can be as simple as adding water to a clay body to make it liquid. Joining slip works well in conjunction with scoring a clay body to adhere moist pieces of clay together. Joining slip is compositionally the same as a moist clay body with extra water, like the slurry created from throwing on the potter's wheel. Simple slurry or slip made this way can be applied on freshly-thrown clay as a surface because their moisture levels are similar.

**Below:** Cary Esser, *Untitled Set 3*, 2014. Slipcast fritware cone 01. (*Image credit: E.G.Schempf*)

Shrinkage is related to the clay content in a slip recipe because clay shrinks as it dries. Since slips are applied after the clay form has already undergone some shrinkage, they need to be formulated to shrink less. To reduce shrinkage, slips generally contain lower clay content than moist clay bodies. Formulating a slip with an increased amount of non-plastic materials such as fluxes or fillers will reduce their rate of shrinkage. Likewise, substituting a portion of the clay in the recipe with another clay that shrinks less or calcining a portion of the clay in the recipe will reduce shrinkage. Calcining involves heating the clay to a sufficient temperature to remove the chemically combined water. Finally, a deflocculated slip shrinks less because it contains less water.

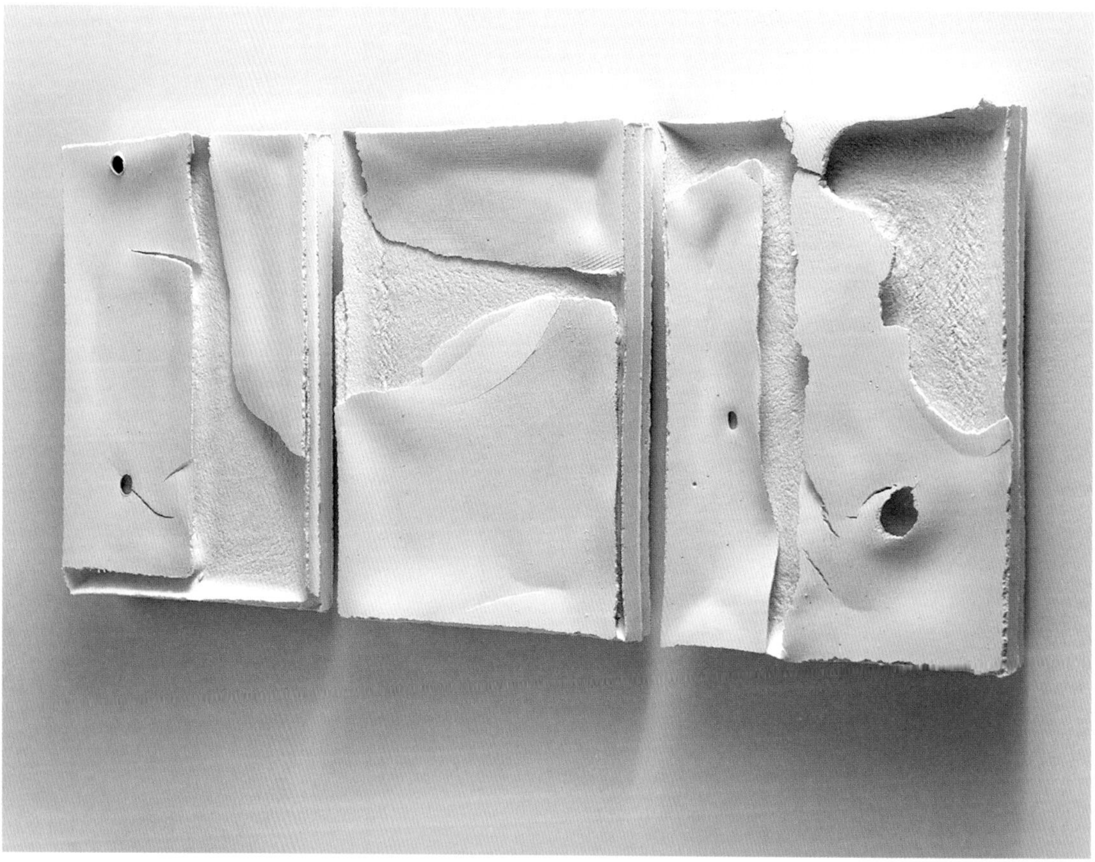

## DEFLOCCULATION

Viscosity is the resistance of a liquid, or in this case slip, to flow. An effective way to reduce the amount of water necessary to make clay liquid and less viscous is through deflocculation. Deflocculation refers to an altered liquid state in which clay particles repel one another while in solution. The natural state of clay is flocculation where clay particles are attracted to each other and stick together in clumps, or flocs. This inherent attraction comes from a dual electrostatic charge on clay platelets where the edges of clay exhibit a positive charge and the faces a negative charge. The positive and negative forces cling together through electrostatic force. In moist clay bodies, the clay particles aggregate in bunches or stacks.

Deflocculation in clay slips can be achieved with the introduction of a deflocculant into the mixture of clay and water. Another term for deflocculant is dispersal agent, or electrolyte. These substances act to disperse the clay particles by altering their electrical affinity. Deflocculants work by changing the electrical charge on the clay platelet edges from positive to negative. With both edges and faces negatively charged, the clay particles separate or deflocculate. This results in a state of particle avoidance such that the clay particles move with less water.

**Below and overleaf:** Jean-Pierre Larocque, *Small Horse*, 2018. (*Image credit: Bertrand Carrières*)

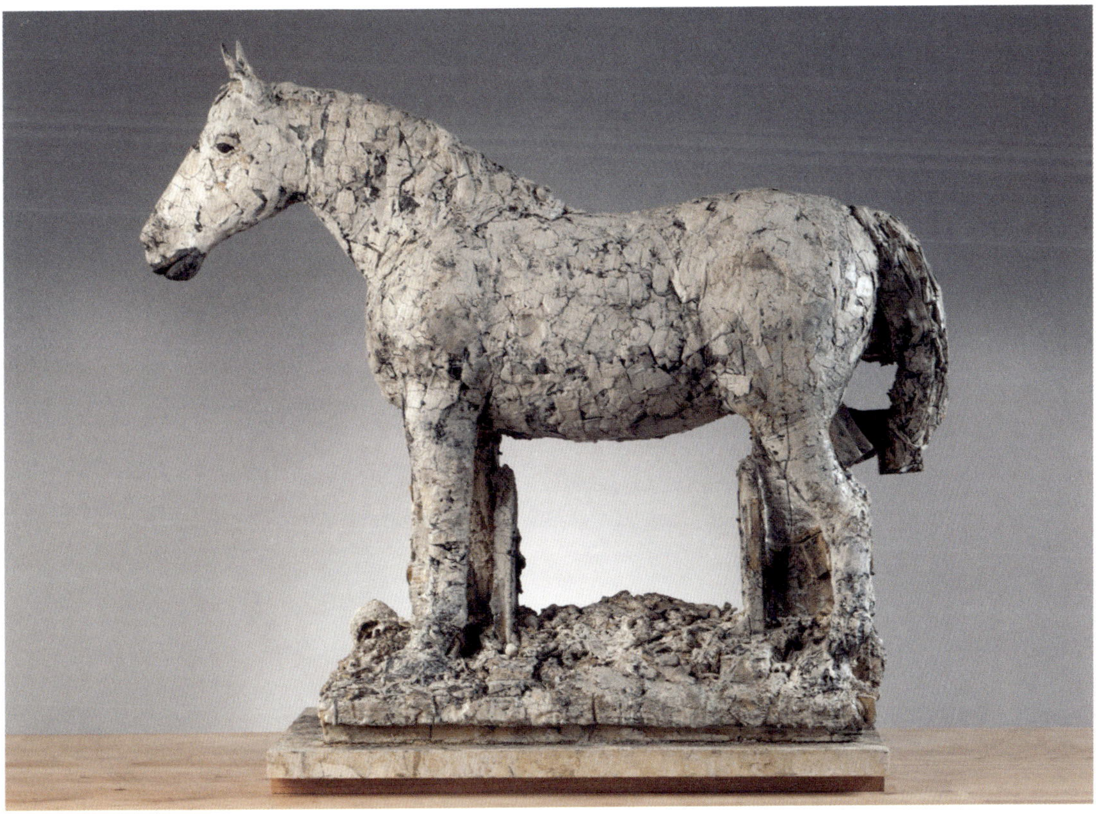

Deflocculation can be accomplished by lowering the acidity of the water mixed with clay with additions of alkaline materials. Soluble alkaline materials like sodium silicate can be used as a deflocculant. Not all slips are deflocculated, but the advantage of a deflocculated slip is that a smaller quantity of water is required to achieve a liquid state. Less water in a slip means reduced shrinkage and a lower incidence of cracking.

Deflocculant amounts should be calculated accurately since mixing excess deflocculant in a slip results in over-deflocculation, a quick increase in the viscosity of a slip and a re-solidification of the mixture. Sodium-based deflocculants have a strong effect on clay suspensions and include sodium silicate, sodium carbonate (soda ash), and sodium phosphate (Calgon). Polyelectrolyte deflocculants including Dispex (ammonium polyacrylate) and Darvan (sodium polyacrylate) have a gentler effect. Soda ash is used in small amounts as a secondary deflocculant with sodium silicate to help deflocculate materials with high organic content like some talcs but is not used in porcelain.

Flocculating liquid clay is the thickening of the mixture which results in increased viscosity. Acidity encourages flocculation and various acidic ingredients can be added to create temporary or longer lasting flocculating effects. Adding vinegar can flocculate a mixture for up to two weeks and gelatine for a few hours, whereas magnesium sulphate (Epsom salts) or calcium sulphate (gypsum) will create a longer lasting effect.

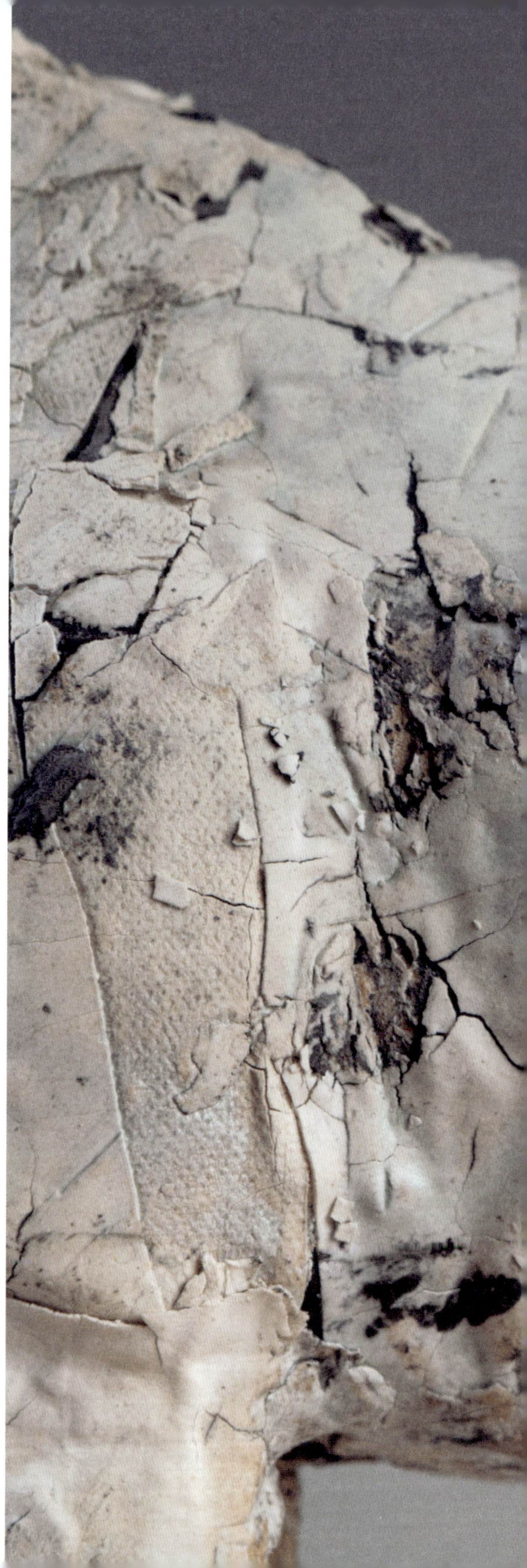

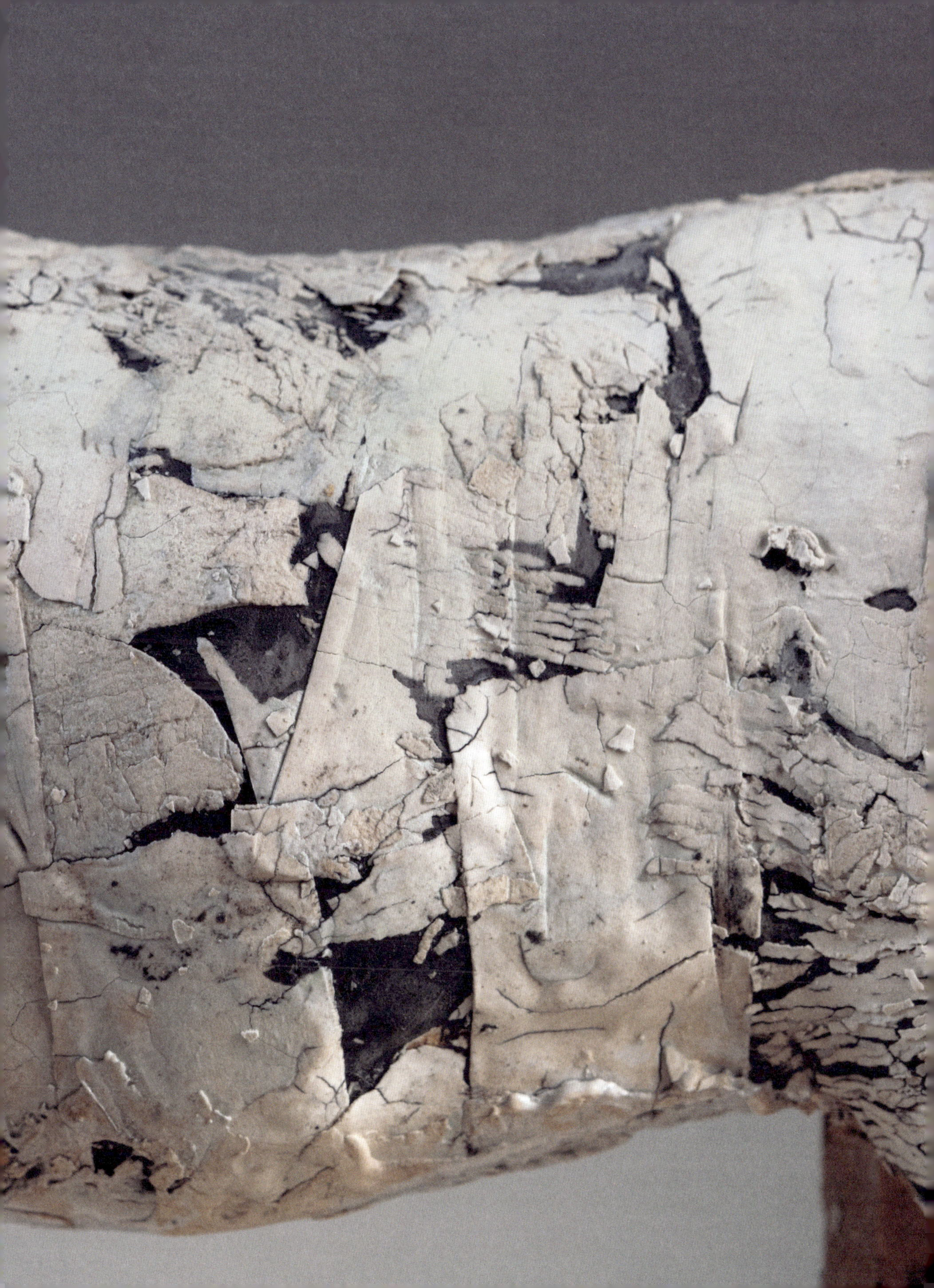

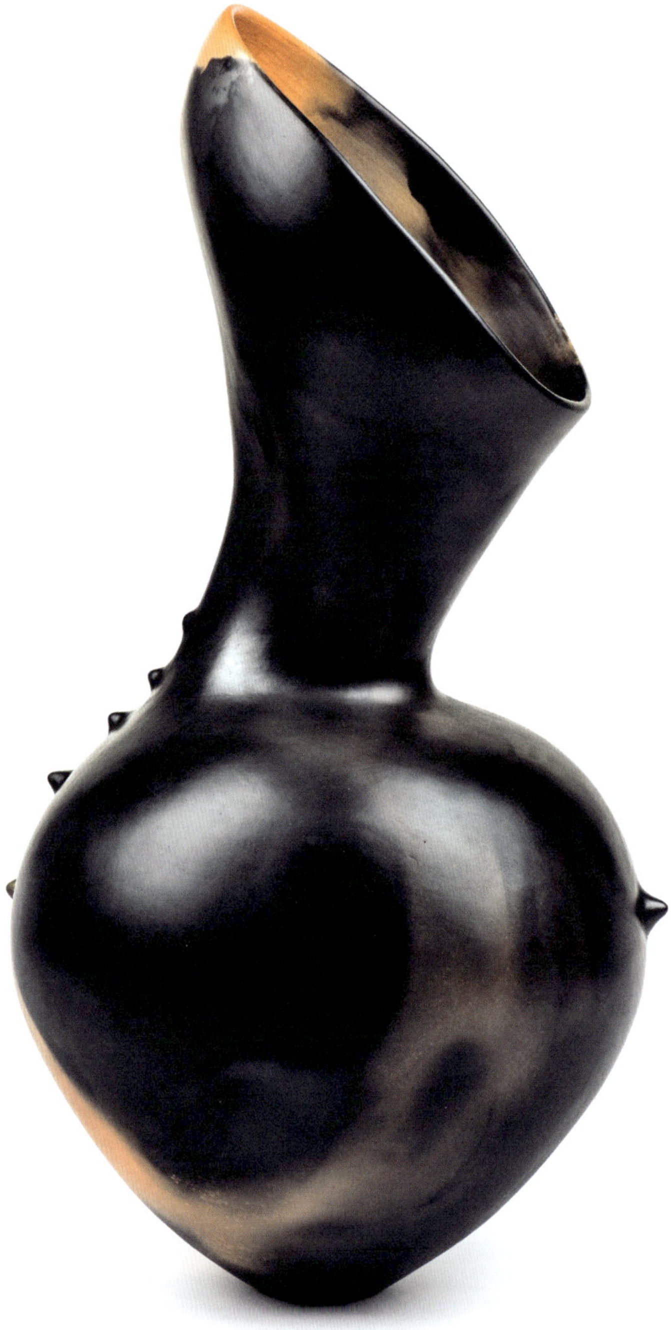

**Above:** Magdalene Odundo, *Untitled*, 2021. The effects of reduction are revealed in the terra sigillata on the surface of this vessel fired in a saggar in a gas kiln. (*Image credit: David Westwood Photography, © Magdalene Odundo*)

## TERRA SIGILLATA

Terra sigillata is prized for its smooth, waxy sheen resulting from the alignment of the very finest particles of clay. Originating in ancient times, this slip is made from clay, water and deflocculant. This simple substance relies on the proper proportion of these three ingredients to achieve deflocculation with excess water. This solution allows the finest clay particles (less than one micron) to float while the heavier clay particles sink with gravity. The distinction between the two layers becomes apparent over time as the heavy particles sink into a layer of sludge that is somewhat darker, thick and grainy in texture at the bottom of the container. The terra sigillata solution, or terra sig, is the top layer of suspended fine clay. The terra sigillata can be decanted from the top by pouring it off or siphoning it. About twice the amount of water is used to make terra sigillata than casting slip. This excess water encourages the heavier particles to settle whereas in casting slip all particles remain suspended.

Terra sigillata is an excellent surface treatment for revealing fine details, since the fine clay bonds closely to the clay body rather than pooling between details as thicker slips or glazes tend to do. Clays of various colours can be used to make terra sigillata and additional colouring oxides can be added to increase the colour range. Terra sigillata surfaces can also be scratched through with sgraffito, as on Greek vessels, to reveal the clay body underneath. Terra sigillata works optimally at low temperatures in the range cone 08–03. Above cone 1, terra sigillata may lose its sheen as it bonds further with the clay body underneath. Unlike other slips, glaze is not generally applied to a terra sigillata surface because it has difficulty adhering to the smooth and sealed surface.

**Notes on terra sigillata composition**

| | |
|---|---|
| **Clay** | Total clay 100%<br>Ball clay for light colours<br>Earthenware for darker colours |
| **Water** | Two times clay weight |
| **Deflocculant** | White clays 0.3%<br>Red clays 0.5%<br>Based on clay weight |
| **Colourants** | Oxides and stains 0.5-10% |

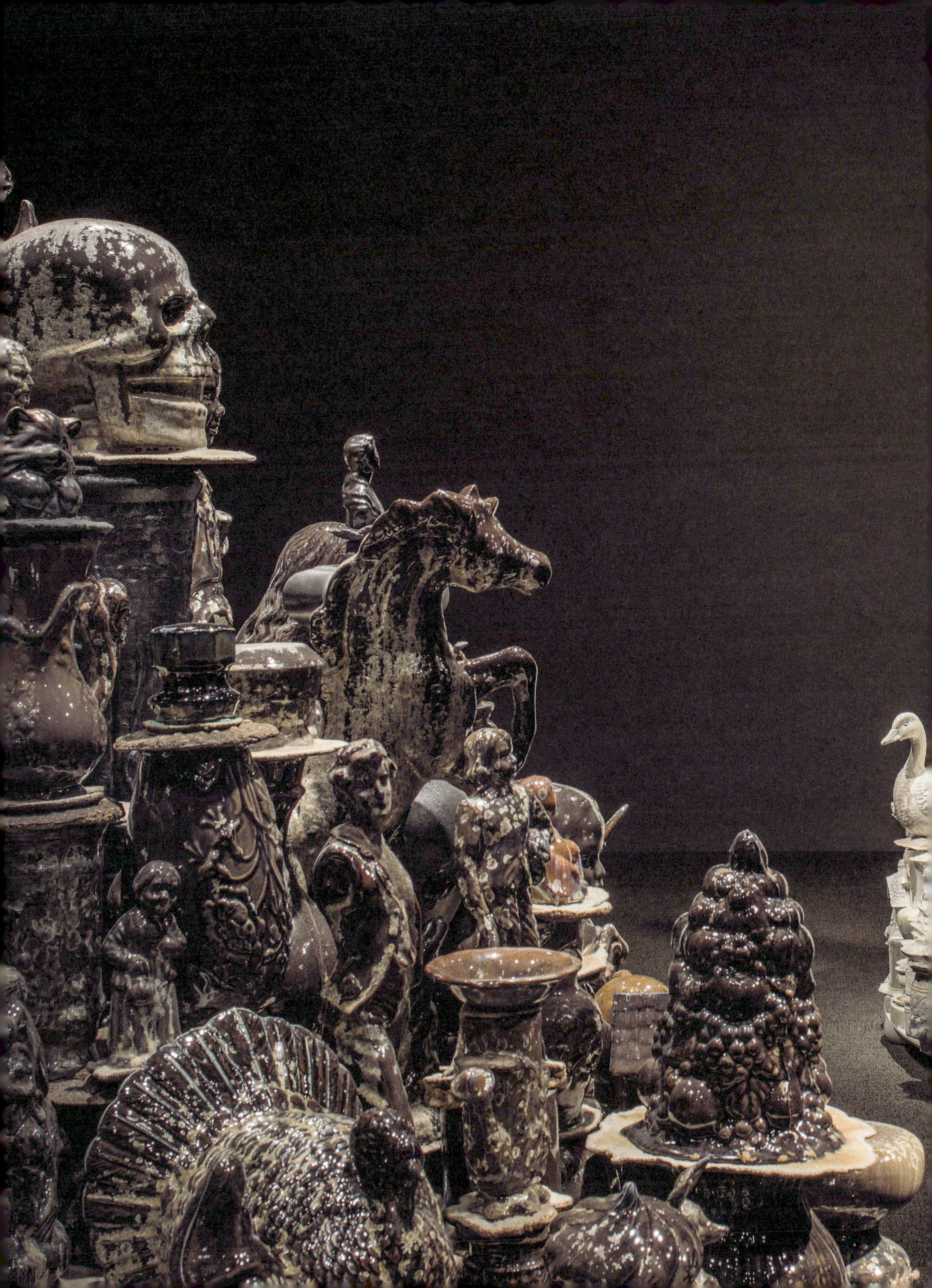

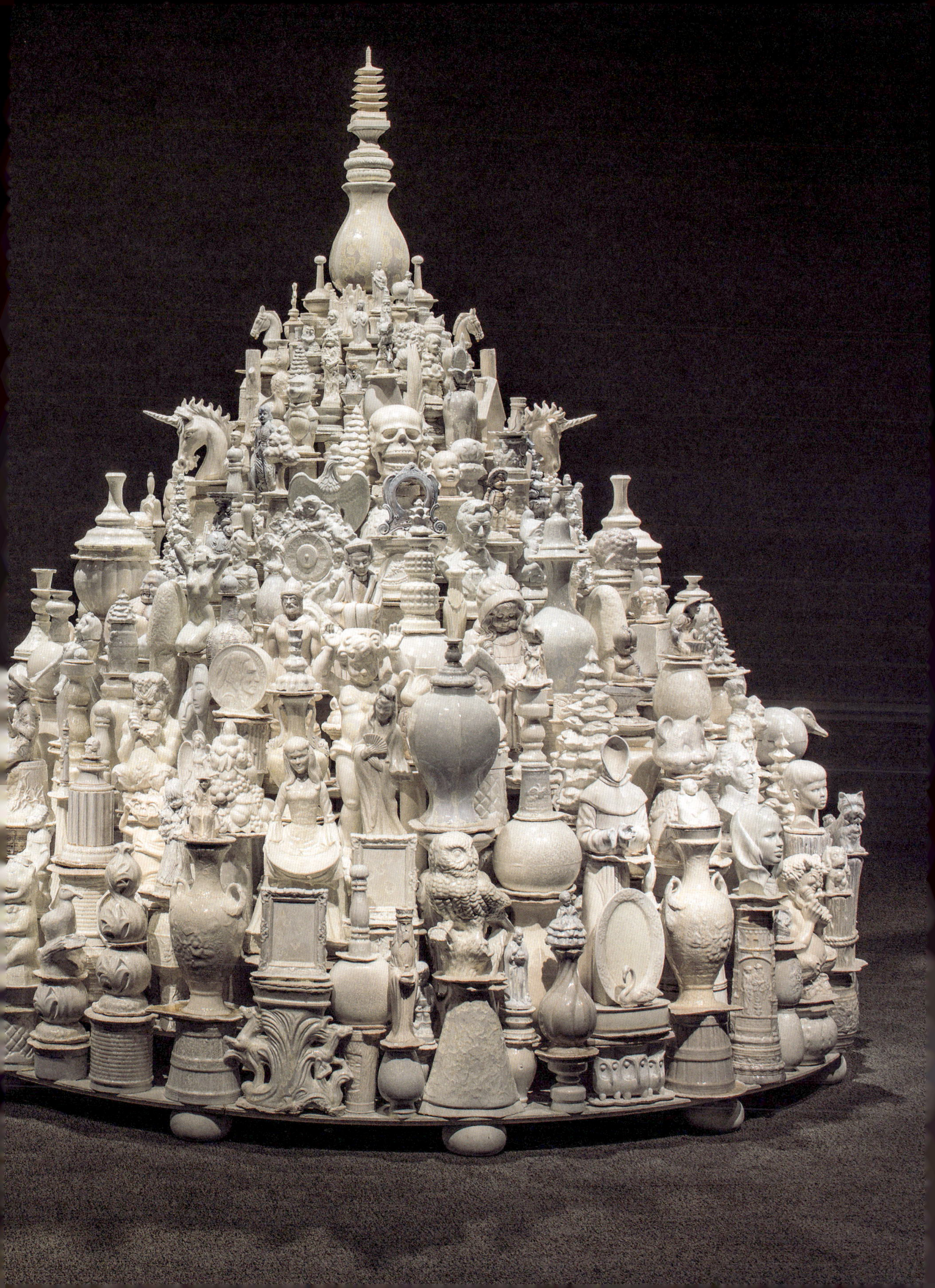

## CASTING SLIP

Casting slips are clay bodies formulated as liquid slips that can be poured into a plaster mould to create clay forms. Slip casting is possible due to the symbiotic relationship between wet clay and the drying effect of plaster which wicks away enough moisture for the clay to stiffen where it touches the mould. When slip casting, the plaster acts like a rigid sponge that absorbs water from the slip through its tiny pores, passing it through small capillaries to the outside of the mould where it evaporates. Once the slip thickens, the mould is emptied of the remaining slip and a cast shell of clay remains. As the clay continues to dry, the clay form shrinks slightly and releases from the plaster mould.

Deflocculation is the key to formulating a casting slip because less water is required for the slip to become fluid. Deflocculation combats the tendency of the clay particles to stick together, and in combination with just enough water will make the casting slip liquid. Moist clay bodies and casting slips usually have about the same amount of water, 40–50%, with somewhat similar shrinkage rates. In most cases, clay bodies can be converted into a casting slip simply by remixing the recipe with deflocculants (see Appendix 7).

Casting slips are composed of clays, fluxes, glass formers, fillers, and colourants, like other clay bodies. Slip casting bodies do not require plasticity, and therefore less plastic clays are more desirable when formulating a casting slip. Bentonites should be avoided in casting slips because they cause the slip to remain wet and soft in the mould, making it difficult to remove. Casting slips generally contain about 50% clay and 50% non-plastic ingredients like fluxes, glass formers, and fillers.

**Notes on casting slip composition**

| | |
|---|---|
| **Clay** | Total clay 50%<br>Kaolin or ball clay for white<br>Other clays for colour |
| **Fluxes** | Total flux up to 35%<br>High fire: potash feldspar<br>Mid-range: feldspar and frit<br>Low fire: frit or talc |
| **Glass former** | Silica/flint up to 25% |
| **Deflocculant** | 0.25–0.5% |
| **Water** | 40–50% of dry batch weight |

**Previous:** Walter McConnell, *A Theory of Everything: Dark Stupa & White Stupa*, 2016. Cast porcelain takes on the precise details of a myriad of commercial mould forms over which a zinc crystalline glaze plays on the surfaces. *(Image credit: Maxwell Makenzie)*

**Overleaf:** Jean-Pierre Larocque, *Small Heads Series*, 2018. Stoneware with engobes. *(Image credit: Bertrand Carrières)*

## SLIP DENSITY

A slip consists of solid particles of ceramic materials suspended in water. The more solid particles are in the suspension, the greater its density. It is useful to be able to measure the density of a slip to diagnose problems with deflocculation or consistently make the same mixture.

The density of a material is calculated by dividing the mass (g) by its volume (ml). For water the density is 1 g/ml. A thin suspension would have a density closer to water at 1 g/ml, and a thicker almost solid suspension would have a density nearing 2.5 g/ml. The density of a slip relative to that of water is often expressed as its specific gravity which is the number of times it is denser than water.

The specific gravity of a slip can be measured by weighing a small sample of slip and performing a simple calculation. Measure 100 ml of slip in a graduated cylinder and weigh the contents. Divide the weight of the contents by 100 to find the specific gravity. For example, if the slip weighs 170 g then dividing 170 by 100 gives us a specific gravity of 1.7.

Knowing the specific gravity of a casting slip can help to remedy many issues. Most casting slips work well with a specific gravity between 1.7–1.8. If your slip is below 1.7 add more dry materials and deflocculant to thicken the density of the casting slip. If the slip is above 1.8 add more water, not deflocculant, to thin the casting slip. Two signs of over-deflocculation in a casting slip are the quick formation of a skin on the surface of the wet slip before or during casting and becoming thixotropic or quickly gelling from the liquid state.

## SLIP OR ENGOBE

An engobe is a thin coating of clay slip layered on a clay surface to change its colour, texture or to apply a design. The terms slip and engobe are often used interchangeably. The term engobe is more commonly used in North America and denotes a more specialized slip formulated for application on wet, leather hard, dry, or bisque clay. In general, slips are made with 50% or more clay and engobes with 50% or less clay.

At various stages in the drying process, different painted and drawn effects are possible with slips and engobes using different methods of application, removal, and alteration. These effects include slip trailing, sgraffito, inlay or mishima, feathering, or mocha diffusion. Complete coverage with a slip or engobe can mask the clay body underneath. Slips with high clay content are opaque, while engobes with low clay content require an added opacifier such as Zircopax to reach opacity.

An engobe is more stable than a glaze, with its composition closer to that of a clay body than a glaze. Slips and engobes have a high viscosity due to their relatively high clay content and will not move, flow, or blur during firing as glazes do. Engobes usually have a very wide tolerance in composition, application, and firing range, although recipes may need adjustment between the extremes of low and high fire. Engobes should cover the underlying clay body with a suitably dense coating of the desired colour or texture, cling to the ware during shrinkage from drying and firing, harden at or below the maturing temperature of the clay form, and survive under a glaze coating without becoming distorted.

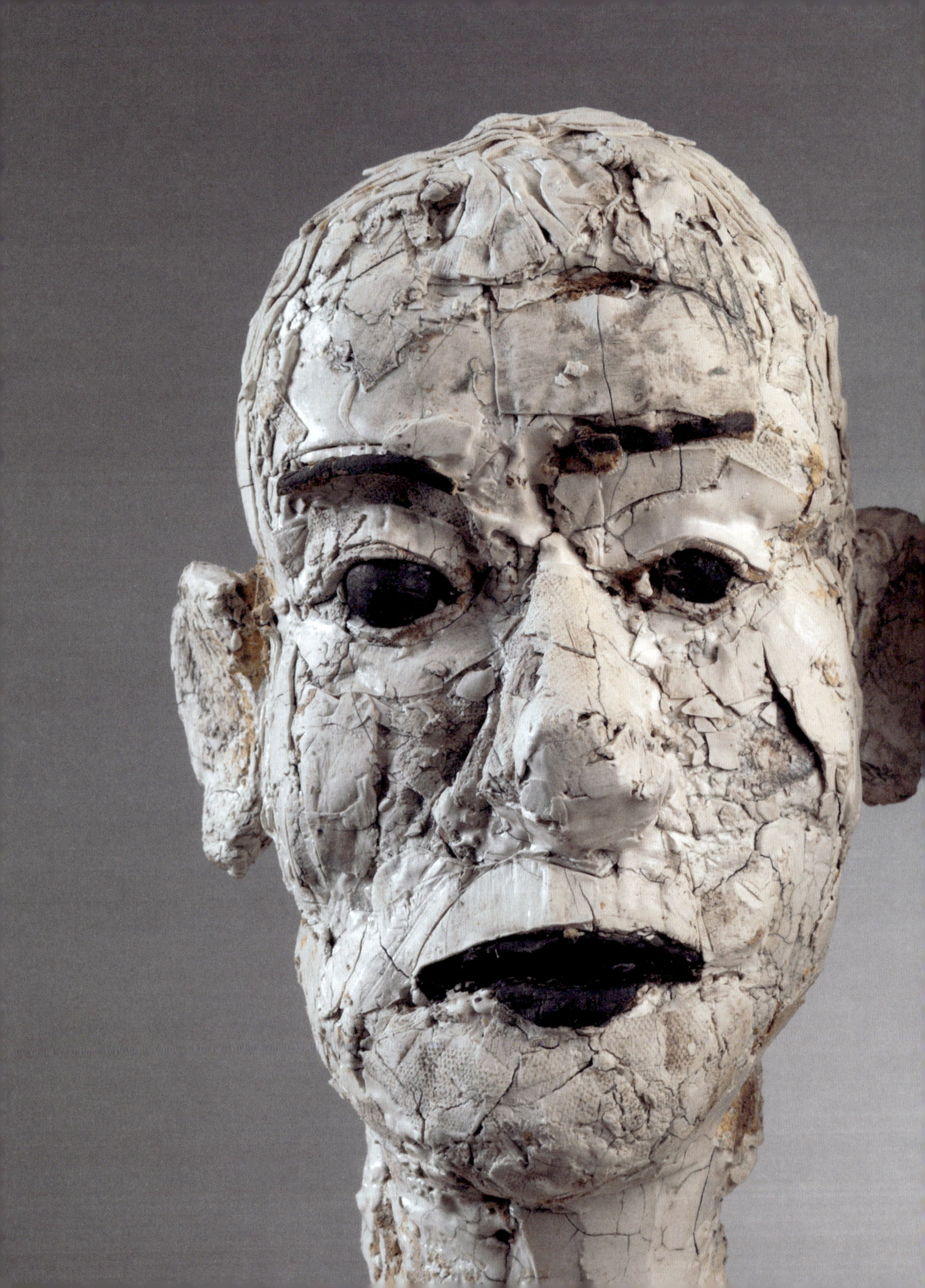

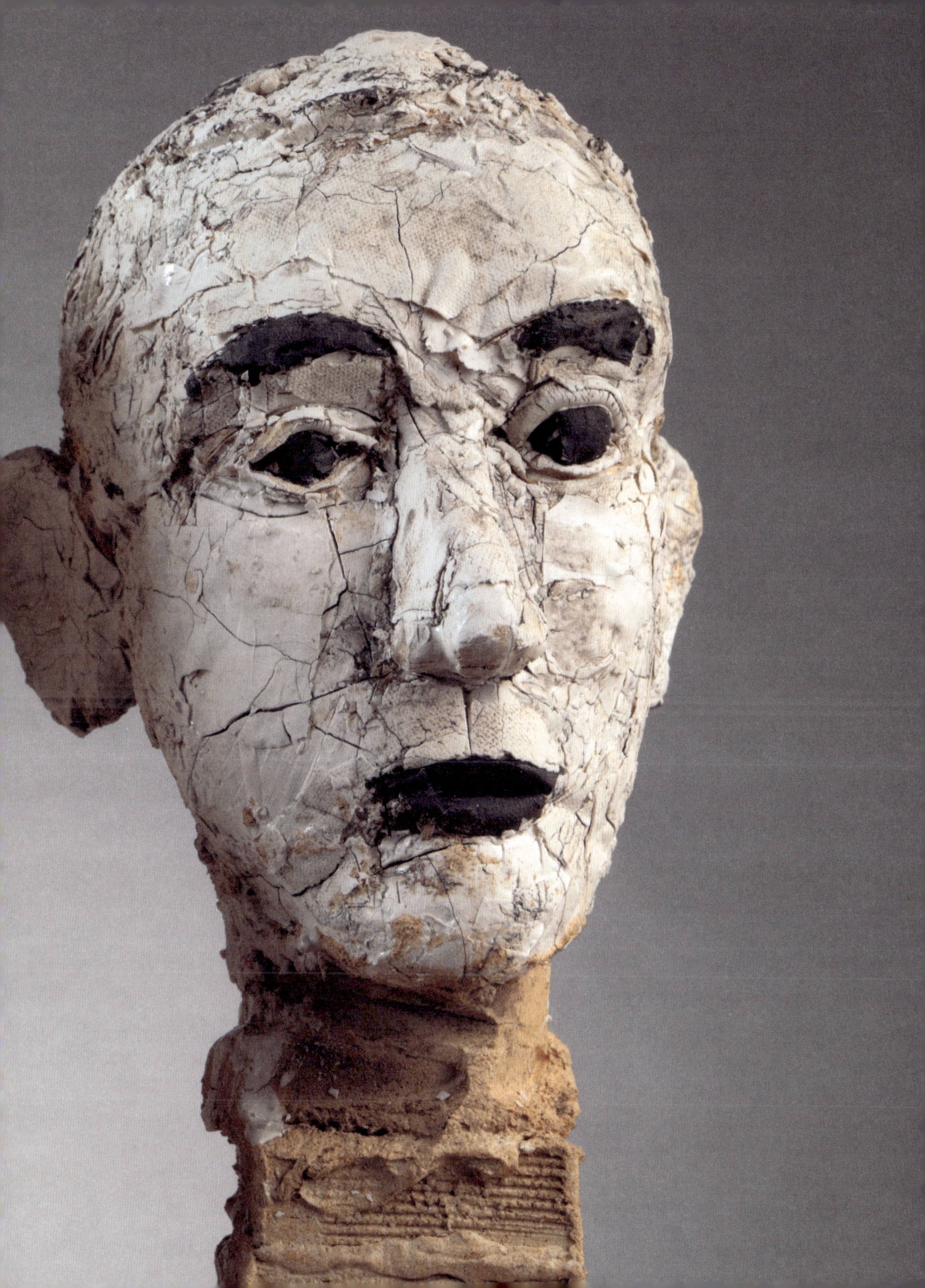

Engobes are formulated based on the timing of application in relation to the clay form. As a clay body dries, it shrinks. Usually applied to a partially dry, leather hard, or even a bisque-fired clay base which has already shrunk, an engobe must be reformulated with less shrinkage to match the shrinkage of the clay body underneath. To reduce the shrinkage of an engobe, the amount of clay or plastic ingredients is reduced, and the amount of non-plastic ingredients is increased. Small amounts of borax or soda ash can be added to slightly flux, harden, and deflocculate an engobe.

Clay is the plastic ingredient in an engobe recipe and the quantity of clay determines the overall shrinkage. An effective way to adjust an engobe for less shrinkage would be to replace up to 75% of the plastic clay content with calcined clay. Calcining clay is the process by which the chemical water in clay is driven off with heat. It is possible to calcine clay oneself by firing powdered clay to cone 010–08. Calcined kaolin may also be purchased from ceramic suppliers. Calcined clay can also be used to reduce crawling.

Apply thin coats of engobe just thick enough to cover the clay. Thick coats may flake, crack, or lift off. Engobes are applied to leather-hard

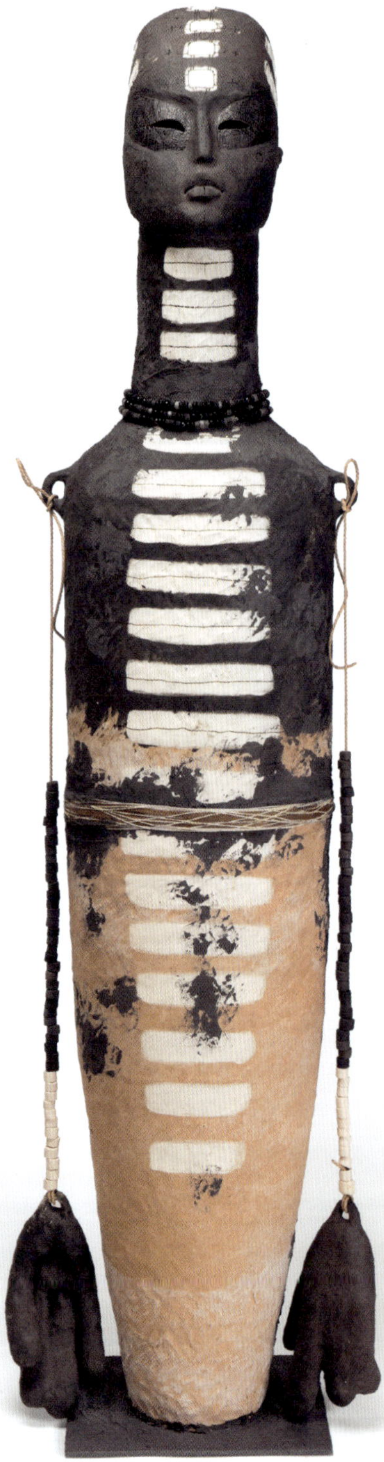

**Left:** Rose B Simpson, *Copilot A*, 2023. Painted with contrasting slips, this figurative sculpture addresses identity, feminism, the legacy of colonialism, racism, and capitalism. (*Image courtesy of the artist, Jessica Silverman, San Francisco and Jack Shainman Gallery, New York. Image credit: Addison Doty*)

clay by brushing, dipping, spraying, sponging or slip trailing with a squeeze applicator. While the surface of an engobe can be slightly more vitreous than a clay body, engobes tend to remain dry and matt. Engobes with enough clay content do not melt and can be used to cover the entire clay form including the base. Glaze may be applied over an engobe to seal the surface.

Engobes are an inexpensive alternative to colouring an entire clay body since they require a smaller amount of colourant. Textured engobes can be formulated with coarse clays or fillers such as sand. Engobe layers can be carved through or applied with stencils to expose the underlying clay or another engobe layer and can be pronounced by utilizing contrasting colours or textures.

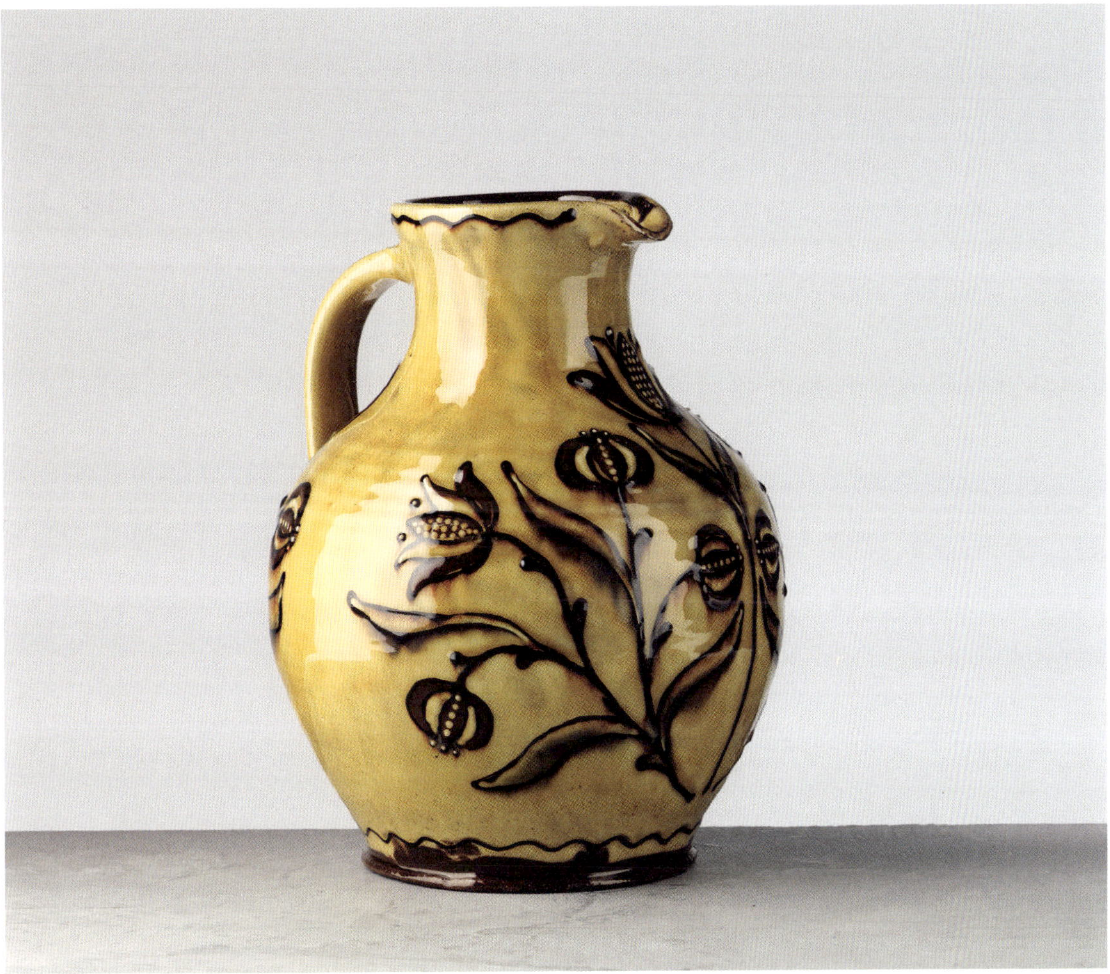

**Above:** Hannah McAndrew, *Fat-bellied Tulip Jug*, 2018. Traditional slip trailing techniques in a tulip design accentuate this fat-bellied red earthenware jug covered in white slip and lead glaze. (*Image credit: Shannon Tofts*)

**Overleaf:** Sanam Emami, *Patterned Tile*, 2021. Opaque engobes and glazes applied with stencilled designs create intriguing windows through which more layers are partially revealed. (*Image credit: Del Harrow*)

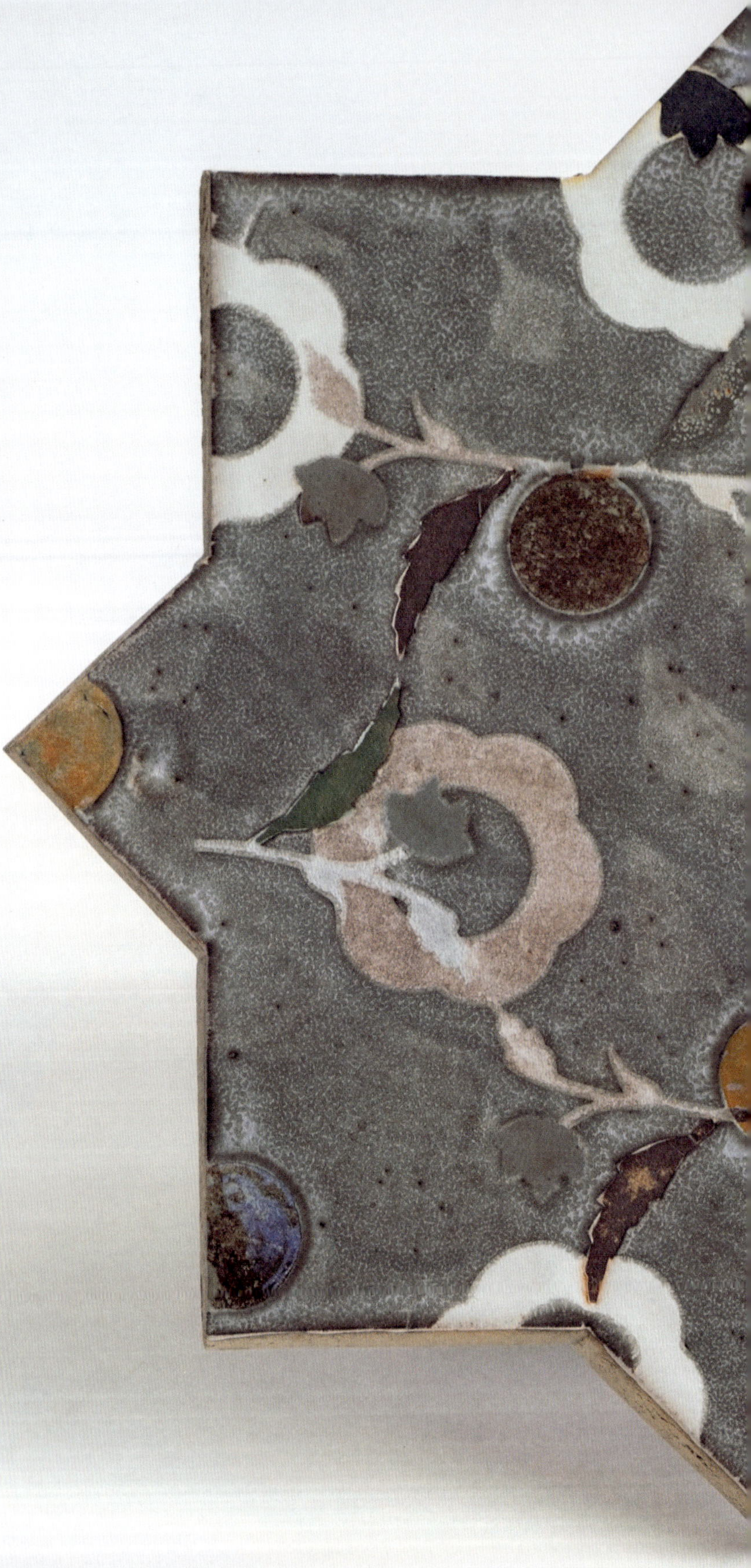

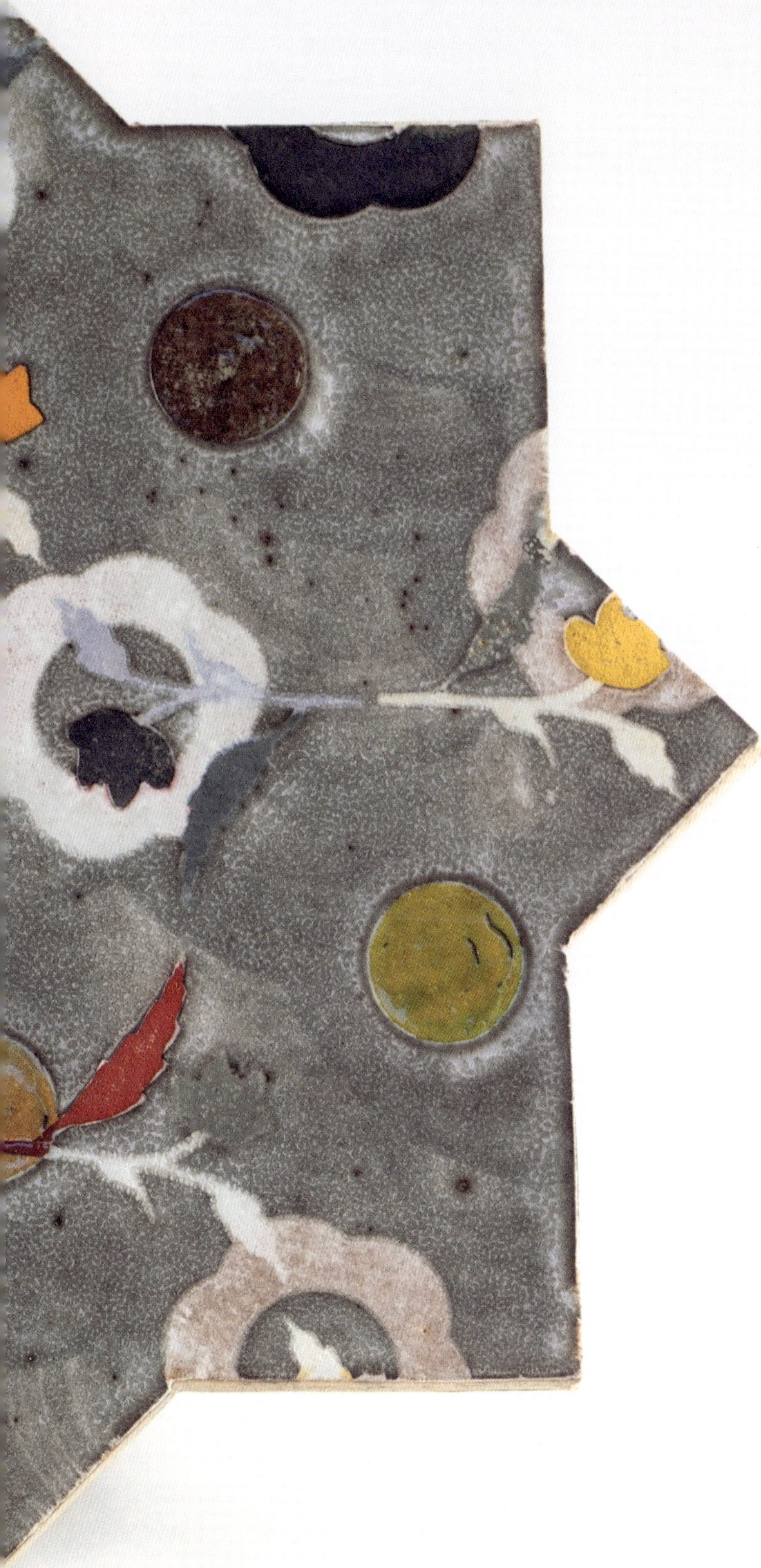

**Notes on engobe composition**

| | |
|---|---|
| **Clay** | Total clay typically up to 50%<br>Ball clay or Kaolin for whiteness<br>Ball clay higher shrinkage<br>Kaolin lower shrinkage<br><br>Decrease clay content for vitreous engobes |
| **Flux** | Total flux up to 40%<br>High fire: feldspar 20%<br>Mid-range: feldspar, frit and talc 30%<br>Low fire: frit and talc 40%<br><br>Fluxing increased in higher amounts with added silica for vitreous engobes |
| **Glass former** | Silica/flint up to 30% |
| **Hardener** | Borax up to 5%<br>Titanium 5-10% |
| **Opacifier** | Tin 5%<br>Titantium 5–10%<br>Zirconium 10–15% |
| **Colourants** | Oxides/Stains 0.5–10% |
| **Texture fillers** | Sand, grog, organics, etc. |
| **Deflocculant** | Dispex, Darvan, etc.<br>Low amounts reduce shrinkage |
| **Flocculant** | Vinegar or Epsom salts<br>Low amounts thicken slip |

**Adi Avidani engobe recipe:**

| | |
|---|---|
| China clay | 40 |
| Ball clay | 50 |
| Potash feldspar | 5 |
| Flint | 15 |
| + | |
| Stain: | 0.5–5% |

Underglazes are commercially-made engobes containing an organic brushing medium. Underglazes can be applied to greenware or onto bisqueware.

Flashing slips are sensitive to atmospheric conditions and respond with variations in colour to the flame path, and reduction in wood and salt kilns. Usually applied to greenware, flashing slips contain 50% or more kaolin clay. Flashing slips are also known as alumina slips with high alumina coming from the high kaolin content. Flashing slips produce interesting colour effects in the presence of sodium from the kiln or from flux. Sodium will produce orange to red with iron in the clay, calcium makes yellow-green, titanium in the presence of iron leads to opalescent effects, zinc with iron can make greens, and the presence of magnesium softens all colours.

**Right:** Roberto Lugo, *Prisoner Sequence I, Orange and Black series Amphora*, 2022. Black underglaze decorative motifs depict stories of contemporary social and political oppression. *(Image credit, Tiffany Smith. Courtesy of R & Company)*

**Overleaf:** Adi Avidani, *Dissolving Lines*, 2024. The slip cast stoneware elements are smoothed, dry sanded, and sprayed with a few layers of engobe before bisque firing to cone 04, then resprayed and fired to cone 5. *(Image courtesy of the artist)*

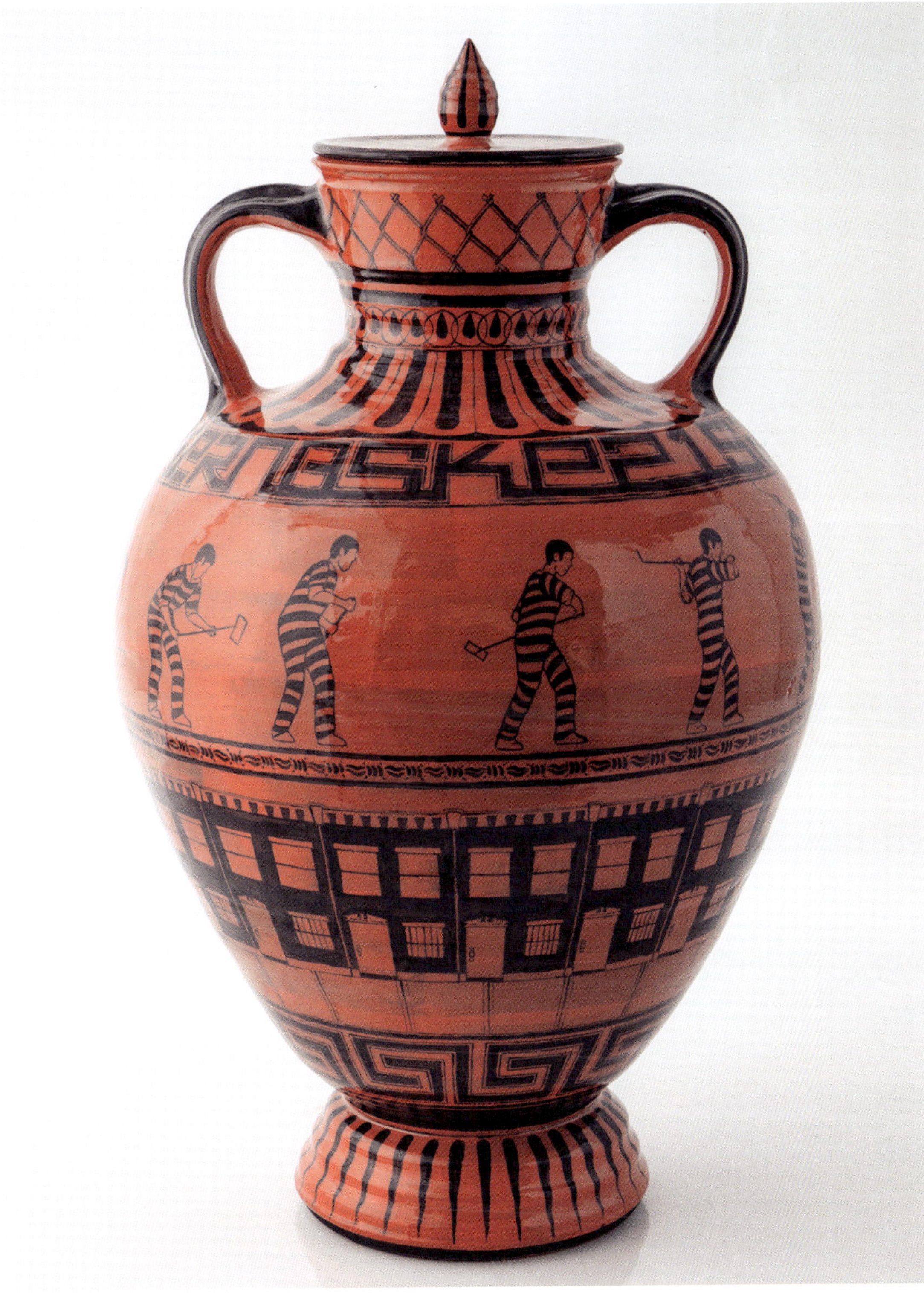

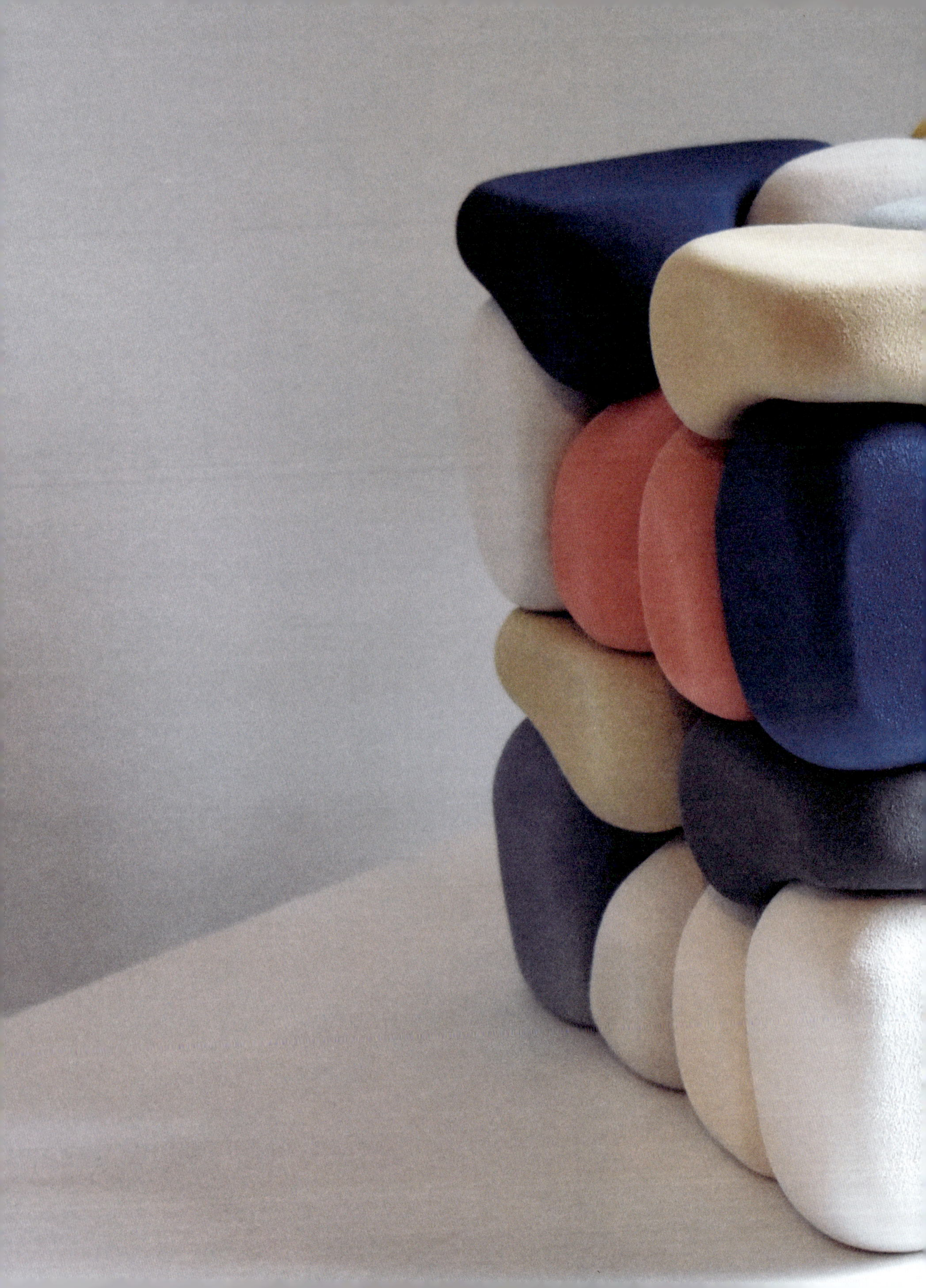

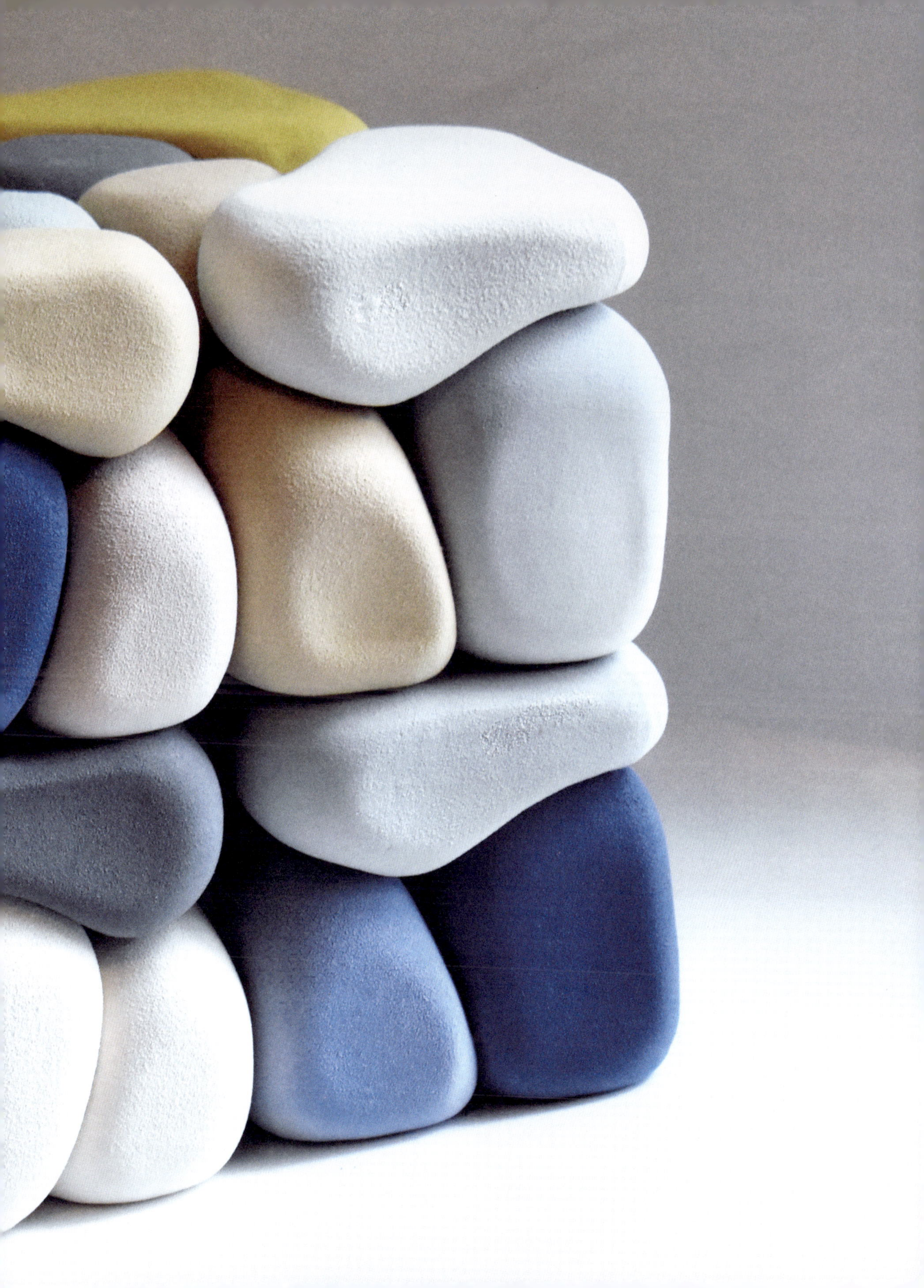

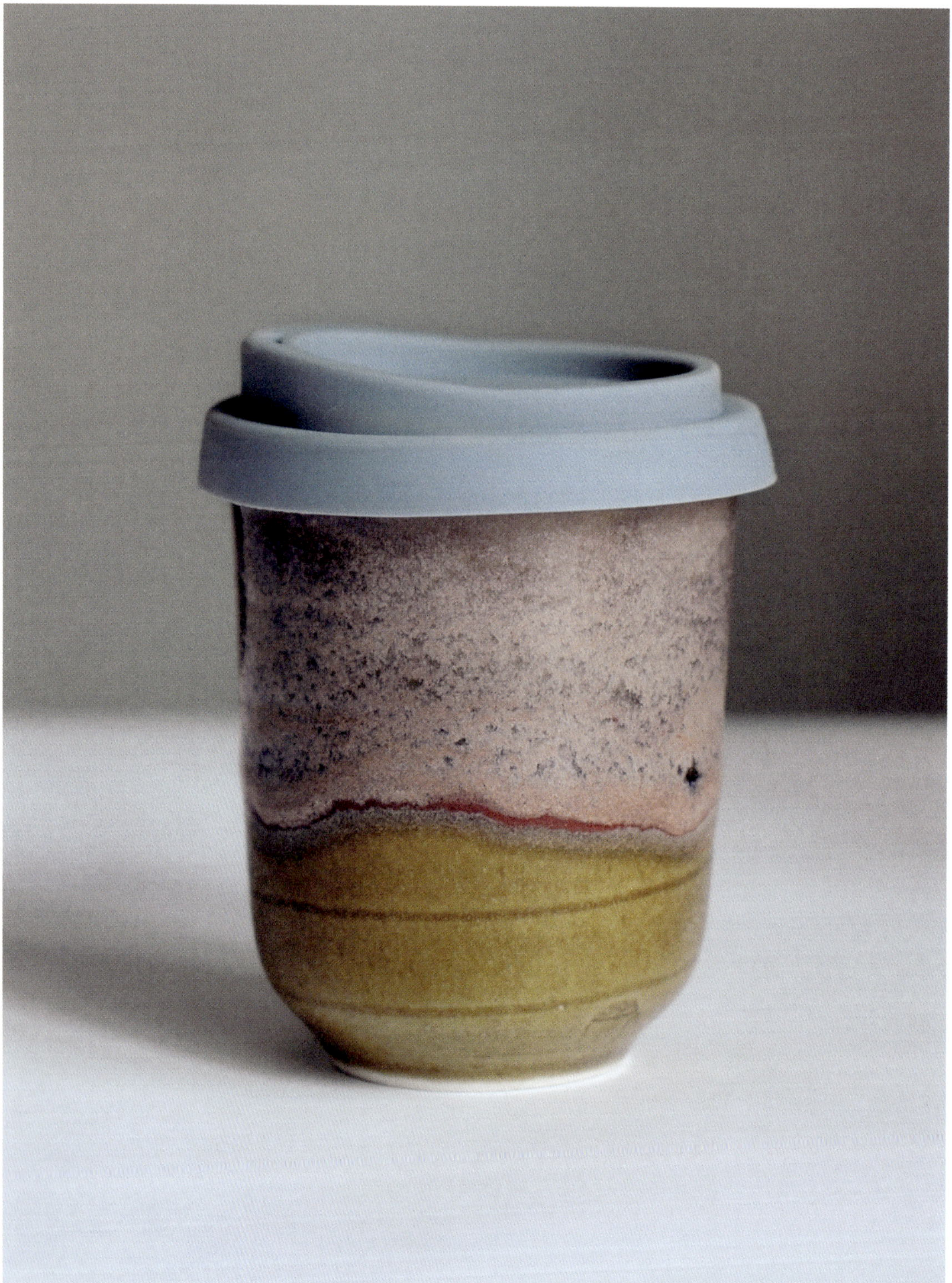

## LOCAL CLAY

Making slip out of local clay, is usually the first step of processing it into a workable substance. Once water is added to a local clay to create a slip or slurry, it can be easily poured through screens or sieves to remove larger particles and organic matter. Afterwards, the slip can be dried out to the proper consistency to make slips, clay bodies and glazes (see Appendix 3).

Because harvesting local clay is a laborious undertaking, one way to make the most of it is as a surface treatment. This way, a small quantity of local clay will extend further than the larger volume required to make a clay body. In addition, some natural clays that are unusable as clay bodies due to lack of plasticity or from impurities, may work perfectly well as a slip. As a surface, the naturally occurring variations in local clays can add interest and a unique character to each piece. To begin to test local clay as a slip, sieve the slip through a 60 mesh sieve then dip a greenware tile in the slip and fire. Modify with flux and glass formers to adjust the fit to the clay body as needed.

Local clays can also create excellent terra sigillata though their deflocculant requirements vary (see Appendix 7). Plastic clays with fine particles will yield more terra sigillata than coarse clays. Ball milling the initial mixture to further reduce clay particle size may yield a larger quantity. Red clays require a higher amount of deflocculant than white clays. To determine the proper amount of deflocculant, make a test batch of several hundred grams. To determine the amount of deflocculant for any particular local clay, add deflocculant gradually to the slip in small increments, until one of the mixtures creates more terra sigillata than the others. Natural clays are highly variable, and each individual batch may have slightly different requirements.

Certain clays are given the designation of a slip clay because they naturally contain a high quantity of fluxes or colourants that act as fluxes and reduce their melting temperature. These clays such as Albany slip melt into a glaze when fired to high temperatures and can easily be used as as glaze for single firings.

**Opposite:** Yuliya Makliuk, 2023. Cup with wild clay slip and commercial glaze. *(Image credit: Yuliya Makliuk)*

## EGYPTIAN PASTE AND SOLUBILITY

Egyptian paste is a self-glazing clay body that contains more flux than typical clay bodies. It is mixed as a liquid slip and dried to the consistency of a mouldable paste. Egyptian paste recipes are low in clay content. This creates an open, porous structure that permits the migration of soluble ingredients to the surface. The bulk of ingredients in an Egyptian paste recipe are fluxes high in alkaline (sodium, potassium, and lithium) oxides. For the self-glazing effect to happen, soluble salts, such as soda ash, sodium bicarbonate, potassium carbonate and lithium carbonate, need to be dissolved in warm water and mixed with small amounts of clay and silica. As the mixture dries, the salts migrate to the surface to create a powdery, salty scum on the surface. The scum melts during firing to create a glassy surface over a lightweight core. Because there is little clay, the mixture lacks plasticity and forms made of Egyptian paste are usually simple and often made in moulds. The usual temperature range for firing Egyptian paste is between cone 012 and cone 04 in oxidation. Egyptian paste becomes brightly coloured with the addition of colourants and the iconic aqua blue is created with copper oxide or copper carbonate.

Egyptian paste is a good example of how soluble fluxes can alter materials, as the soluble salts move out of the clay body to the surface. The glassy coating that forms on the surface of Egyptian paste is a result of the salts recombining with silica during firing. Whenever soluble ingredients are used in either a clay body or glaze, their composition may change as migrating salts can lead to unpredictable results. Solubility may cause uneven drying and fluxing with too much flux in some areas and not enough in others. Because of this, soluble materials tend to be avoided, except when such effects are desired. Issues of solubility are eliminated when the same material is used in a non-soluble form such as in frits.

**Notes on Egyptian paste composition**

| | |
|---|---|
| **Clay** | Total clay up to 20% – white clays best<br>Bentonite up to 3% |
| **Flux** | Total flux up to 50%<br>Nepheline syenite, feldspar, frit<br>Minimum 10% soluble salts, typically sodium bicarbonate and soda ash |
| **Glass former** | Silica/flint up to 40% |
| **Filler** | Silica sand optional |
| **Colourants** | Oxides/stains 0.1–10% |

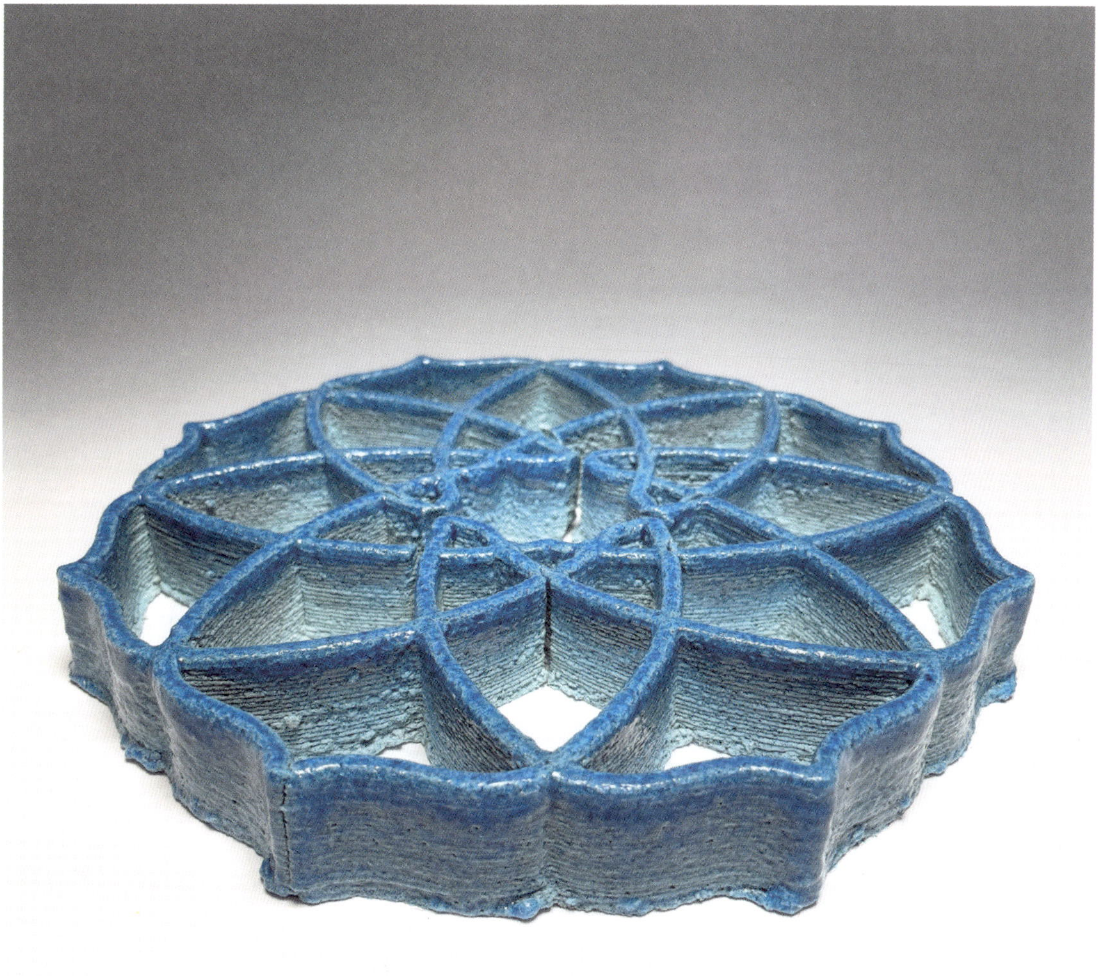

**Above:** Lesley Baker, *Crossover*, 2023. Copper with sodium creates the characteristic aqua blue colour of Egyptian paste. Utilizing 3D printing technology may bypass some of the difficulty in forming this non-plastic material. (*Image credit: Lesley Baker*)

CHAPTER FIVE

# GLAZE

A glaze is a thin, glassy coating used to cover the clay surface. In its unfired state a glaze is a liquid suspension of mineral particles in water. The glaze is applied to the dry or biscuit-fired clay piece, which absorbs the water and leaves a layer of powdery glaze on the surface. During firing, the glaze melts into a molten, glassy substance. Once fired, the glaze is similar in composition to glass with the addition of clay to make the glaze less fluid in the melt.

There are several benefits that the addition of clay brings to a glaze:

1. Clay brings in alumina which acts as a stabilizer/ stiffener to keep the glaze from flowing off the piece when it melts in the hot kiln.

2. Clay keeps the glaze minerals suspended in water in the glaze bucket before the glaze is applied.

3. Clay contains silica and alumina in combination. Ingredients that already contain multiple materials, rather than purified ones, melt more readily and combine with the other glaze materials.

**Opposite:** Noelle Hoover, Single Tooth Productions, stoneware bowl with underglaze and drippy glaze. *(Image credit: Noelle Hoover)*

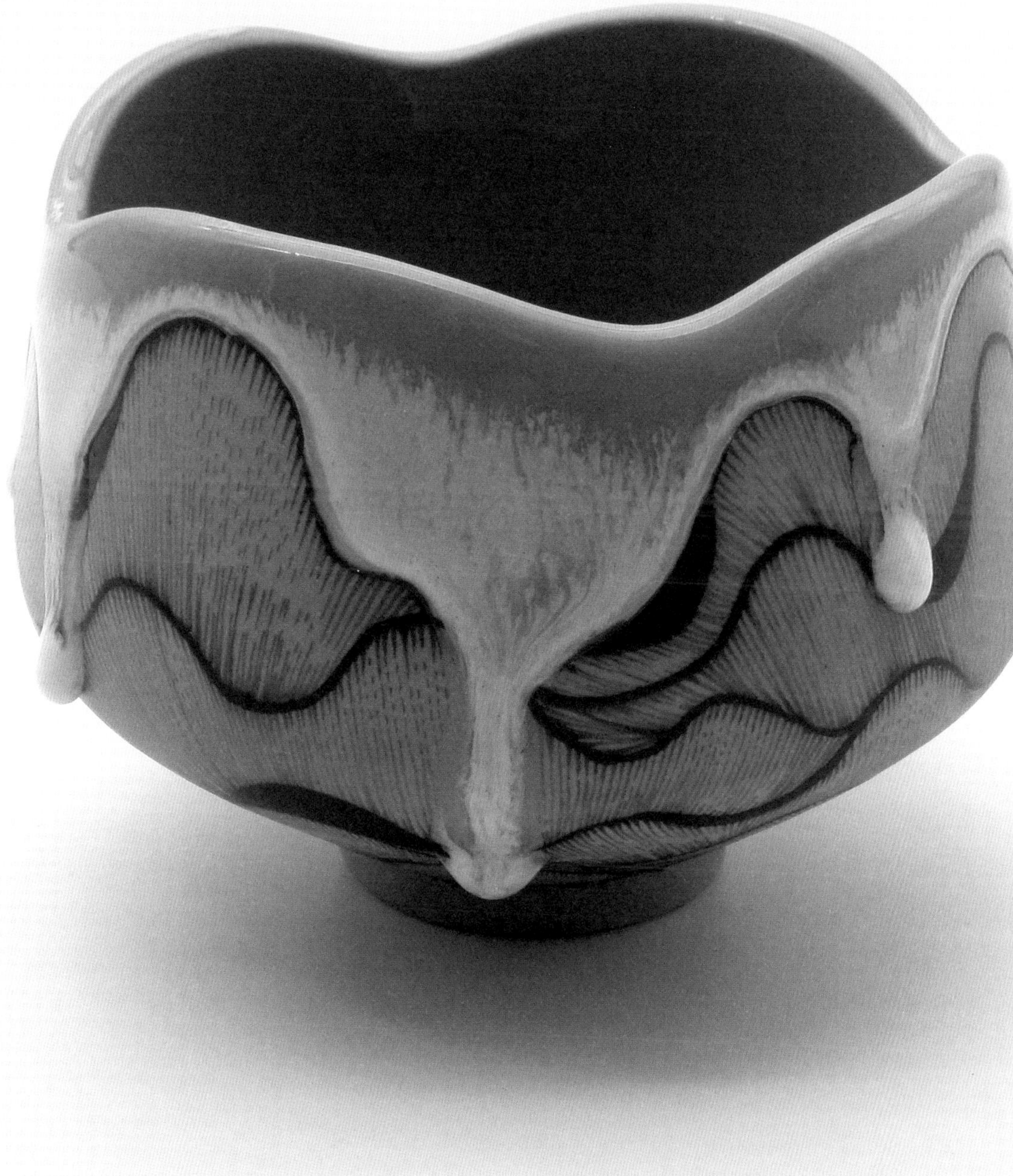

**CLAY BODY GLAZE**

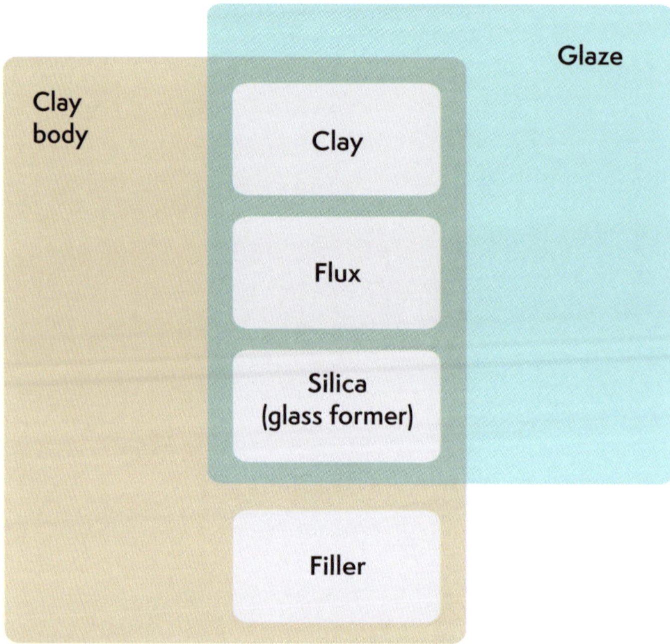


## CLAY VERSUS GLAZE

Clay bodies and glazes are related chemically. They are made from the same ingredients; clay, flux and silica, combined in different proportions. Clay bodies have more clay, while glazes contain more flux. In addition, clay bodies can contain fillers to aid even drying and prevent cracking.

As shown in the diagram above, a glaze is made up of clay, flux and silica. The clay component is further made up of silica and alumina so glazes are comprised of three components: silica, alumina and flux. These can be categorized according to their function; silica is a glass former, alumina is a stabilizer and fluxes help to melt the glaze.

Glass former – some silica comes from the clay added to the glaze but more silica is needed to create glass.

Stabilizer – alumina comes from clay; we rarely use pure alumina in glaze.

Flux – various minerals help melt alumina and silica to form glaze.

Silica and alumina have very high melting points. However, when they are combined in the right proportions, together with fluxes, the melting point is much lower. The ratio of two or more materials that has the lowest melting point is known as a eutectic. This is an important concept in glaze

PHASE DIAGRAM

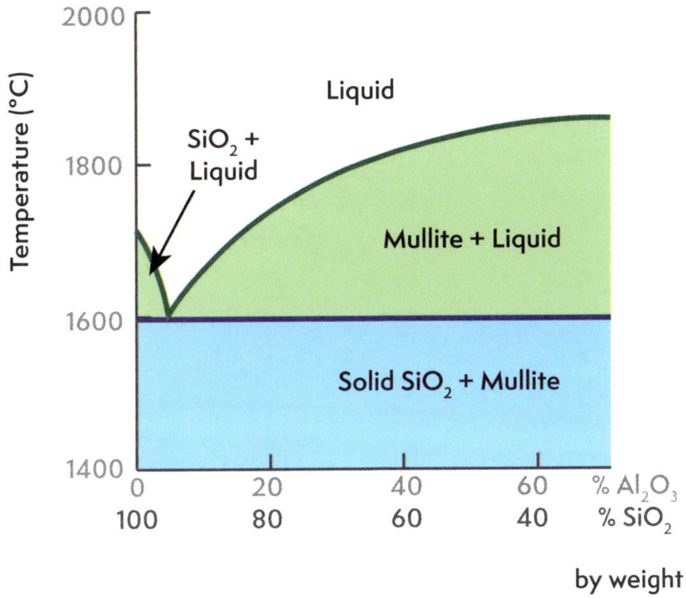


**Above:** Phase diagram for silica and alumina. We can see the eutectic ratio (lowest melting point). The green areas are mixtures of liquid plus solid crystals. The blue area is solid silica plus alumina in the form of mullite. Mullite is a crystalline compound of alumina and silica: $3Al_2O_3.2SiO_2$. At the eutectic composition the solid melts completely at the lowest possible temperature. Either side of this eutectic point melting takes place more gradually over a range of temperatures.

composition. It means that by combining two or more materials in the right proportions, the melting point can be lowered. Most glazes are made by combining silica, alumina and flux in the proportions at, or close to, this eutectic point. Silica melts at 1710°C (3110°F) while alumina melts at the even higher temperature of 2050°C (3722°F). However, if they are combined in the ratio 1 alumina to 8 silica (by number of molecules in the molecular formula, see Chapter 7), the melting temperature is reduced to 1545°C (2813°F). This particular ratio 1:8 is the eutectic, or lowest-melting combination of silica and alumina. Adding fluxes will reduce the melting temperature still further as the fluxes introduce eutectic mixtures of three or more materials that melt at even lower temperatures.

In the next section we will look at the fundamental oxides that make up glazes and how they are categorized by their function as glass former, stabilizer or flux. These basic oxides are supplied by various powdered rocks and minerals, each of which may contain several oxides.

## GLAZE OXIDES

Most glazes contain silica, alumina and at least two flux oxides, an alkali metal such as sodium and an alkaline earth such as calcium, which together bring down the melting temperature and combine with the silica to make the glaze chemically stable. The alkali metal oxides have the greatest fluxing power, while the alkaline earth oxides have a greater effect on the colour and surface texture of the glaze. There are two other glass formers besides silica: boron and phosphorus. These are always used in combination with silica.

13 basic oxides that make up a glaze:

Alumina

Silica
Boron oxide
Phosphorus oxide

Sodium oxide
Potassium oxide
Lithium oxide

Calcium oxide
Magnesium oxide
Barium oxide
Strontium oxide

Zinc oxide
Lead oxide

This list of oxides used in glazes can be divided into three groups: stabilizer (alumina), glass former (silica), and fluxes.

### 1. Stabilizers $R_2O_3$

Alumina ($Al_2O_3$)

Stabilizers have the general formula $R_2O_3$ where R is the metal and O is oxygen. The exception is the oxide of boron which has the formula $B_2O_3$ but is a low-temperature glass former, not a stabilizer.

### 2. Glass formers $RO_2$

Silica ($SiO_2$)

Glass formers have the general formula $RO_2$ where R is the semi-metal and O is oxygen. Exceptions to this rule:

Boric oxide ($B_2O_3$)

Boron is an exceptional oxide that is both a glass former and flux.
Boron is usually added in the form of a frit to bring the glaze melting temperature down from stoneware to mid and low temperatures.

Phosphoric oxide ($P_2O_5$)

Phosphoric oxide is a less commonly used glass former.
Phosphorus is added in the form of bone ash or wood ash as a flux in glazes and to make bone china.

### 3. Fluxes $R_2O$, RO

The fluxes are subdivided into the primary fluxes – the reactive alkali metals that have the chemical formula $R_2O$ – and the secondary fluxes, the less reactive alkaline earths with the formula RO, where R is the metal and O is oxygen. Zinc and lead also act as fluxes and are known as metallic oxides, also with the formula RO.

| **Alkali metal oxides $R_2O$** | **Alkaline earths RO** |
|---|---|
| $Li_2O$ Lithium oxide | MgO Magnesium oxide |
| $Na_2O$ Sodium oxide | CaO Calcium oxide |
| $K_2O$ Potassium oxide | SrO Strontium oxide |
| | BaO Barium oxide |

| **Metallic RO** |
|---|
| ZnO Zinc oxide |
| PbO Lead oxide |

In glazes, insoluble materials are preferred, as soluble ones will dissolve in water and be absorbed into the clay body, causing problems. Each oxide used in glazes has a range of properties which affect the final glaze:

Melting point: this gives an indication of how a material will melt in the glaze, although its eutectic point with silica is a more accurate indication of the actual melting temperature of the glaze.

Volatility: some materials are volatile at high temperatures and the fumes may spread to other pieces in the kiln.

Thermal expansion: the thermal expansion of a glaze material affects how well the glaze fits the clay body.

Viscosity: the viscosity is the measure of the fluidity of the glaze in the melt.

Surface tension: this affects the thickness of the glaze layer once melted. High surface tension causes glaze droplets to form spherical beads.

Hardness: in glazes for functional ware, hardness and scratch resistance are useful properties.

Toxicity: an important consideration, both during use of the raw materials and leaching from the fired glaze.

Colour response: the oxides in the glaze, particularly alumina and zinc can affect the glaze colour.

## ALUMINA

**Sources:**

- Clay
- Kaolin, china clay
- Ball clay
- Bentonite
- Alumina hydrate
- Aluminium oxide
- Kyanite
- Pyrophyllite
- Feldspar
- Frit

Alumina melts at the extremely high temperature of 2050°C (3722°F) and is known as a refractory, a material that withstands melting in the kiln, although it will melt at lower temperatures as part of eutectic mixtures. Alumina makes a glaze more viscous in the melt and prevents it from running off the pot. It is usually added to the glaze in the form of clay (kaolin, ball clay or bentonite) which has the additional function of keeping the other materials suspended in water in the glaze bucket. Alumina contributes high surface tension which makes it an essential ingredient of shino and gloop glazes. Owing to its high drying shrinkage, adding excess alumina in the form of clay can cause crawling in glazes, which can be corrected by calcining half the clay. Adding excess alumina will cause the glaze to become matt, with subdued colour. Alumina hydrate is a source of pure alumina used to make batt wash and wadding for wood and soda firings, but it is not used in glazes as it does not enter into the melt in its pure form.

**Traditional shino, cone 10, 1300°C (2372°F) reduction**

| | |
|---|---|
| Soda feldspar | 33 |
| Nepheline syenite | 33 |
| Ball clay | 33 |

**Opposite:** Simone Leigh, *Village Series*, 2023–24. Stoneware 145 x 114 x 114 cm. *(Image credit: Courtesy of Matthew Marks Gallery, © Simone Leigh)*

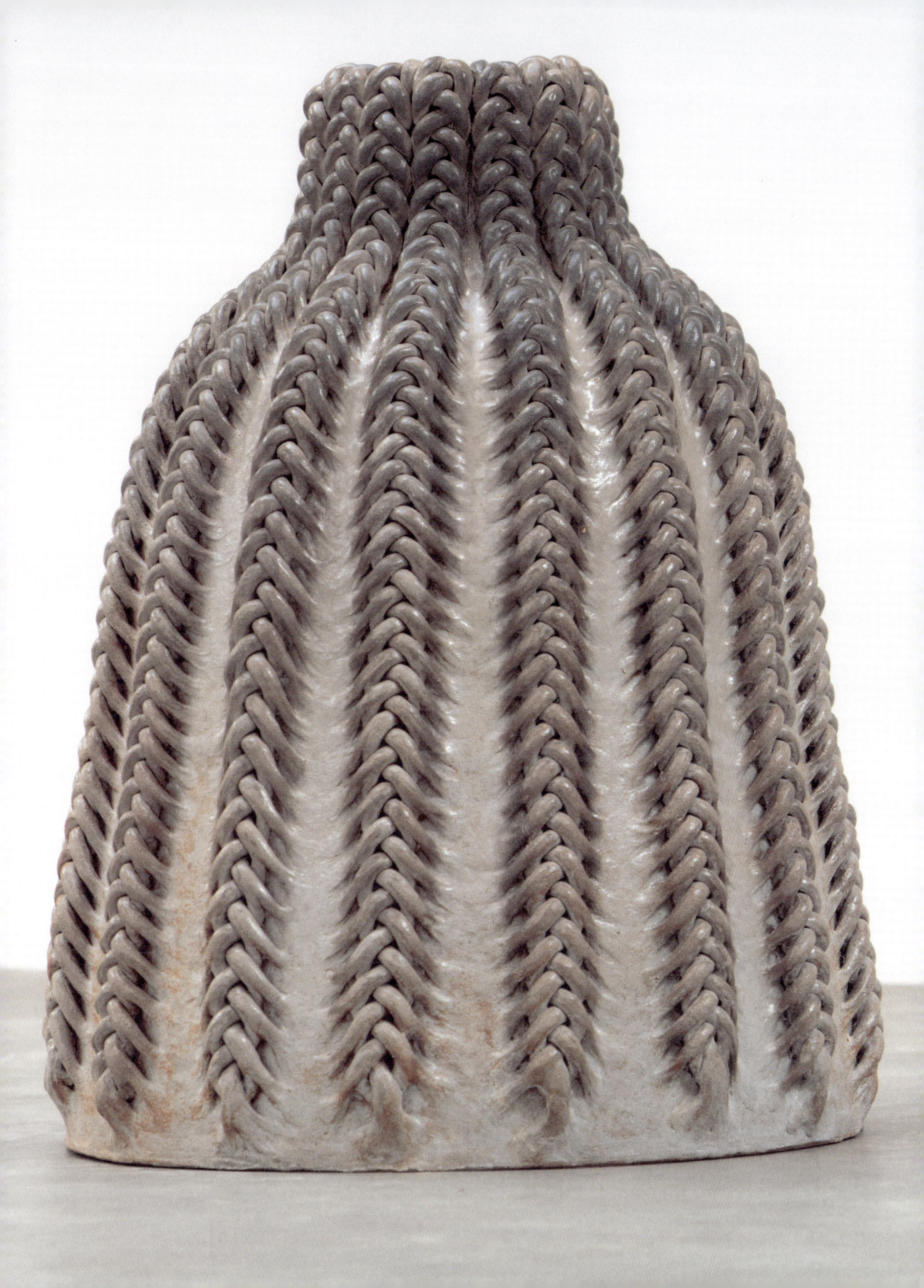

## SILICA

**Sources:**
Silica
Quartz
Flint
Clay
Feldspar
Wollastonite
Talc
Wood ash
Zirconium silicate
Frit

Silica is the main glass former in a glaze and melts at a temperature of 1710°C (3110°F). If combined with fluxes and alumina in a glaze it will melt at lower temperatures such as those attainable in a kiln (1100–1300°C (2012–2372°F)). Silica is usually sourced from the minerals quartz and flint but is also present in other glaze materials including clay and feldspar. Quartz is a crystalline form of silica, while flint is an organic form derived from sedimentary rocks. They are both pure silica and either can be used in glazes. Traditionally, flint was used as it was easier to grind after calcining in a kiln.

**Transparent glaze, cone 9, 1280°C (2336°F)**

| | |
|---|---|
| Potash feldspar | 27 |
| Silica | 32 |
| Whiting | 21 |
| Kaolin | 20 |

**Right:** John Roloff, *Night Ship/Mined Hull/The Frozen Sea*, 1986. Treated with stains and a fused silica glaze. (*Image credit: Lee Fatheree*)

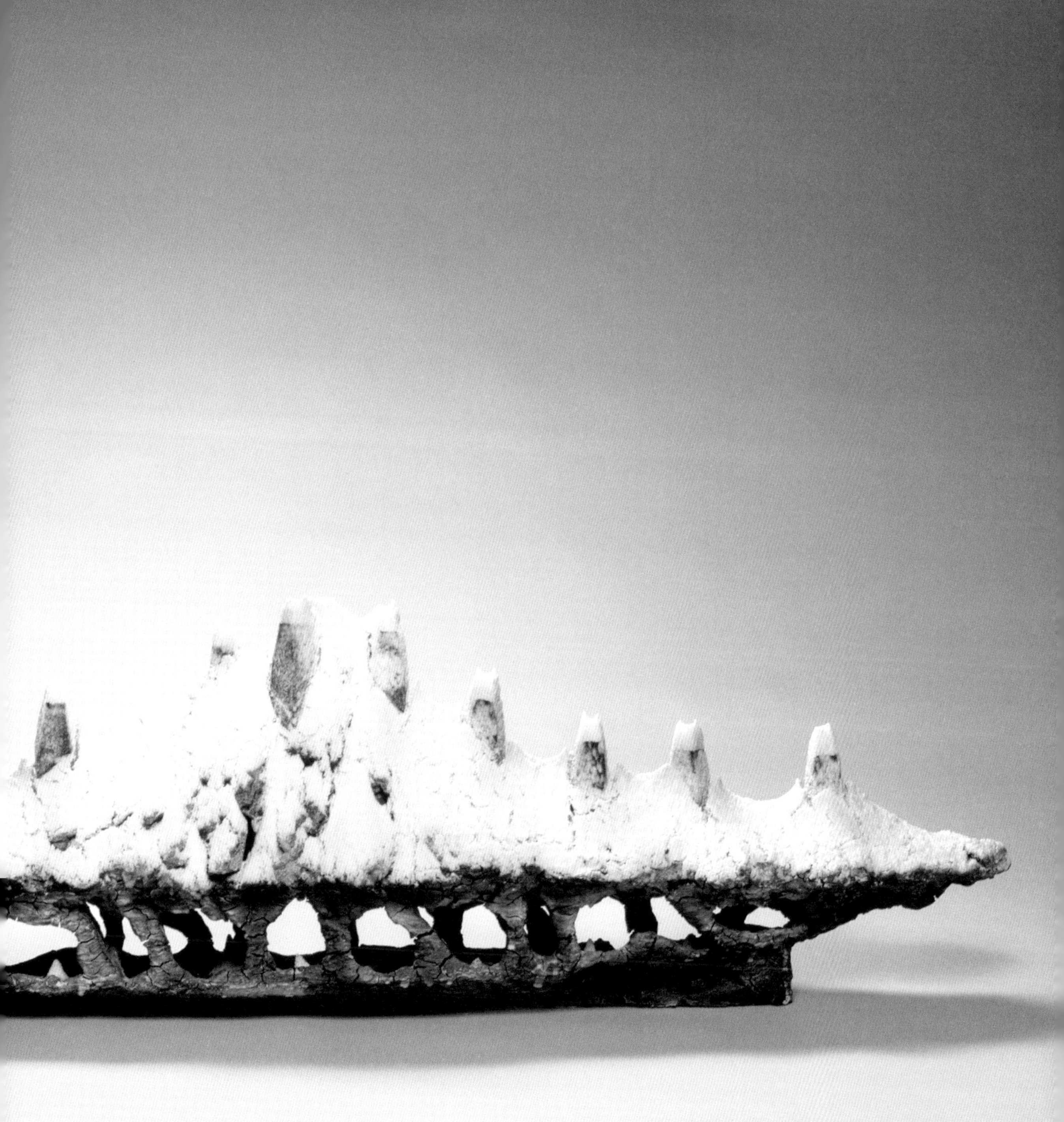

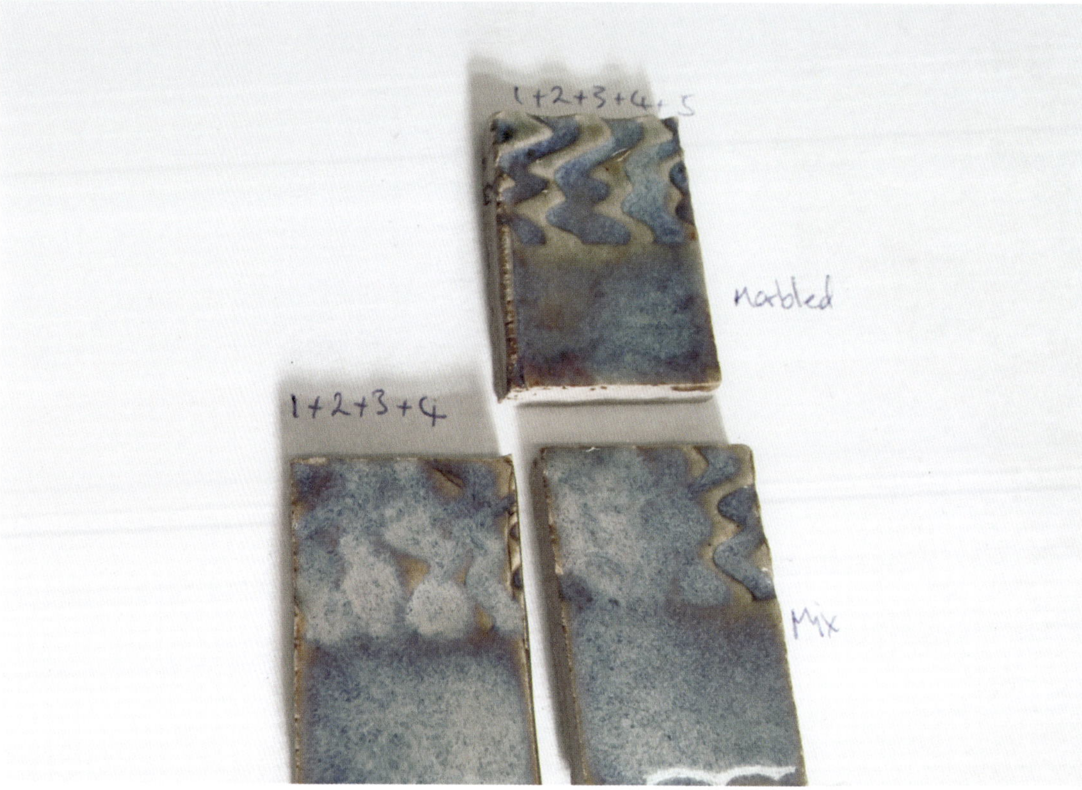
1+2+3+4+5
Marbled
1+2+3+4
Mix

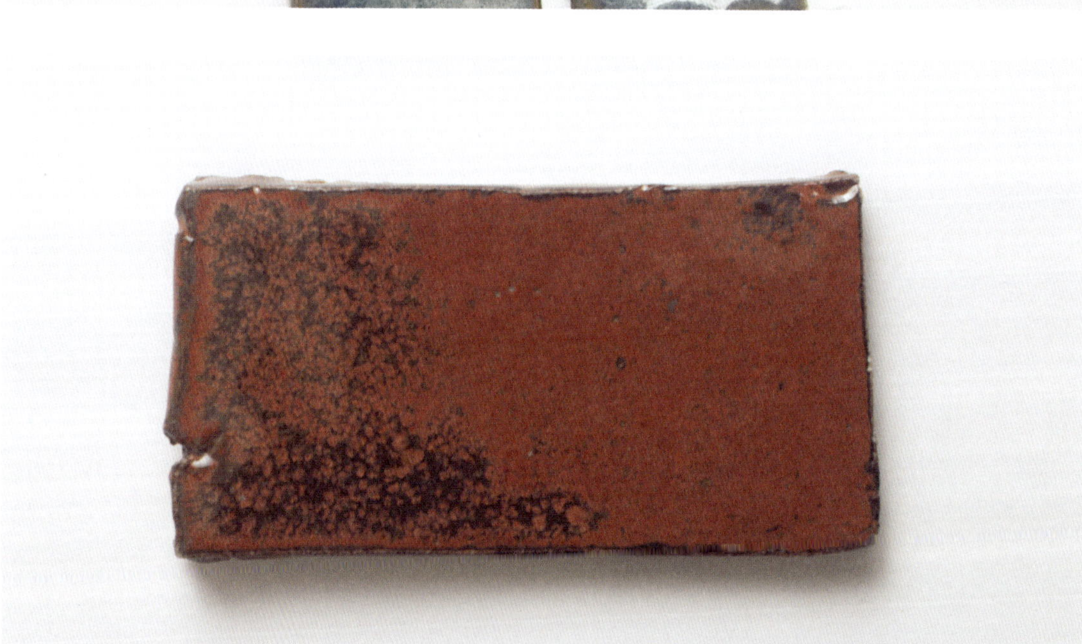

## BORON

**Sources:**
Gerstley borate
Colemanite
Frits (Ferro 3124, 3134)
Borax

Boron oxide is a glass former as well as a flux, used in mid and low temperature glazes to enable them to melt fully. It has a low expansion and prevents crazing. It is less fluid than lead oxide and when combined with calcium it can sometimes cause cloudiness in transparent earthenware glazes. It tends to give cool, bright colours with colouring oxides. It can be sourced from the naturally occurring, slightly soluble Gerstley borate, colemanite and commercially made frits. Its soluble form, borax (sodium borate) is not generally used in glazes as it can be absorbed into the biscuit ware and causes fluxing problems.

**Glossy transparent glaze base, cone 8, 1250°C (2282°F) (High boron stoneware glaze)**

| | |
|---|---|
| Potash feldspar | 34 |
| Calcium borate frit | 14 |
| Whiting | 11 |
| China clay | 13 |
| Quartz | 23 |
| Dolomite | 5 |

**Opposite top:** High boron with added rutile 5% and colouring oxides copper 0.2, chromium 0.1, iron 0.4 and manganese 1%

**Opposite bottom:** Iron red glaze test tile

## PHOSPHORUS

**Sources:**
Bone ash (tricalcium phosphate)
Wood ash

Phosphorus is an additional glass former like silica and boron. It can be introduced in small amounts into a glaze from wood ash or bone ash. It provides opacity and crystallization in certain glazes such as iron red and chun blue. It can also be used to make crawl and foam glazes. Phosphorus is a non-metal with an acidic oxide which is slightly soluble in water, and can cause the glaze to flocculate and thicken over time. This can be counteracted by adding a few drops of a solution of sodium carbonate or sodium silicate to the glaze.

**Iron red glaze, cone 6–8, 1240–1260°C (2264–2300°F)**

| | |
|---|---|
| Potash feldspar | 47 |
| Bone ash | 15 |
| Lithium carbonate | 4 |
| Talc | 17 |
| Quartz | 11 |
| China clay | 6 |
| + | |
| Red iron oxide | 11.5 |

## SODIUM

**Sources:**
Soda feldspar (Minspar)
Nepheline syenite
Gerstley borate
Borax
Soda ash (sodium carbonate)
Wood ash
Frits

Sodium oxide acts as a flux, beginning its fluxing action at a relatively low temperature of around 800°C (1472°F) and once melted is very fluid. It is volatile above 1200°C (2192°F), a property that makes it useful in soda firing, where sodium carbonate is introduced into the hot kiln, where it vaporizes and reacts with the surface of clay to form a glaze. In most stoneware glazes, sodium is introduced in its insoluble form as soda feldspar or nepheline syenite (although these can become slightly soluble in water over time). In lower-melting earthenware glazes, sodium is introduced as a frit, a man-made combination of fluxes and silica. Sodium gives a bright colour response with colouring oxides such as copper and cobalt oxide. It has a high thermal expansion and can cause crazing (crackle) in a glaze. Sodium oxide has low surface tension and makes very fluid glazes.

**Opposite top:** Linda Swanson, *Spectre 1111202* (detail), 2020. The high sodium content in the centre glaze creates crazing patterns that contrast with surrounding balanced glaze

**Opposite bottom:** Kuan crackle glaze test tiles, base glaze and with added chromium 0.25, copper oxide 1, borax frit 10, red iron oxide 1, cobalt 0.25 and rutile 2

A glaze high in sodium:

**Kuan crackle glaze, cone 6–8, 1240–1260°C (2264–2300°F)**
Apply very thickly.

| | |
|---|---|
| Nepheline syenite | 80 |
| Whiting | 5 |
| Borax frit | 15 |
| + | |
| Bentonite | 2 |

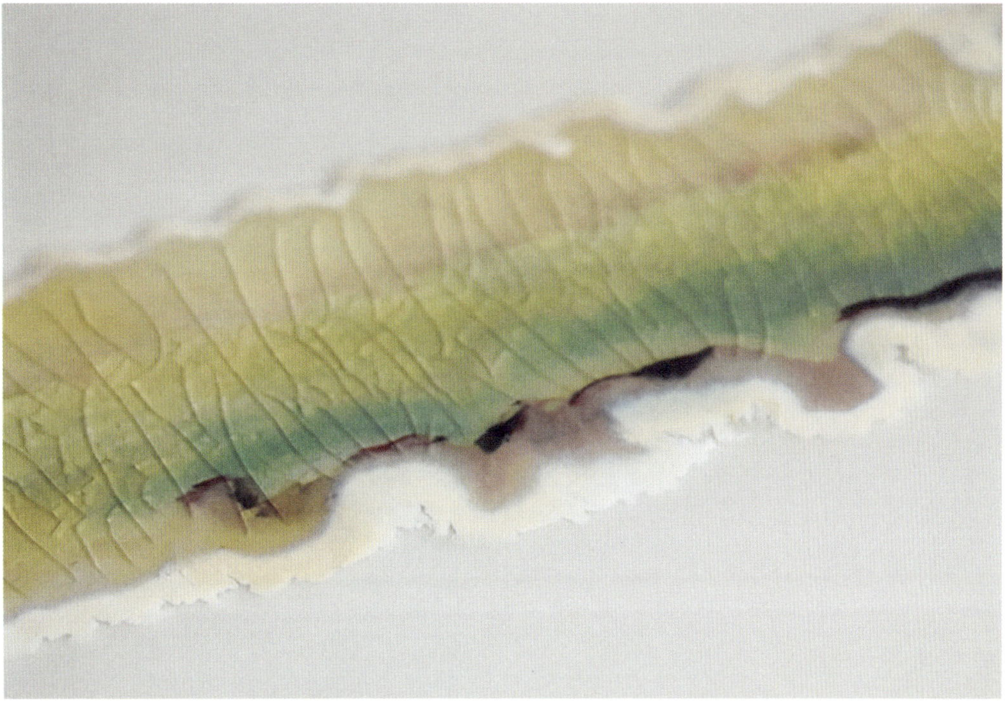

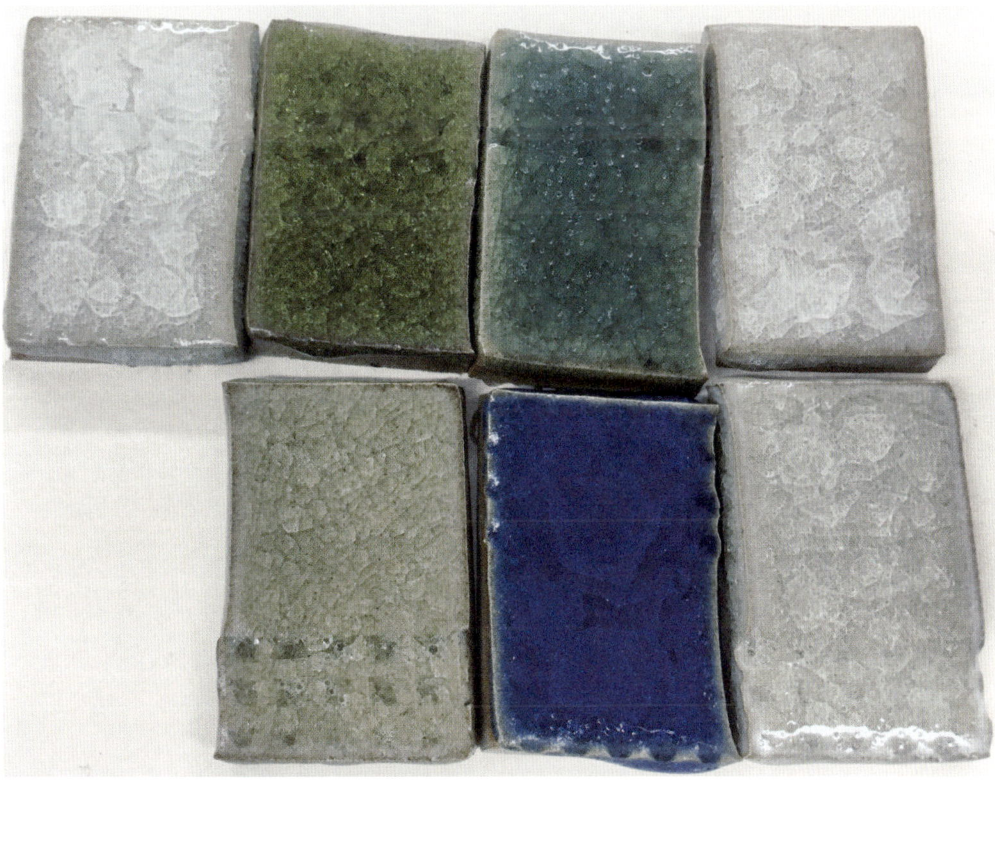

**Above:** Ash glaze test tile

## POTASSIUM

**Sources:**
Potash feldspar (Custer, G200, Mahavir)
Cornwall stone
Granite, pumice, volcanic ash
Wood ash
Pearl ash (potassium carbonate)

Potassium oxide begins its fluxing action at a slightly lower temperature than sodium, 750°C (1382°F) and has a wider firing range. It produces a stronger and more scratch-resistant glaze than sodium. However, the glaze will be less fluid than a sodium-based glaze. The insoluble source of potassium used in glazes is potash feldspar. Most feldspars contain both potassium and sodium and are named after whichever is the predominant oxide. However, potash feldspars are less soluble than those high in sodium and will cause fewer problems in the glaze bucket such as hard panning. The colour response to potassium is similar to sodium and in addition it achieves a higher brilliance owing to an increased refractive index. Potassium glazes have a very low surface tension and are very fluid.

**Transparent ash glaze, cone 9, 1280°C (2336°F)**

| | |
|---|---|
| Potash feldspar | 40 |
| Wood ash | 40 |
| China clay | 20 |

## LITHIUM

**Sources:**
Lithium carbonate
Spodumene
Petalite
Lepidolite
Amblygonite
Some frits (Fusion F493)

Lithium oxide melts at a slightly higher temperature than sodium or potassium but a small addition of lithium carbonate will significantly reduce the melting temperature of a glaze. Lithium is the lightest metal and its oxide has the highest fluxing power, producing a fluid melt with a bright colour response. It starts its fluxing action at 800°C (1472°F) and is less volatile than sodium oxide above 1200°C (2192°F). Lithium used in small amounts in combination with the other fluxes will produce glazes that are hard and scratch resistant. Lithium oxide has a lower expansion than the other alkali metals and is used to make clay bodies that can withstand thermal shock. It can cause shivering in glazes and can be used to correct crazing if it is substituted for some of the sodium and potassium in a glaze. Lithium can be sourced from lithium feldspars such as petalite and spodumene, as well as the slightly soluble lithium carbonate, which is a much more concentrated form of lithium than the feldspars. For example spodumene has one fifth and petalite has one tenth of the concentration of lithium carbonate by weight.

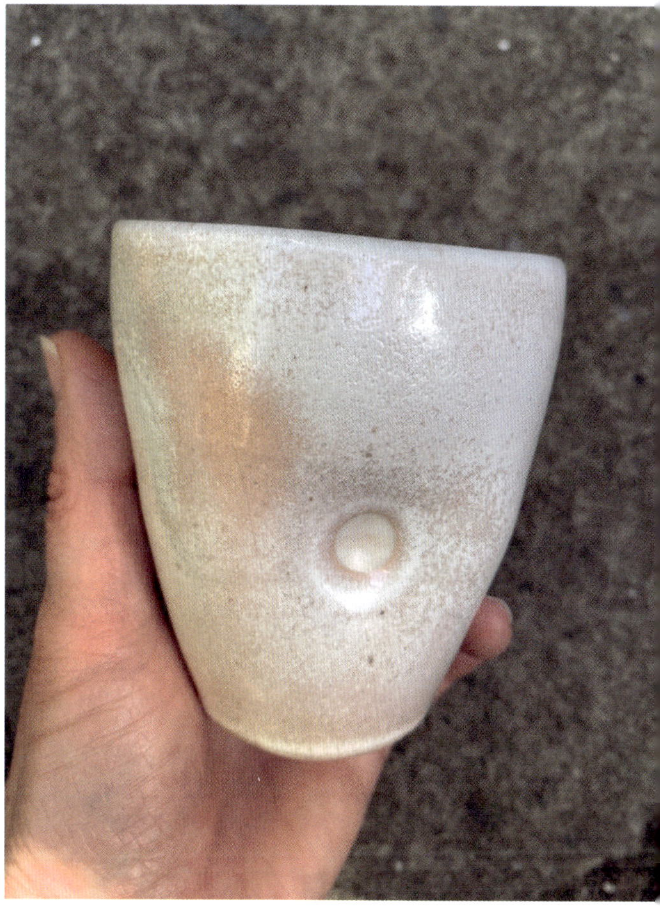

**Spotted shino, cone 9–12, 1280–1320°C (2336–2408°F)**

| | |
|---|---|
| Nepheline syenite | 41 |
| Spodumene | 31 |
| Kaolin | 24 |
| Soda ash | 4 |

**Above:** Linda Bloomfield, dimpled cup, porcelain, spotted shino glaze, anagama fired at Guldagergaard, Denmark 2022

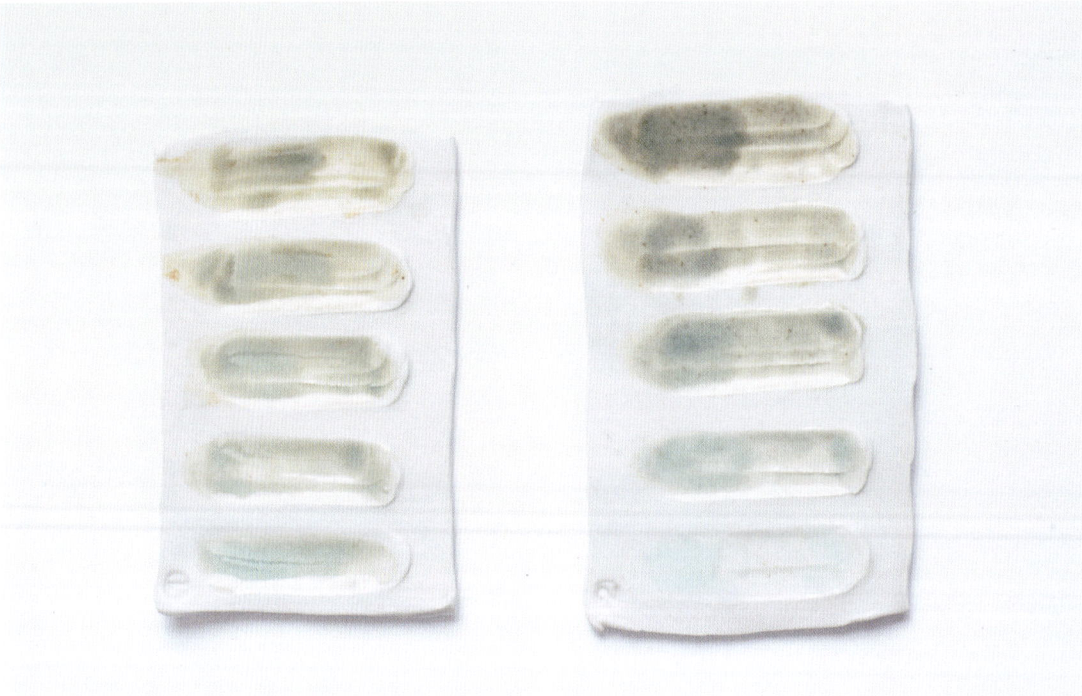

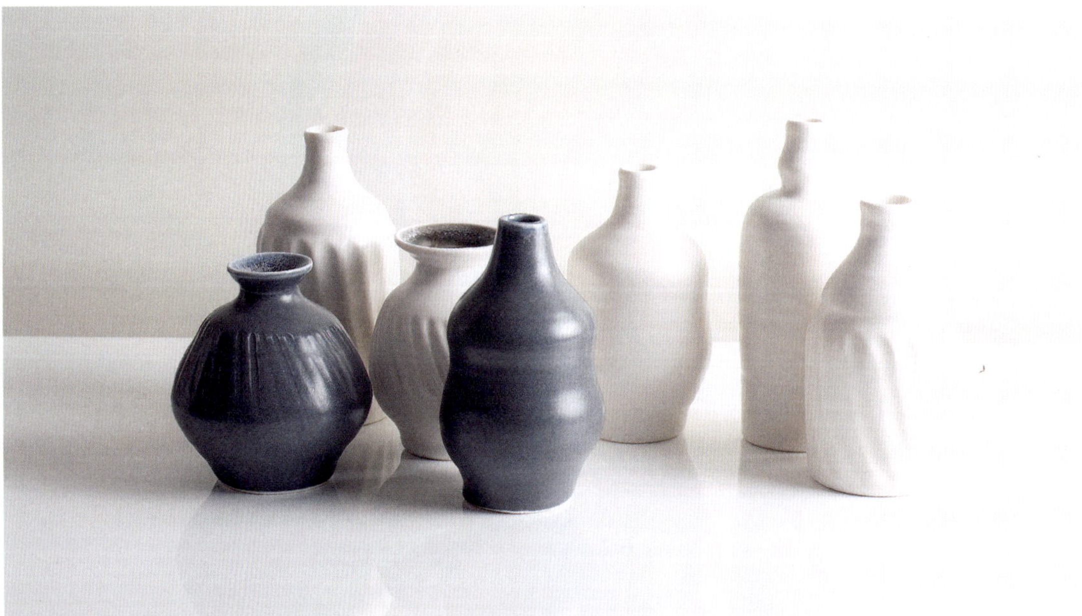

**Above top:** Glossy grey-green celadon test tile with 0.5, 0.8, 1.1, 1.5 and 2% red iron oxide
**Above bottom:** Linda Bloomfield, *Porcelain bottles and vases*, magnesium satin matt glazes, fired to cone 8

## CALCIUM

**Sources:**
Whiting (calcium carbonate)
Wollastonite (calcium silicate)
Dolomite
Gerstley borate
Colemanite (calcium borate)
Bone ash (calcium phosphate)
Wood ash
Some feldspars (Cornwall stone)
Frits

Calcium oxide is a less active flux with a higher melting point than the alkali metal oxides (oxides of lithium, sodium and potassium) but in combination with those primary fluxes it forms a more chemically stable and durable glaze. Its fluxing action begins at around 1100°C (2012°F). It is commonly sourced from whiting (calcium carbonate) and also from wollastonite (calcium silicate). It can also be sourced from bone ash (calcium phosphate) or wood ash. It has a good colour response with colouring oxides, and is used as the main alkaline earth flux in chrome-tin pink, celadon and chun blue glazes. In large amounts it can form crystalline matt glazes which usually have more subdued colours.

**Glossy grey-green celadon, cone 9, 1280°C (2336°F) reduction**

| | |
|---|---|
| Wollastonite | 33 |
| China Clay | 27 |
| Flint | 21 |
| Potash Feldspar | 19 |
| Red Iron oxide | 0.5 |

## MAGNESIUM

**Sources:**
Magnesium carbonate
Dolomite
Talc
Some frits (Ferro 3249)

Magnesium oxide has a high melting point and is used as a flux in stoneware glazes above 1170°C (2138°F). It makes glazes more viscous and in large amounts contributes opacity and a satin matt texture. It affects the colours of some colouring oxides giving pale colours, particularly with cobalt which it turns purple and nickel which it turns pale green. Its high viscosity and high surface tension make it a useful addition to crawl glazes and oilspot glazes. It has very low thermal expansion and is useful in correcting crazing. The main sources of magnesium used in glazes are dolomite (calcium magnesium carbonate) and talc (magnesium silicate). It is usually used in combination with calcium. Light magnesium carbonate (hydromagnesite) is used to make crawl, or lichen-effect glazes.

**Magnesium satin matt glaze, cone 8, 1250°C (2282°F)**

| | |
|---|---|
| Potash feldspar | 33 |
| Talc | 21 |
| Whiting | 12 |
| China clay | 15 |
| Quartz | 16 |
| Zinc oxide | 3 |

## BARIUM

**Sources:**
Barium carbonate
Some frits (Fusion F-403)

Barium carbonate is used in stoneware glazes above 1175°C (2147°F) to make brightly coloured, stony matt glazes. It has a bright colour response to colouring oxides, giving turquoise and blue with copper, chartreuse green with chromium, purple and pink with nickel and manganese. Its high viscosity in the melt makes it suitable for use in volcanic glazes. Barium carbonate is toxic.

**Nickel pink matt, not food safe, cone 8, 1260°C (2300°F) oxidation**

| | |
|---|---|
| Potash feldspar | 33 |
| Barium carbonate | 40 |
| Zinc oxide | 16 |
| China clay | 5 |
| Quartz | 6 |
| + | |
| Nickel oxide | 1.5 |

## STRONTIUM

**Sources:**
Strontium carbonate
Some frits (Fusion F-581)

Strontium carbonate is used in stoneware glazes above 1090°C (1994°F) to make coloured matt glazes. Strontium carbonate can be used as a non-toxic substitute for barium carbonate. It is used at mid stoneware temperatures and begins its fluxing action above 1090°C (1994°F). Its colour response and stony matt texture is similar to that obtained using barium carbonate but slightly less bright. It also has a high viscosity in the melt and can be used in both matt and glossy glazes.

**Strontium matt turquoise, not food safe, cone 8, 1250°C (2282°F)**

| | |
|---|---|
| Nepheline syenite | 55 |
| Strontium carbonate | 19 |
| Lithium carbonate | 2 |
| China clay | 6 |
| Flint | 8 |
| Calcium borate frit | 2 |
| + | |
| Copper oxide | 2 |

**Opposite top:** Nickel pink barium test tile
**Opposite bottom:** Strontium matt glaze test tile
**Overleaf:** Linda Bloomfield, porcelain bowl with vanadium slip and strontium matt turquoise glaze, fired to cone 8

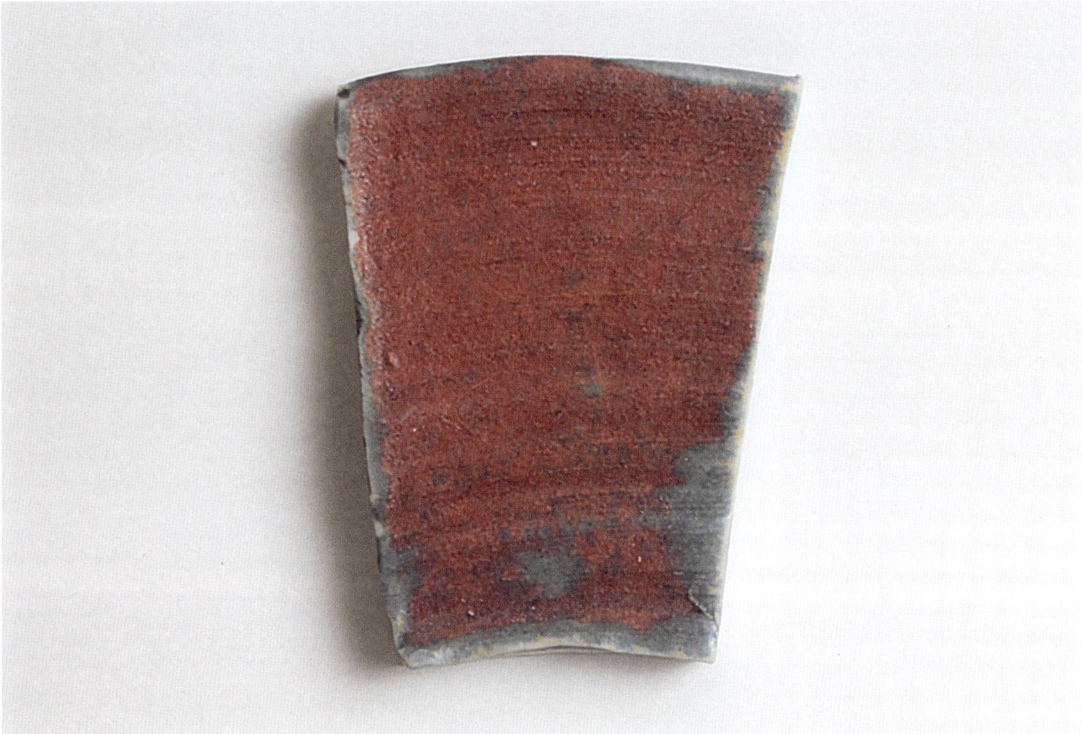

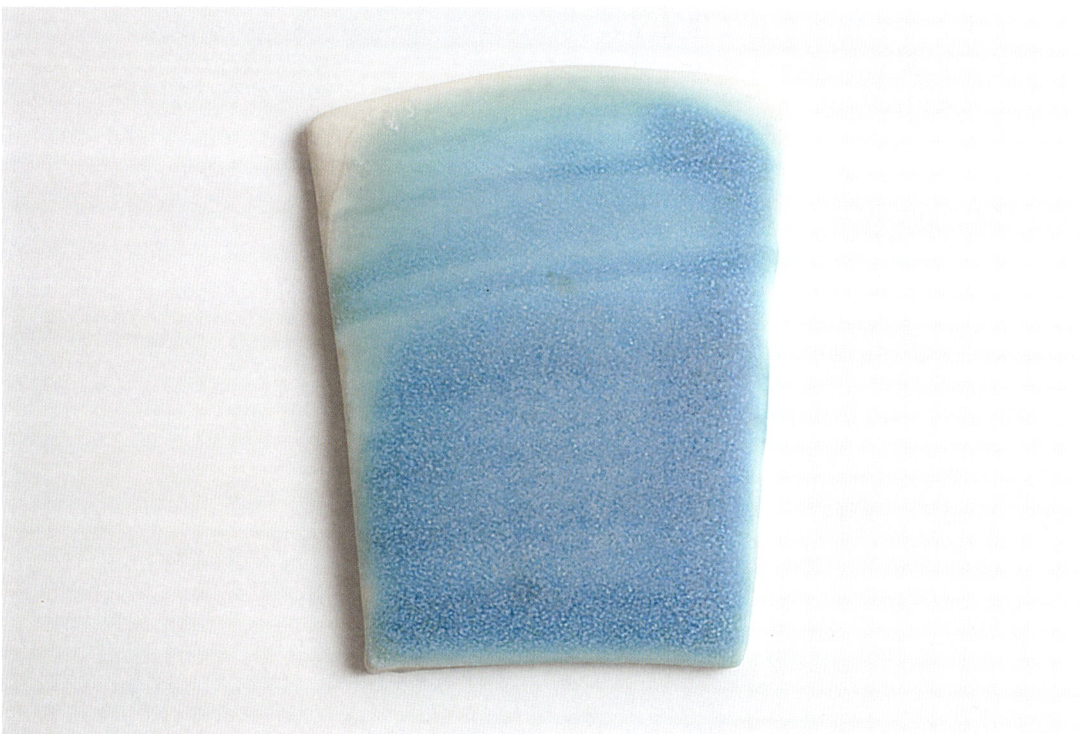

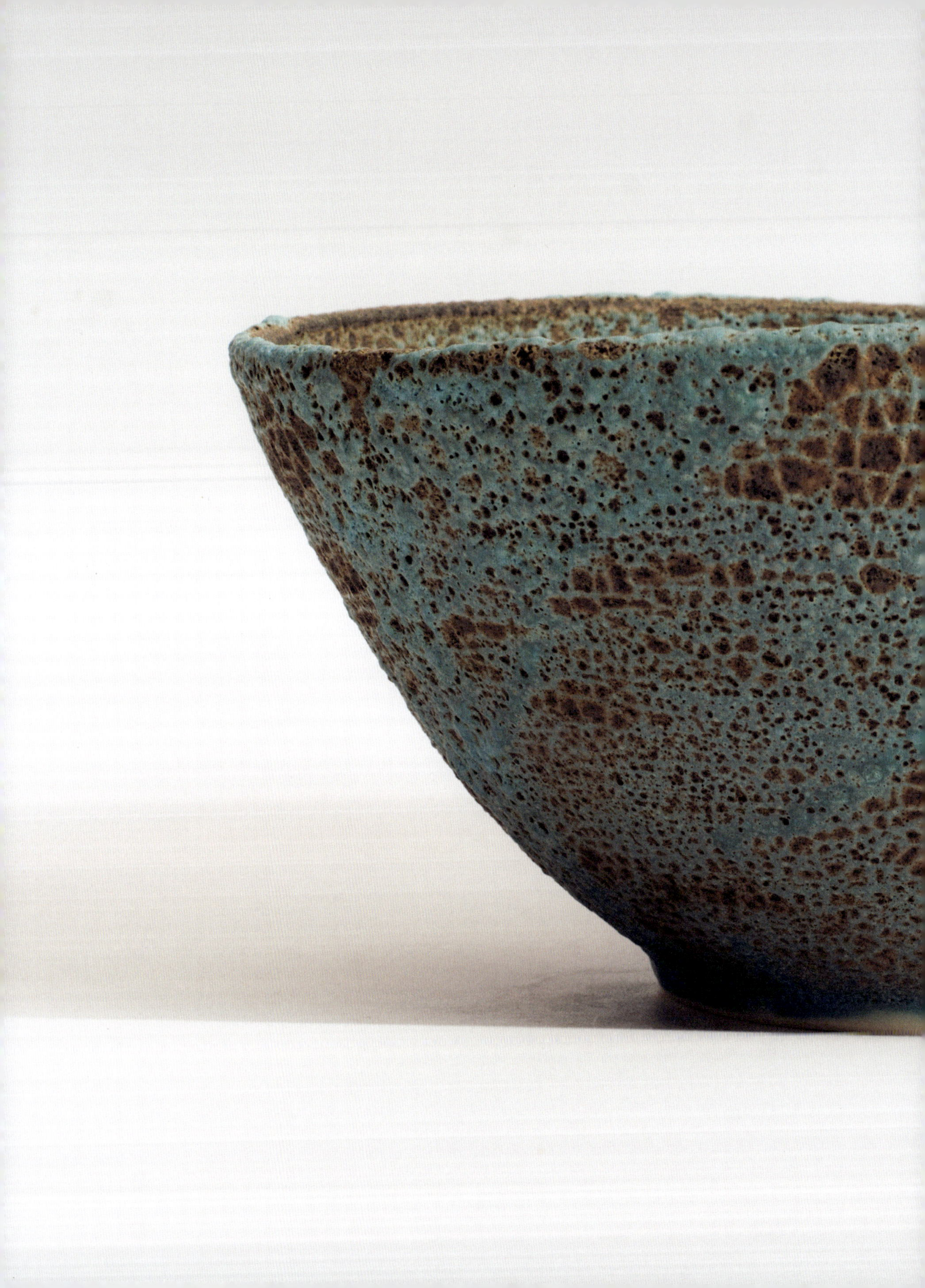

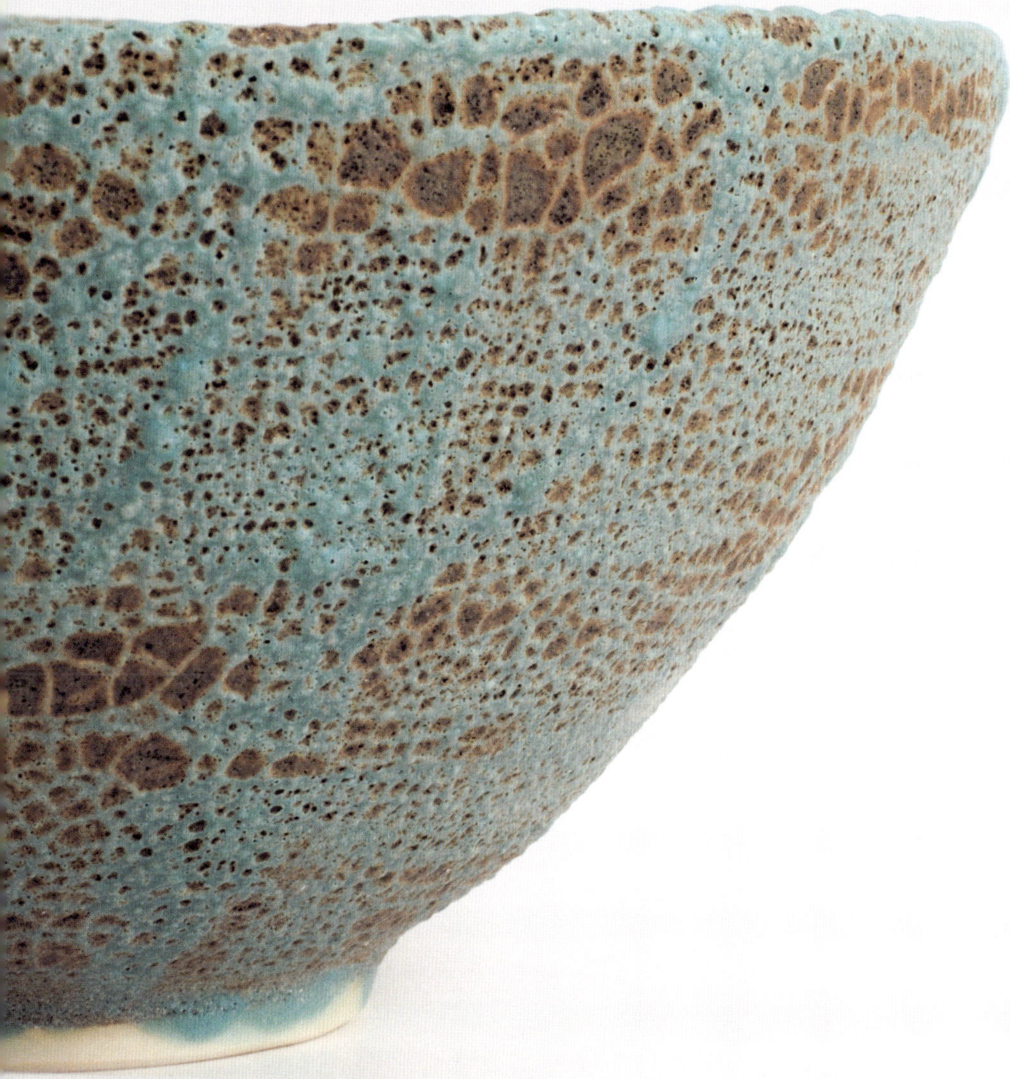

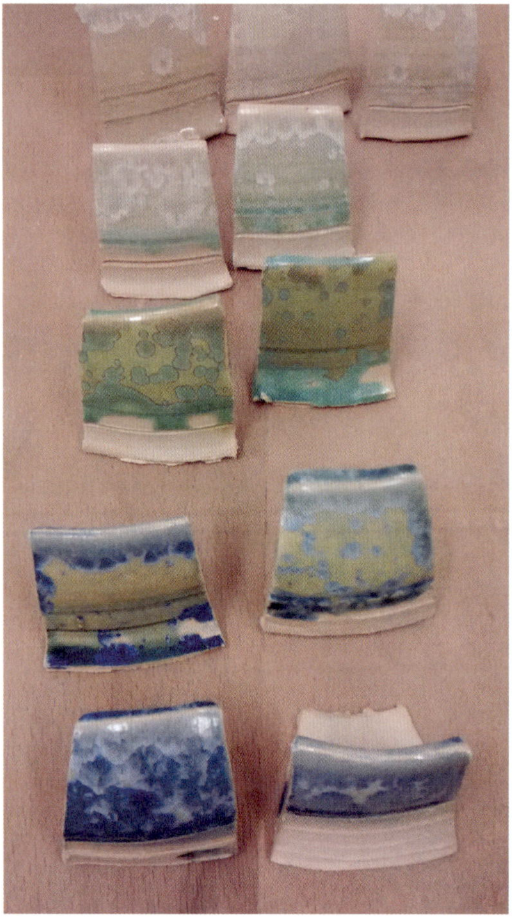

## ZINC

**Sources:**
Zinc oxide
Some frits (Fusion FZ-16)

Zinc oxide is used as a flux in mid-temperature glazes. It forms a eutectic together with calcium oxide and silica that melts in the range 1100–1200°C (2012–2192°F) known as a Bristol glaze. It is always used in conjunction with other fluxes and will also form eutectics with the other alkaline earth fluxes, magnesium, strontium and barium. A small amount around 5% will increase the melt, while larger amounts up to 25% will make crystalline glazes where it forms crystals of zinc silicate. It has a low thermal expansion and can be used to correct crazing. However, zinc oxide has high shrinkage on drying as well as high surface tension, so can cause crawling. To prevent shrinkage, the oxide can be calcined at around 815°C (1500°F). It has an adverse effect on the colours of some colouring oxides and stains, particularly those containing chromium, which it turns brown or pink.

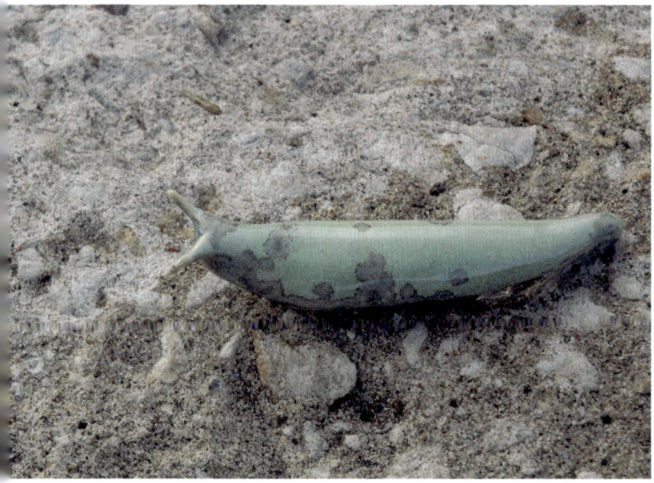

**Zinc crystalline glaze, cone 8, 1260°C (2300°F)**
This glaze has large crystals if cooled slowly below 1100°C (2012°F)

| | |
|---|---|
| Ferro frit 3110 | 47 |
| Zinc oxide | 23 |
| China clay | 3 |
| Quartz | 23 |
| Titanium dioxide | 4 |

## LEAD

**Sources:**
Lead frits, lead bisilicate,
lead sesquisilicate
China paints

Lead is still used by some traditional slipware potters as it gives a warm colour response with colouring oxides, such as honey yellow from iron oxide. It acts as a flux at earthenware temperatures to give clear, transparent glazes that are very fluid and resistant to crazing. However, lead is toxic and accumulates in the body so it is no longer used widely in glazes. Boron can be used as an effective substitute for lead.

**Earthenware glaze, transparent, cone 2, 1100°C (2012°F)**

| | |
|---|---|
| Lead bisilicate | 75 |
| Ball clay | 25 |

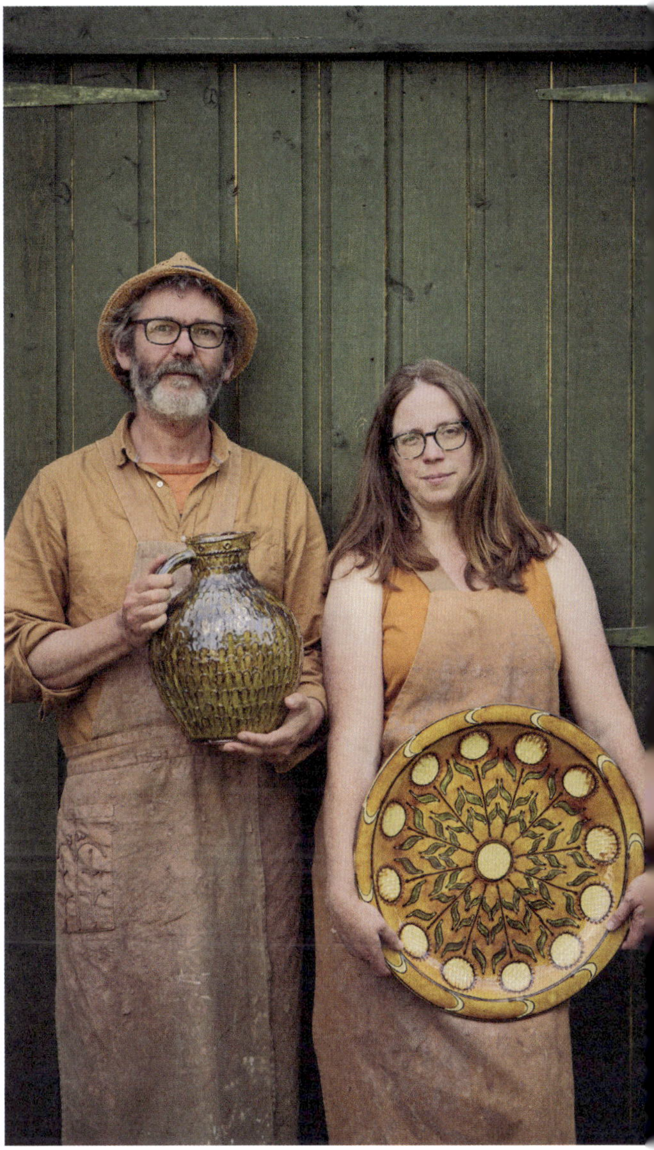

**Opposite top:** Zinc crystalline glaze, base glaze, with added copper oxide, nickel oxide and cobalt oxide

**Opposite bottom:** Linda Swanson, *Infestation*, 2011. Zinc crystalline glaze with copper

**Above:** Doug Fitch and Hannah McAndrew, 2023. Large jug and slip-trailed platter made of red earthenware with white slip and lead glaze. *(Image credit: Shannon Tofts)*

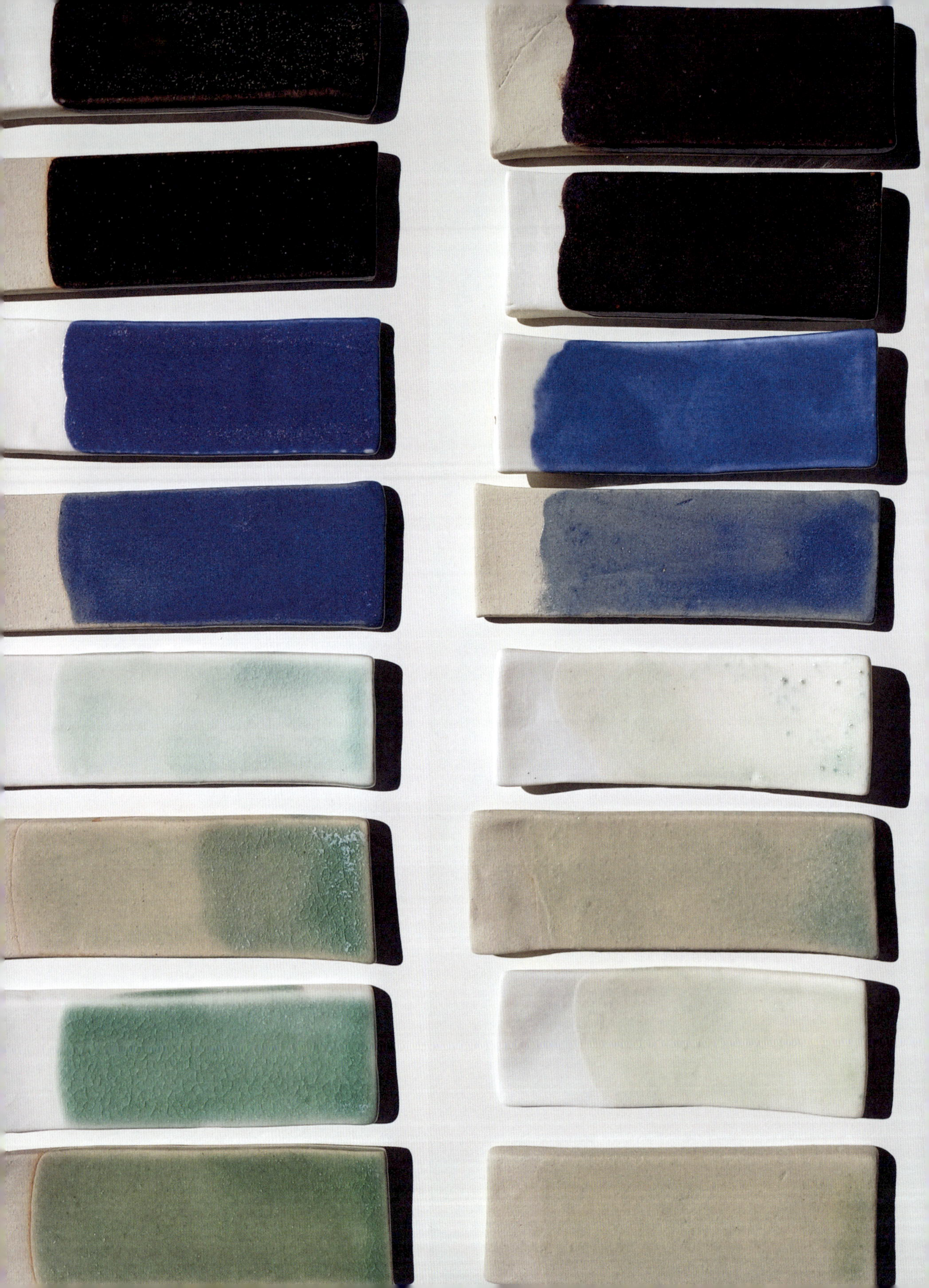

CHAPTER SIX

# GLAZE MATERIALS

## MATERIALS AND THE OXIDES THEY SUPPLY

The oxides in glazes are not usually found in nature in their pure states. They are usually supplied as powdered rocks and minerals such as feldspar and dolomite. Many glaze materials like feldspar contain several oxides: soda, potassia, silica and alumina. These mixtures of oxides tend to melt better than single oxide materials such as alumina (as discussed earlier in the section on eutectics, p. 100). However, sometimes only pure sources are available (such as barium carbonate) or fluxes may occur in combination with other fluxes as a mineral (such as dolomite, a carbonate of calcium and magnesium). These carbonates convert to oxides during firing, when carbon dioxide is given off.

Minerals are defined by a chemical formula, such as whiting, calcium carbonate ($CaCO_3$) while rocks can be far more variable, as they are made up of mixtures of minerals in varying proportions. Some materials such as barium carbonate are prepared chemically by precipitation from barium sulphate. Other materials are supplied to the ceramics industry as rocks, including nepheline syenite and limestone. Their compositions vary from different sources; for example, one limestone source may be pure calcium carbonate, while another may also contain magnesium carbonate which might affect the colour of the glaze. Some materials are kept constant by combining rocks from different parts of the mine or even from different mines. The compositions of materials also vary over time as old mines close and new mines open. Rocks and minerals supplied to the ceramics industry are usually ground to a 200 mesh powder or finer.

- Feldspar is composed of multiple glaze oxides: silica, alumina, sodium, potassium or lithium
- Frit is composed of multiple glaze oxides: silica, boron, various fluxes

**Opposite**: Sigurlina Osuala. Eggshells substituted for whiting create slightly different colour reactions in the same recipe: 40 potash feldspar, 30 quartz, 10 dolomite, 10 china clay, 10 calcium carbonate (left whiting and right eggshells). *(Image credit: Sigurlina Osuala)*

## FELDSPAR ($KNaO.Al_2O_3.6SiO_2$)

Feldspar is a mineral found in granite rock containing potassium (K) and sodium (Na) as well as alumina ($Al_2O_3$) and silica ($SiO_2$). Feldspars are variable materials and contain both potassium and sodium, but are named potash or soda feldspar depending on which is present in the higher proportion. In North America, they usually have brand names, such as Custer Feldspar, a potash feldspar that is no longer mined, or Minspar, a soda feldspar from Spruce Pine, North Carolina. Cornwall stone is a type of decomposed granite containing potash feldspar. As it is no longer mined, it is usually substituted with other feldspars and silica. Granite rock and volcanic ash can be used in glazes instead of feldspars. Some feldspars, such as petalite, contain lithium instead of potassium or sodium.

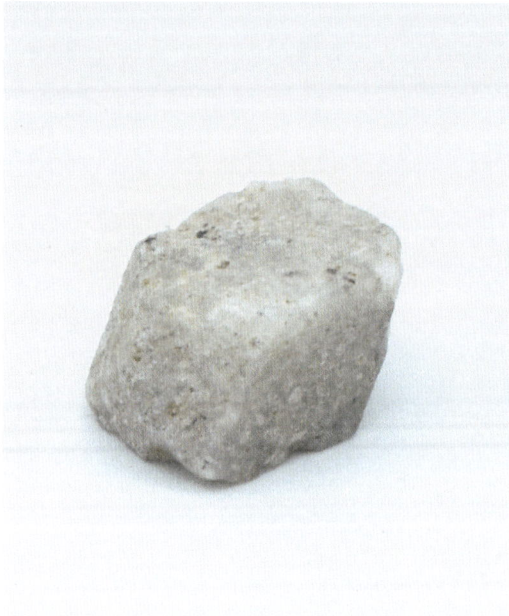

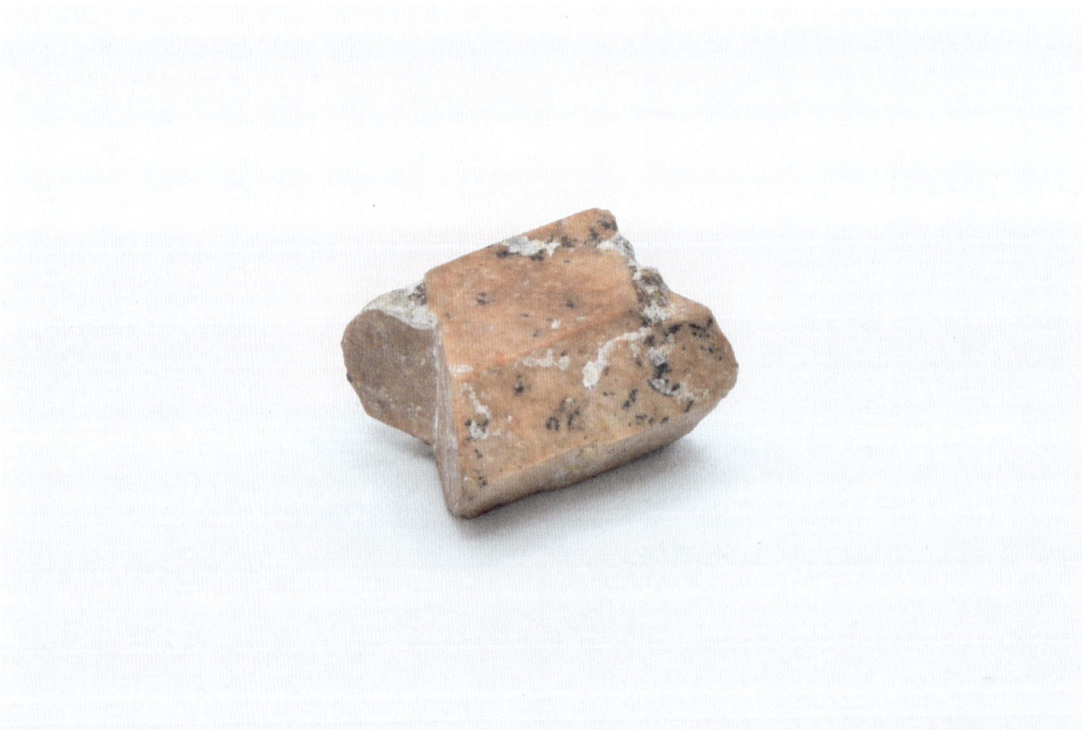

## FRIT

Frits are like man-made feldspars, made from flux materials such as borax, that would otherwise be soluble in the glaze. The fluxes are melted together with silica and ground to a powder. Since they have already been melted into a glass, they melt more easily than raw minerals. The most commonly used frits are those containing boron (US Ferro frit 3134, 3124 and UK standard borax frit), which is needed to melt low fire and mid-temperature glazes. Some frits such as Ferro frit 3195 are almost a complete glaze and only require the addition of clay. Glass can also be ground and used as a frit. There are two main types of glass; soda-lime glass and borosilicate glass. Glass has a different composition to frit, but it can be used as a substitute for alkaline frit or borax frit.

**Opposite top:** Orthoclase, potash feldspar
**Opposite bottom:** Feldspar
**Below:** Galena, lead sulphide

- Ferro frit 3134: a borax frit containing calcium, sodium, boron and silica used to replace Gerstley borate in mid-temperature glazes. This frit contains relatively high boron.

- Ferro frit 3124: a borax frit containing calcium, sodium, boron, silica, similar to 3134 but with alumina.

- Ferro frit 3195: almost a complete low fire glaze, requiring only the addition of 10–15% clay.

- Ferro frit 3110: an alkaline frit similar to feldspar, containing sodium, calcium and silica, used to make crystalline glazes in combination with zinc oxide. This frit has high expansion.

- Ferro frit 3249: a low expansion frit containing boron and magnesium.

- Lead bisilicate: a frit containing lead and silica. It has a low solubility and is safer to use than lead oxide, which is highly toxic.

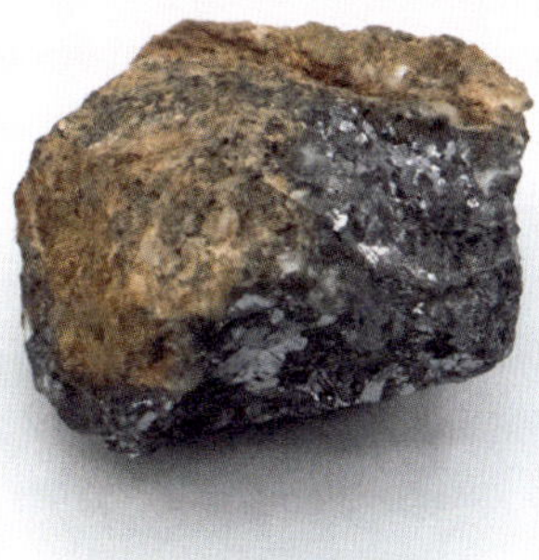

## BARIUM CARBONATE – BARIUM (Ba), $BaCO_3$

Barium carbonate is manufactured from the mineral barite, or barium sulphate, which, although it is non toxic, is not used in glazes as it releases sulphur dioxide gas during firing. Barium carbonate is used to make brightly-coloured matt glazes and is toxic. Barium carbonate is also used in small amounts to prevent soluble salt scumming in casting slips. It breaks down and gives off carbon dioxide at high temperatures.

## BENTONITE – SILICA (Si), ALUMINA (Al), $Al_2O_3.4SiO_2.H_2O$

A type of highly plastic ball clay used in small amounts for suspending glaze materials in water. Found in the Black Hills in Wyoming, sodium bentonite is the type of bentonite used in porcelain clay bodies and glazes, while calcium bentonite has other uses as an adsorbent (it bonds to materials such as fats and oils). Only 2–3% is needed in a glaze; adding too much can cause it to thicken and gel.

## BONE ASH – CALCIUM (Ca), PHOSPHORUS (P), $Ca_3(PO_4)2$

Bone ash is a natural form of calcium phosphate made from calcined animal bones. The synthetic form of bone ash is called tricalcium phosphate. It is used in small amounts to make chun blue glazes and up to 15% to make red iron glazes. It may cause crawling and blistering and in larger amounts and can be used to make foam glazes. It is used as a flux in bone china.

## BORAX – BORON (B), SODIUM (Na), $Na_2O.B_2O_3.10\ H_2O$

Borax is a soluble mineral composed of hydrated sodium borate. It comes from evaporated lake deposits in California and Turkey. For use in glazes, it is usually combined with silica and made into a frit to prevent it dissolving in the glaze bucket and precipitating out over time. It is used to make Ferro frits 3124, 3134 and 3195.

## CLAY – SILICA (Si), ALUMINA (Al), $Al_2O_3.2SiO_2.2H_2O$

Clay is used as the main source of alumina in glazes and to suspend the other glaze materials in water. Kaolin, ball clay or bentonite can be used in glazes. Kaolin is a pure, white clay, while ball clay has a finer particle size but contains impurities such as iron oxide. Bentonite is used in small amounts (1–2%) to suspend glazes that are otherwise low in clay. Clays often have brand names such as EPK, (Edgar Plastic Kaolin), OM 4, (Old Mine 4 ball clay) and Hyplas71, an English ball clay that is high in silica. Locally found clays can be used in glazes but testing is essential.

## COLEMANITE – CALCIUM (Ca), BORON (B), $2CaO.3B_2O_3.5H_2O$

Colemanite is hydrated calcium borate found in California, Chile and Turkey. It can be used in glazes but the initial heating in the kiln must be slow as water is given off and can cause the glaze to spit off. It is better to use a calcium borate frit such as Ferro frit 3134.

## CORNWALL STONE – POTASSIUM (K), $K_2O.Al_2O_3.8SiO_2$

Cornwall stone is a potash feldspar higher in silica than other potash feldspars. It also contains a small amount of calcium. It was originally mined in Cornwall, England, before that source became uneconomical to mine. A substitute material is made by combining potash feldspar with silica and whiting.

## CUSTER FELDSPAR – POTASSIUM (K), $K_2O$ $Al_2O_3.6SiO_2$

Custer feldspar is an American potash feldspar that was mined in South Dakota from 1928–2023. It had a slightly lower potassium content than G-200 and Mahavir potash feldspar.

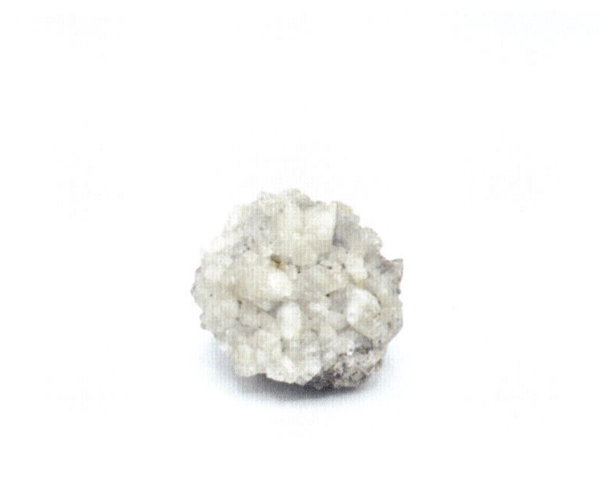

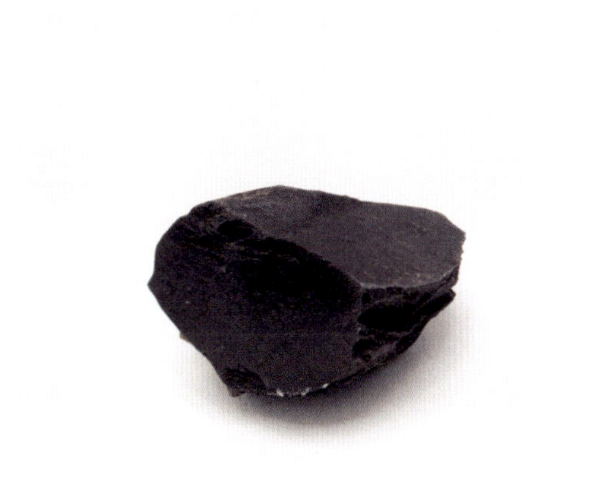

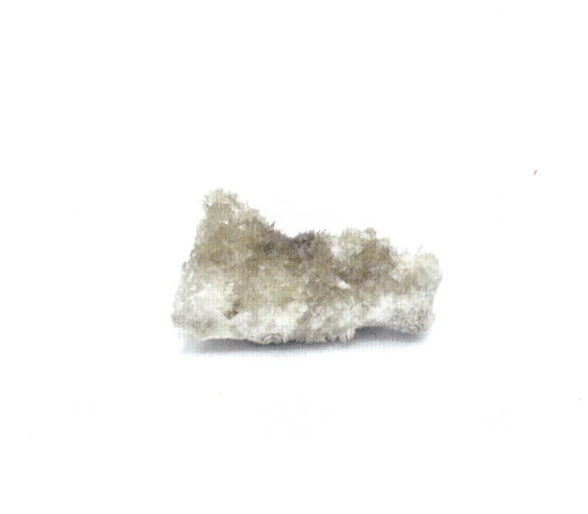

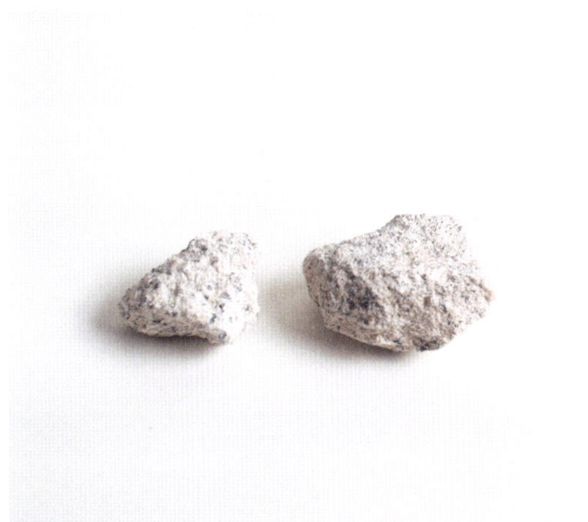

**Above top left:** Barite
**Above bottom left:** Colemanite

**Above top right:** Shale
**Above bottom right:** Cornish stone

## DOLOMITE – MAGNESIUM (Mg), CALCIUM (Ca), $MgCO_3.CaCO_3$

Dolomite is calcium magnesium carbonate, named after the range of limestone mountains in the Italian Alps. It is used to make matt glazes. Its low expansion can help to correct crazing. Dolomite gives off carbon dioxide above 600°C (1112°F).

## G-200 FELDSPAR POTASH FELDSPAR – POTASSIUM (K), $K_2O.Al_2O_3.6SiO_2$

G-200 Feldspar is a discontinued American potash feldspar with high potassium that was mined in Georgia. Use another potash feldspar such as G-200 EU or Mahavir feldspar as a replacement.

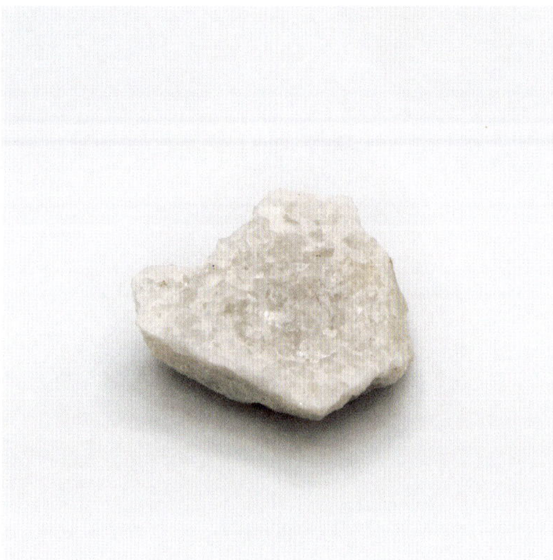

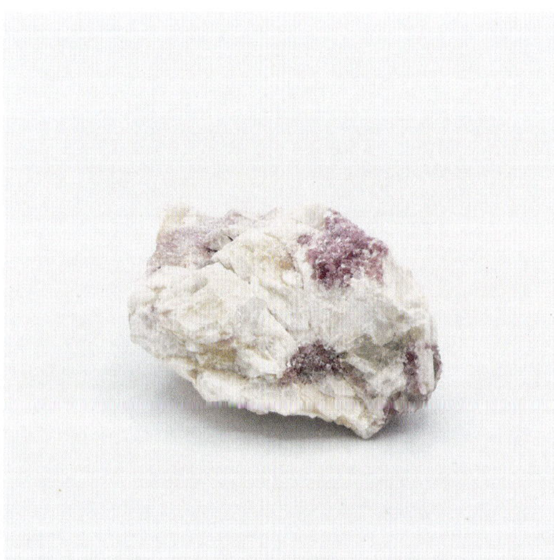

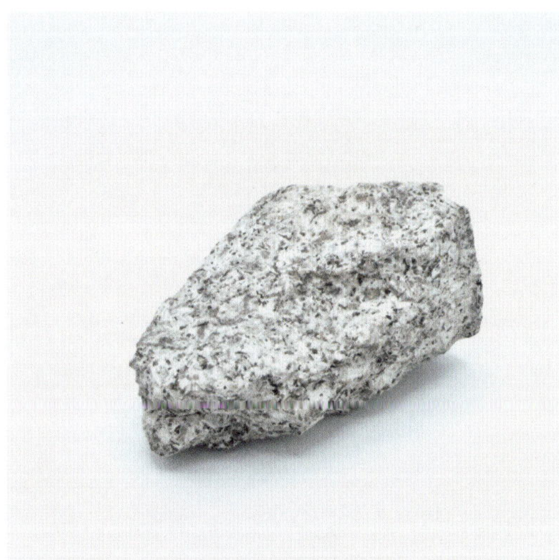

## GERSTLEY BORATE – BORON (B), CALCIUM (Ca), SODIUM (Na)

Gerstley borate is a mineral that was mined in Boron, California. It contains ulexite, colemanite and clay. As it is no longer mined, suppliers now make up borates of a similar composition by combining these materials. It often makes the glaze thicken and gel over time, so US Ferro frits 3134 and 3124 or UK standard borax frit can be used instead.

## KAOLIN – SILICA (Si), ALUMINA (Al), $Al_2O_3.2SiO_2.2H_2O$

Kaolin is a pure, white clay, used as the main source of alumina in glazes and to suspend the other glaze materials in water. Kaolins often have brand names such as EPK, (Edgar Plastic Kaolin) from Florida, Tile number 6 from Georgia or Grolleg from Cornwall, also known in the UK as china clay.

## LITHIUM CARBONATE – LITHIUM (Li), $Li_2CO_3$

Lithium carbonate is a concentrated source of lithium, a strong alkali metal flux. It is slightly soluble in water and has recently become very expensive as it is in high demand for use in electric car batteries. It is much higher in lithium than spodumene or petalite, both feldspars which also contain silica and alumina.

**Opposite top left:** Dolomite
**Opposite top right:** Kaolin
**Opposite bottom left:** Lithium source: lithia pegmatite
**Opposite bottom right:** Nepheline syenite

## MAGNESIUM CARBONATE – MAGNESIUM (Mg), $MgCO_3$

Magnesium carbonate (magnesite) is a pure source of magnesium used to make matt glazes, usually in combination with calcium. Light magnesium carbonate is a hydrated variety (hydromagnesite) used to make crawl glazes.

## MINSPAR SODA FELDSPAR – SODIUM (Na), $Na_2O.Al_2O_3.6SiO_2$

Minspar is an American soda feldspar similar to the discontinued NC-4 feldspar from Spruce Pine in North Carolina. It contains both sodium and potassium but is higher in sodium.

## NEPHELINE SYENITE – SODIUM (Na), $K_2O.3Na_2O.4Al_2O_3.9SiO_2$

Nepheline syenite is a feldspathoid rock, similar to feldspar but with less silica. It is mined in Canada and Norway. It is a more concentrated source of sodium than soda feldspar. It melts at a lower temperature than feldspar in the range 1100–1200°C (2012–2192°F) and is often used to make matt glazes owing to its low silica content. It has a variable composition with different levels of silica from different sources, with a molecular formula falling between 1:1:2 (flux:alumina:silica, per Hamer) and 1:1:4.65 (per Rhodes).

## PEARL ASH – POTASSIUM (K), $K_2CO_3$

Pearl ash is potassium carbonate. It is soluble in water so most glazes use insoluble potash feldspar instead. Pearl ash is much higher in potassium than feldspar.

## PETALITE – LITHIUM (Li), $Li_2O.Al_2O_3.8SiO_2$

Petalite is a lithium feldspar which is higher in silica than the other lithium feldspar spodumene. Lithium encourages bright colours and has a lower thermal expansion than sodium or potassium.

## POTASH FELDSPAR – POTASSIUM (K), $K_2O.Al_2O_3.6SiO_2$

Potash feldspar is the generic name for feldspars that are high in potassium. They usually contain both sodium and potassium but the potassium content is higher. Feldspars also contain alumina and silica. Custer feldspar and G-200 were American potash feldspars. European potash feldspars include Norfloat from Norway and FFF from Finland. Mahavir feldspar is a potash feldspar from India.

## SAND – SILICA (Si), $SiO_2$

Silica sand is made of coarse particles of silica, silicon dioxide. Beach sand may contain other materials such as iron oxide which lower the melting temperature. It can be added to clay bodies or placed on the kiln shelf when firing large pieces to help them to shrink without cracking. Do not use in a glaze as its coarse particles will not melt.

## SILICA – SILICA (Si), $SiO_2$

Silica is silicon dioxide, the main glass former in glazes and also used in clay bodies. It is supplied as quartz or flint, finely ground to a mesh size of 325 for use in glazes. Silica is also present in feldspars and clays.

## SODA ASH – SODIUM (Na), $Na_2CO_3$

Soda ash is sodium carbonate. It is soluble in water and most glazes use insoluble soda feldspar instead. However, wood-fired carbon-trap shinos do use soda ash which rises to the surface of the glaze as it dries and traps carbon in the glaze. Soda ash is much higher in sodium than feldspar. Its other main use in pottery is in soda firing where it is introduced into the kiln at high temperature. Soda ash gives off carbon dioxide at high temperatures.

## SODA FELDSPAR – SODIUM (Na), $Na_2O.Al_2O_3.6SiO_2$

Soda feldspar is the generic name for feldspars that are high in sodium. Soda feldspars contain both sodium and potassium but are higher in sodium. Feldspars also contain alumina and silica. Minspar and NC-4 are the brand names of American soda feldspars.

## SPODUMENE – LITHIUM (Li), $Li_2O.Al_2O_3.4SiO_2$

Spodumene is a lithium feldspar lower in silica than petalite. Hence it is more concentrated in lithium than petalite and it is also more expensive.

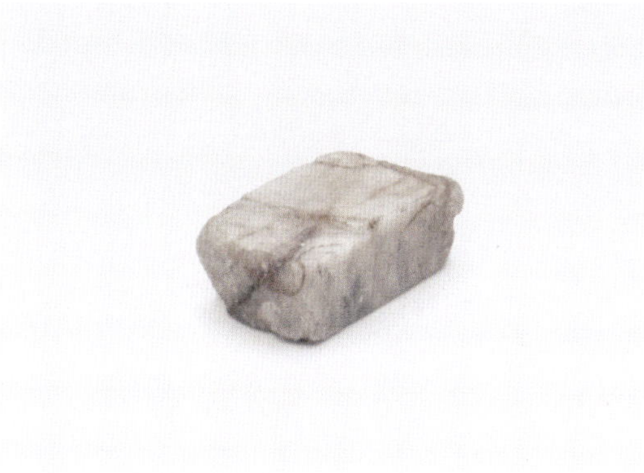

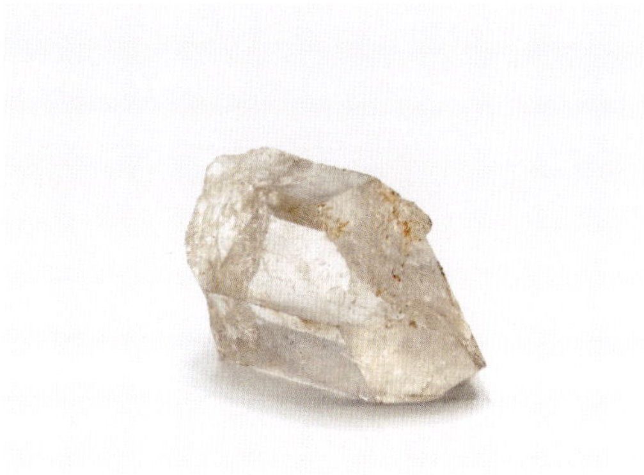

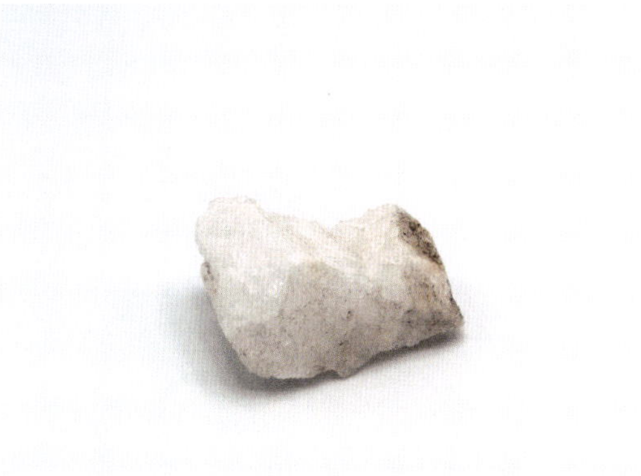

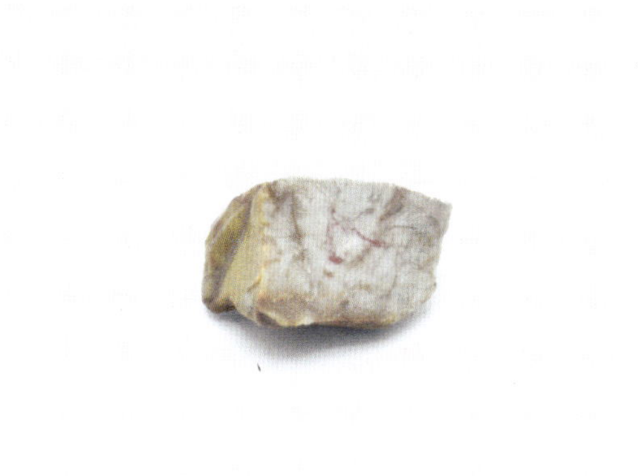

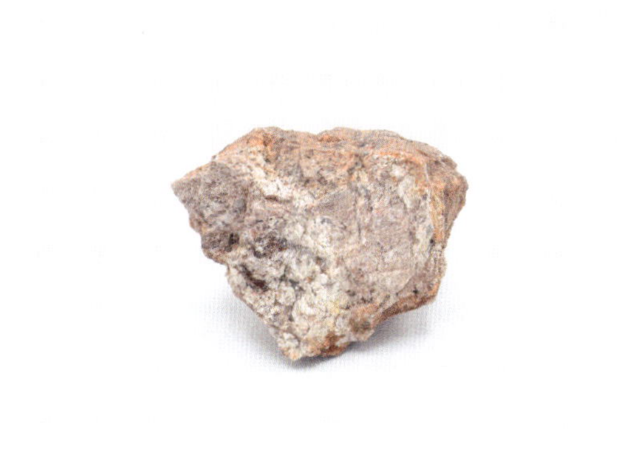

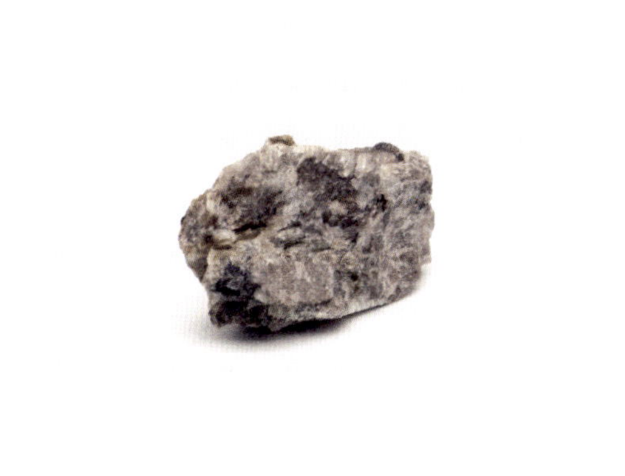

**Above top left:** Feldspar
**Above middle left:** Silica source: quartzite
**Above bottom left:** Sodium feldspar source: pegmatite containing albite

**Above top right:** Silica source: quartz crystal
**Above middle right:** Silica source: flint
**Above bottom right:** Spodumene

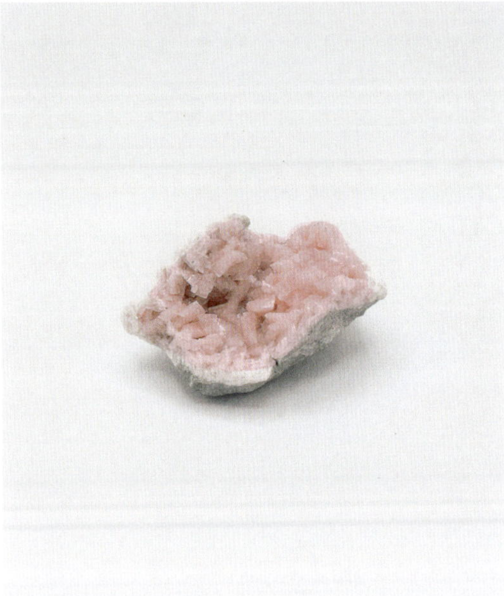

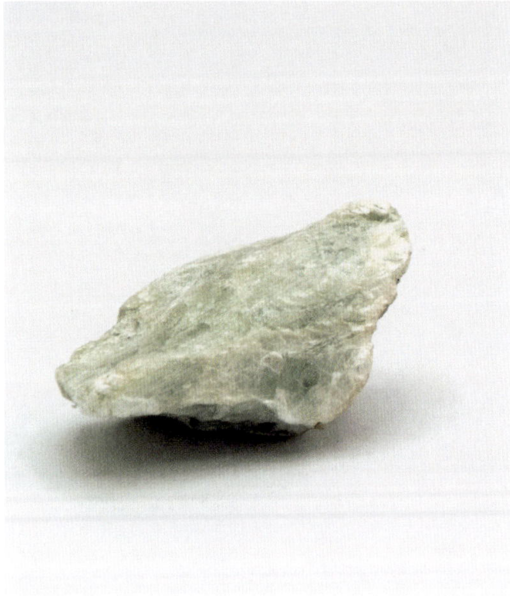

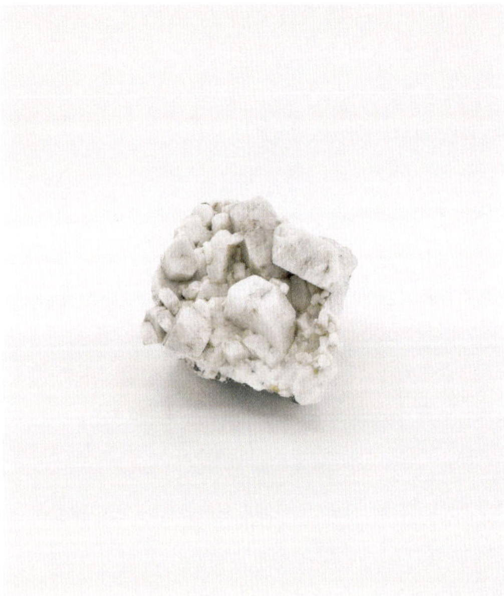

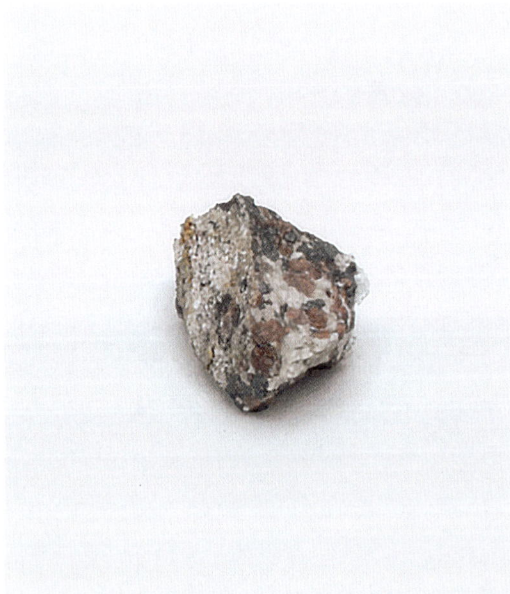

**Above top left:** Strontium source: strontianite
**Above bottom left:** Wollastonite

**Above top right:** Talc
**Above bottom right:** Zinc source: zincite

## STRONTIUM CARBONATE – STRONTIUM (Sr), $SrCO_3$

Strontium carbonate is a non-toxic substitute for barium carbonate, used to achieve brightly coloured matt glazes. It can be used instead of barium in in copper blue glazes, where it will give similar colours, but not in nickel pink, barium-zinc glazes. Strontium carbonate gives off carbon dioxide above 600°C (1112°F).

## TALC – MAGNESIUM (Mg), SILICA (Si), $3Mg.4SiO_2.H_2O$

Talc is magnesium silicate, used to make low fire clay bodies and satin matt glazes. There have recently been problems with sourcing good quality talc. Some talc sources are contaminated with iron oxide and organic matter, while other talc deposits have been associated with asbestos, a magnesium silicate mineral with thin, fibrous crystals that has been found to cause cancer. Talc has low thermal expansion and helps to prevent crazing.

## WHITING – CALCIUM (Ca), $CaCO_3$

Whiting is calcium carbonate, also known as chalk or limestone, used widely as a secondary flux to make glossy and matt glazes. Whiting gives off carbon dioxide above 600°C (1112°F).

## WOLLASTONITE – CALCIUM (Ca), SILICA (Si), $CaSiO_3$

Wollastonite is calcium silicate often used in industry to enable fast firing and avoid bubbles in the glaze. The needle-shaped crystals of wollastonite pack together better than the round particles of whiting, and trap less air in the glaze layer. Wollastonite is mined in north-eastern New York, Finland and Mexico.

## WOOD ASH – CALCIUM (Ca), POTASSIUM (K), SODIUM (Na)

Wood ash contains the minerals absorbed by the growing tree, including calcium, silica, potassium, sodium, magnesium and phosphorus. Different types of tree have different compositions of wood ash. Washing wood ash before adding to a glaze removes some of the soluble alkalis: potassium and sodium salts and makes it less caustic to use.

## ZINC OXIDE – ZINC (Zn), ZnO

Zinc oxide is used to make crystalline glazes. It has high drying shrinkage and is often calcined to prevent crawling, although it can readily absorb water from the air and become lumpy. It has a low thermal expansion and can be used to prevent crazing. However, it has an adverse reaction with some colouring oxides including chromium, which it turns brown or pink.

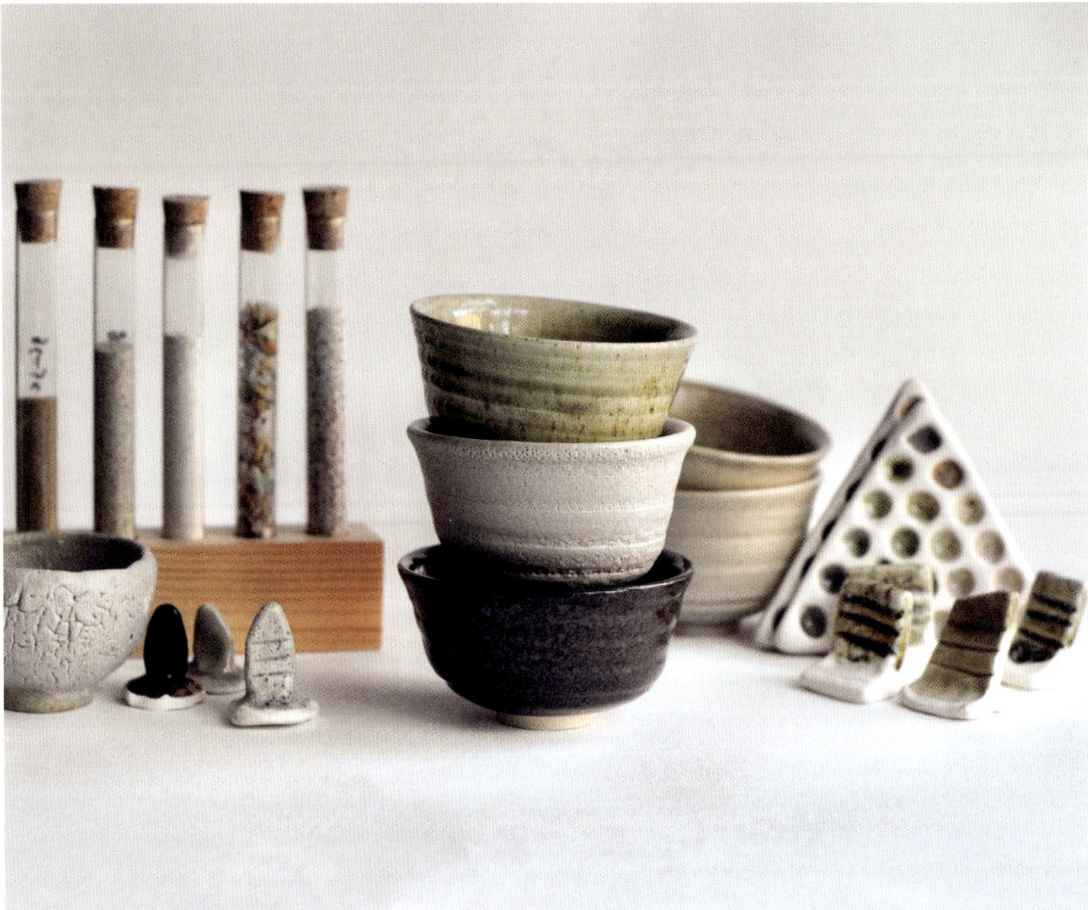

## USING LOCAL AND RECYCLED MATERIALS

Instead of using refined glaze materials bought from a supplier, it is possible to make glazes using locally found materials such as clays, rocks, shells and wood ash. There are four main types of material needed to make a glaze: clay (stabilizer), feldspar (primary flux), calcium (secondary flux) and silica (glass former).

## CLAY

Locally dug clays or slips can be used as decorating slips and glazes. These may contain iron oxide and titanium dioxide as impurities, which will give colour and texture to the slip or glaze. All clays can be recycled before firing and used in slips and glazes, as can waste clay slip from the ceramics studio sink trap. Biscuit ware can also be ground down to use in glazes and as grog in clay bodies.

**Above:** Yuliya Makliuk, *Zero Waste Glazes*, 2024. Material testing of studio waste recycled into glaze on white stoneware, cone 6. *(Image credit: Yulia Makliuk)*

## PRIMARY FLUXES – FELDSPARS AND GRANITE

Volcanic dust or pumice can be used in glazes instead of feldspar. Rock dust from cutting granite can also be used in glazes as a feldspar replacement. If you are using found rocks, biscuit fire and then crush them. To get a fine powder for use in glazes, a ball mill is essential. This is a rotating drum filled with porcelain balls that grinds the glaze mixed with water. Igneous rocks such as granite, rhyolite, andesite and basalt can be used in glazes as a source of alkali metal fluxes, silica and alumina. Other rocks such as limestone and marble can be used as a source of calcium in a glaze, while shales can be crushed and used in the same way as clay.

## SECONDARY FLUXES – CALCIUM SOURCES

Wood ash is high in calcium and also contains potassium, magnesium, silica, phosphorus and iron and is almost a complete glaze on its own, or with the addition of clay to make it less runny. It needs to be sieved before using in glazes. Soak it in water, pour off the soluble alkali and add fresh water, sieve it through a coarse sieve, then leave it to dry out on a porous surface before using in a glaze. Wood ash can also be used in glazes without washing. The soluble alkali in wood ash is very caustic and rubber gloves should be worn when using ash glazes.

Sea shells, egg shells, animal bones and fish bones can all be calcined (biscuit fired) and ground to a powder. Be careful when mixing calcined shells with water as they heat up when the calcium oxide reacts with the water. Bone ash is an ingredient of bone china and can be used in small amounts in glazes as a flux to replace whiting. Another source of calcium is dust from cutting marble or limestone, which can be used instead of whiting to make glossy and matt glazes.

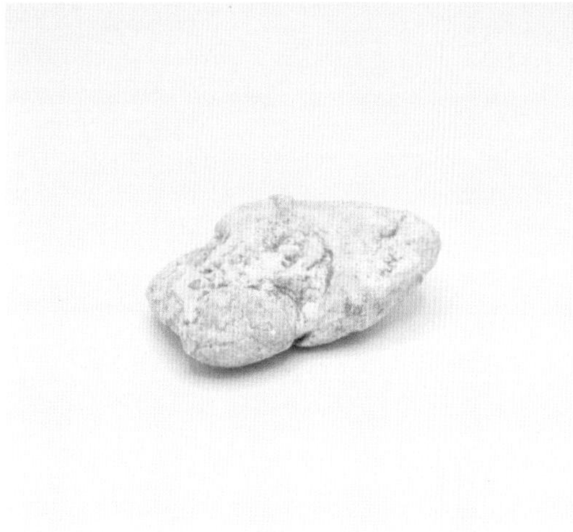

**Wood ash glaze, cone 9, 1280°C (2336°F)**

| | |
|---|---|
| Wood ash | 40 |
| Potash feldspar | 40 |
| Clay | 20 |

**Bone ash clay body, 1250°C (2282°F)**

| | |
|---|---|
| Bone ash | 50 |
| Cornish stone | 25 |
| China clay | 25 |

**Above:** Limestone

**Overleaf:** Carly Breame, 2022. Tableware made using fish bones, oyster shells, wine bottle glass, potato peelings and fruit peelings from a restaurant in Margate, England. *(Image credit: Carly Breame)*

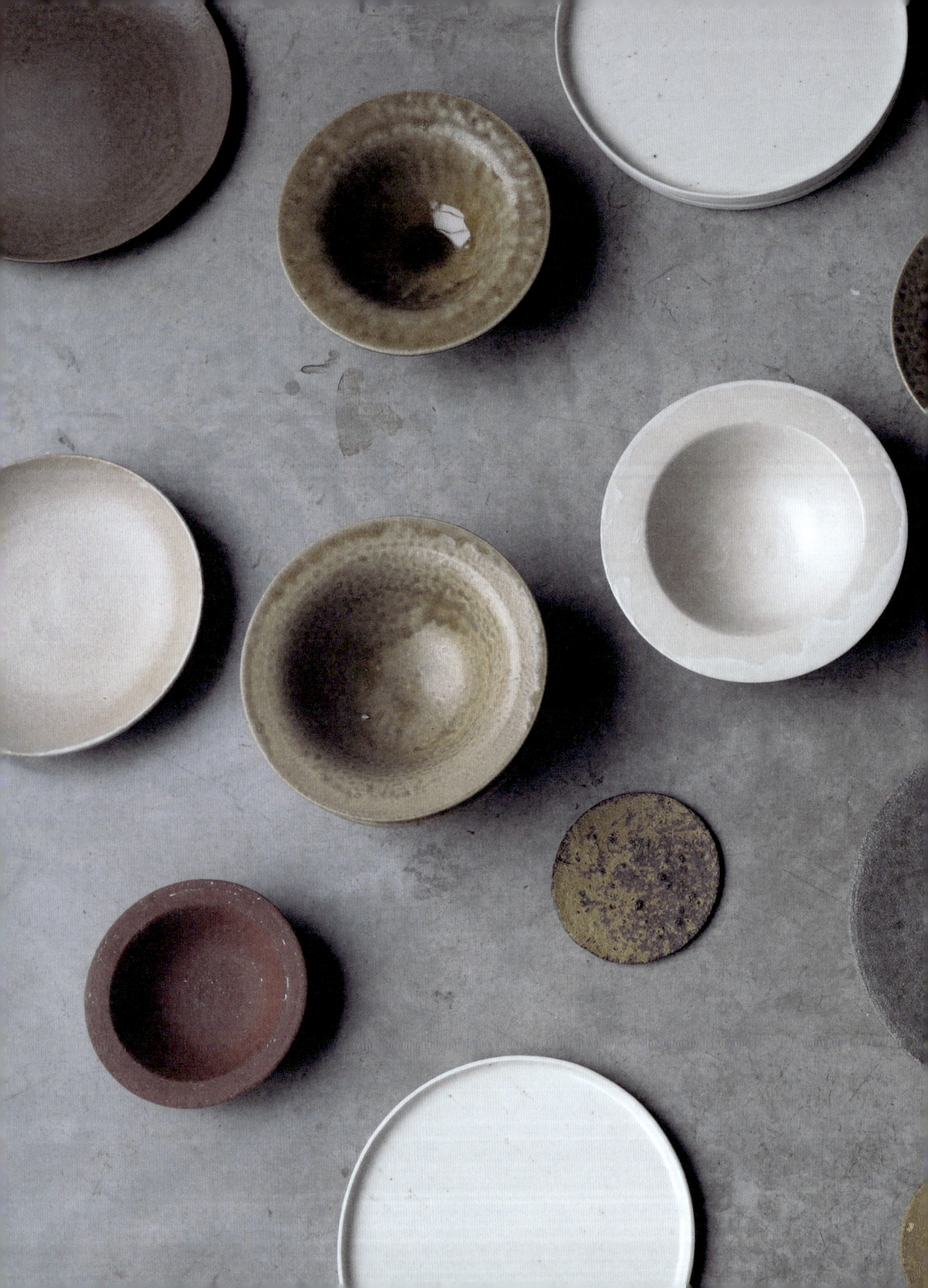

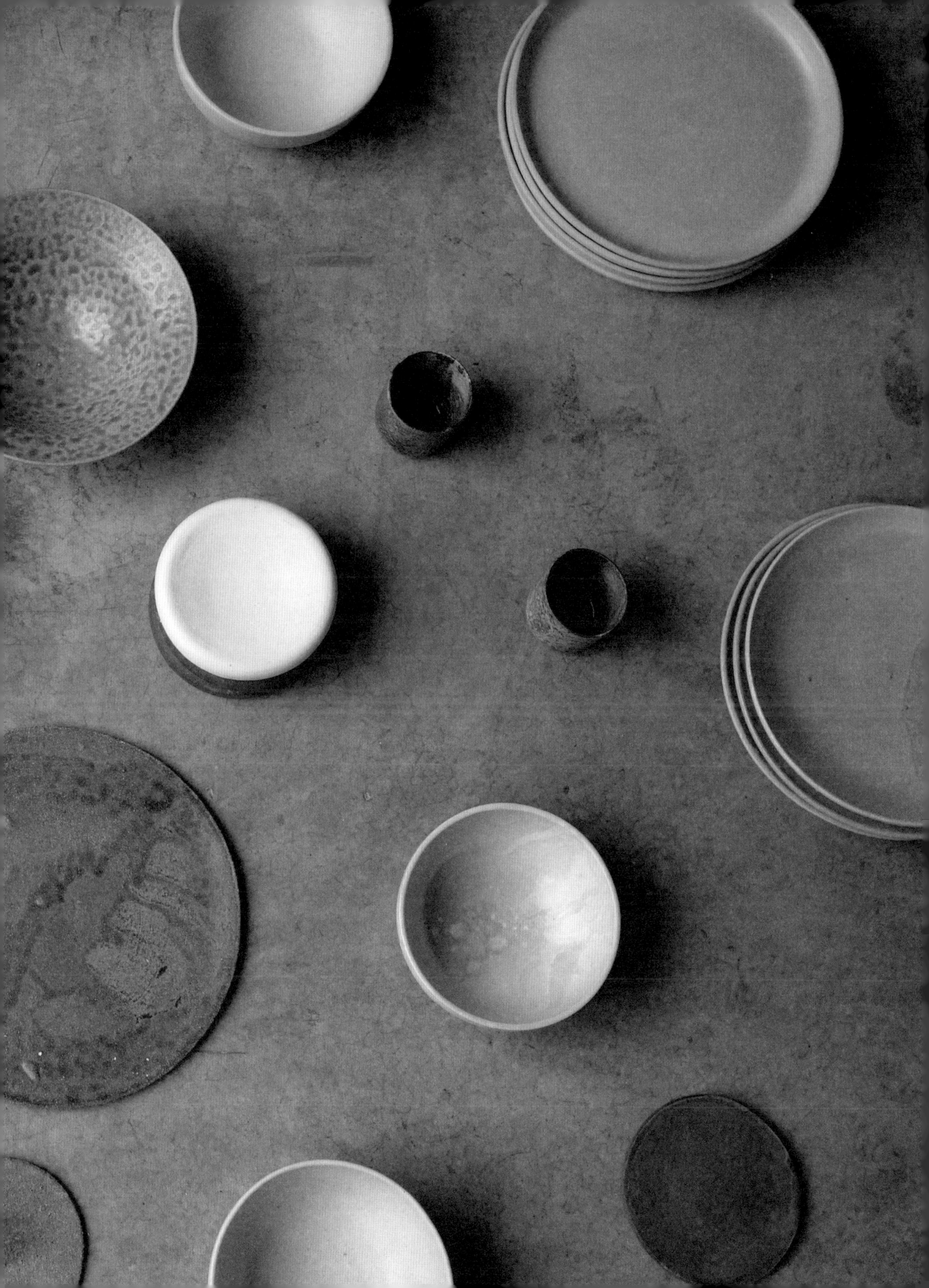

## SILICA

Grass ash and rice hull ash are high in silica and are already in powdered form. Pumice and volcanic ash are also high in silica and can also be obtained as a fine powder used for polishing. Recycled glass from bottles can be ground and used in glazes as a replacement for frits. Glass contains sodium, calcium, silica and colourants and can be a substitute for alkaline frit (Ferro frit 3110). Laboratory glass and Pyrex also contain borax and can be used as a substitute for borax frit (Ferro frit 3134). Blue glass is a good source of cobalt, while green and brown glass contain iron oxide. You can grind glass in a ball mill or an old coffee or spice grinder (but do not use it again for food after grinding glass).

**Glossy transparent glaze, cone 8, 1260°C (2300°F)**

| | |
|---|---|
| Potash feldspar | 34 |
| Borax frit | 14 |
| Whiting | 16 |
| Quartz | 23 |
| China clay | 13 |

**Earthenware glaze, transparent, cone 04, 1060°C (1940°F)**

| | |
|---|---|
| Borax frit (Ferro frit 3195) | 85 |
| Kaolin | 15 |

For feldspar, use ground granite, volcanic ash or pumice
For frit, use ground glass
For whiting, use wood ash or calcined shells
For quartz, use grass or straw ash
For clay, use locally dug or recycled clay

Without knowing the compositions of local materials, you can still make glazes by trial and error, by doing line blends (mixing wood ash and clay in different proportions), biaxial blends (mixing flux materials with quartz and kaolin in different proportions in a grid format) or triaxial blends (mixing three different materials such as wood ash, clay and feldspar). The biaxial grid method is explained by Ian Currie in his book Revealing Glazes and online at www.currie.to. A Currie grid is a 7x5 biaxial grid where the fluxes and colourants remain constant but the clay and silica are varied. The tests are made from four corner glazes: the base fluxes with no clay or added silica, with high clay (40% kaolin), with high silica (50% silica), and with both high clay and high silica (25% kaolin, 40% silica). These four corner glazes are mixed with water and sieved. The rest of the grid is generated by combining the four corner glazes in varying proportions, measuring wet volume using a syringe.

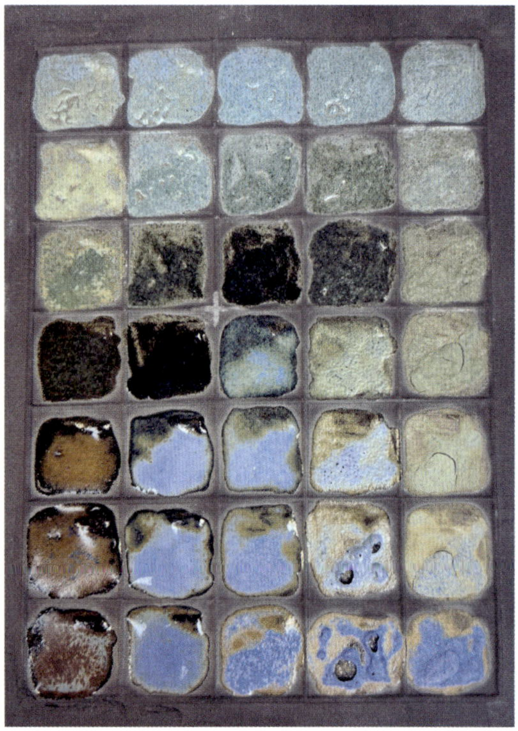

## ANALYSING A BIAXIAL GRID

### A balanced glaze

Glossy transparent glazes are the result of the three parts of the glaze (glass former, stabilizer, flux) being in balance. This means there is enough of each to completely melt and form the glass structure with everything tied up and nothing left 'hanging' in the structure. If we increase/decrease any of these three parts, the appearance of the glaze changes. When there is more of one oxide than necessary to form the glass structure, the excess of this oxide remains suspended in the structure. This causes the structure to be less transparent and the excess oxides may also form crystals which refract light and dampen gloss.

### Excess silica

Glazes with an excess of silica are called silica matts. These surfaces can be harsh and dry or appear rough like sandpaper. As the silica amount in a glaze increases beyond the balanced amount, crazing begins to appear until excess silica inhibits the glaze from melting.

### Excess alumina

Glazes with an excess of alumina tend to be stiff glazes with a tendency to form pinholes, since the glaze is not fluid enough to flow to close holes. As the amount of alumina increases, the glaze surface turns from transparent glossy to satin, then opaque matt and finally becomes a dry matt glaze.

### Excess flux

Glazes with an excess flux oxide may appear glossy and matt in patches where the flux forms crystals. Alternatively, the glaze may become fluid and runny. Alkali metal fluxes are powerful fluxes that in large amounts will cause glazes to run and craze. Alkaline earth fluxes such as calcium, magnesium and barium are weaker and tend to crystallize and form blotchy to matt surfaces in high amounts. Zinc in excess with absence of alumina will form visible crystals.

As an alternative to extensive testing, you can have a material analysed in a laboratory by x-ray fluorescence spectroscopy. This is much less expensive than having a glaze tested for leaching. Once you have the chemical analysis, you can substitute the material in a glaze recipe in glazy.org.

## COLOURING OXIDES

Instead of buying colouring oxides (see Chapter 8), you can grind flakes of rust to make red iron oxide, or use red, yellow or brown clay dug from a stream, road cutting or riverbank, which will contain varying levels of iron oxide. Wood ash, particularly from pine, contains iron oxide and manganese dioxide. Cobalt can be recycled from blue glass, but remember you will also be introducing silica, sodium and calcium from the glass, which may make your glaze more runny. Copper can be obtained from offcuts of electrical wire. Mix the copper wire with a small amount of vinegar, leave for a few days to make verdigris, then scrape off and use as copper carbonate for decoration or in a glaze.

**Opposite:** Camille David-Boudreault, Biaxial test tile of Bristol glaze with cobalt and titanium on black clay, fired at cone 6. Original glaze recipe: Custer feldspar 37, whiting 13, zinc oxide 11, kaolin 16, silica 23, bentonite 2, titanium dioxide 5.5, cobalt carbonate 2

CHAPTER SEVEN

# GLAZE CHEMISTRY AND CALCULATION

It is better to use glaze oxides as they occur naturally (mixed up in combination in a feldspar or clay) because they melt more readily than pure oxides. However, we need to work out the proportions of each oxide in the glaze in order to get the right balance for melting (the lowest melting combination, known as a eutectic; see Chapter 5).

## COMPOSITION OF A GLAZE

It is important to understand which material in a glaze supplies each oxide:

- Silica ($SiO_2$), the glass former, from quartz or flint and a small amount from the clay added to the glaze
- Alumina ($Al_2O_3$), the stabilizer, from the clay added to the glaze
- Fluxes: feldspar supplies potash and soda (KNaO) and whiting supplies calcium (CaO)

A combination of two or three fluxes melts better than one flux alone. It is also important to combine different types of fluxes in the correct proportions to make the glaze chemically stable so that it will not leach or break down in contact with acids or alkalis.

**Opposite:** Mel Arsenault, *Coliadinae*, 2021. Glazes appear quite different in the raw and fired states as revealed in this piece fired to 1223°C (2233°F). *(Image credit: Mel Arsenault)*

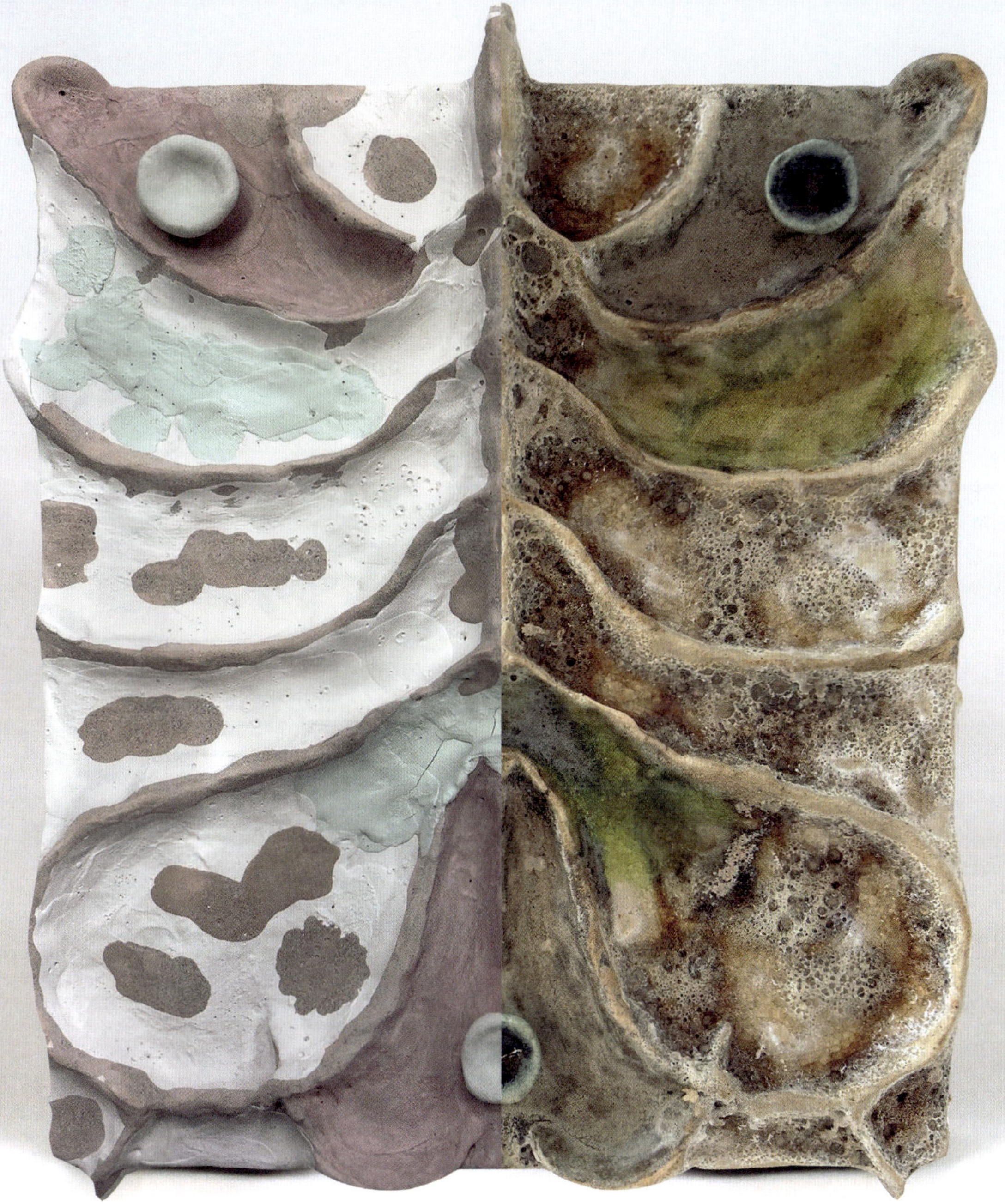

## UNITY MOLECULAR FORMULA

It is useful to calculate the molecular formula of a glaze in order to better understand your glaze and how to adjust it. The molecular formula shows the proportions of the different oxides in the glaze. You can predict whether the glaze will be glossy or matt, and whether it will be chemically stable and safe for use on functional ware.

### CALCULATING THE MOLECULAR FORMULA FROM A GLAZE RECIPE

**Glaze recipe, cone 10, derived from a Tony Hansen glossy transparent**

| | |
|---|---|
| Feldspar | 30 |
| Quartz | 30 |
| Whiting | 20 |
| Kaolin | 20 |

A glaze recipe is usually written by the weight percent of each material in the glaze, adding up to one hundred. To convert into numbers of molecules, each recipe weight is divided by the molecular weight of the material.

### MOLECULAR WEIGHT

Molecular weights are the sum of the atomic weights in the molecule. For example, the molecular weight of whiting, calcium carbonate ($CaCO_3$) is the sum of the atomic weights of calcium, carbon and three oxygens, 40 + 12 + (3 x 16) = 100 (see Appendices 1 and 11). The carbon is lost during firing as carbon dioxide gas, leaving only calcium oxide in the glaze.

To calculate the molecular formula, the chemical formula of each glaze material is first split into its constituent oxides. Below we see the chemical formula of each glaze ingredient:

Potash feldspar $K_2O.Al_2O_3.6SiO_2$
Quartz $SiO_2$
Whiting $CaCO_3$
Kaolin $Al_2O_3.2SiO_2.2H_2O$

We can see in this list the numbers of molecules of each oxide in each glaze ingredient. For example kaolin has one molecule of alumina ($Al_2O_3$) and two molecules of silica ($SiO_2$).

### MOLECULAR EQUIVALENT

Divide the glaze recipe weight percent for each material by the molecular weight to get the molecular equivalent. This is the number of molecules of each oxide in the glaze.

- I used potash (K) feldspar, with a theoretical molecular weight 556 (see Appendix 11). The glaze recipe has 30% feldspar, so the molecular equivalent for feldspar is 30/556 = 0.054 x $K_2O$ and 0.054 x $Al_2O_3$ and 0.054 x $6SiO_2$

- Quartz has a molecular weight of 60, and the glaze recipe is 30% quartz, so 30/60 = 0.5 x $SiO_2$

- Whiting has a molecular weight of 100 and the glaze recipe is 20% whiting, so is 20/100 = 0.2 x $CaCO_3$ (carbon dioxide $CO_2$ is lost during firing to leave CaO)

- Kaolin has a molecular weight of 258 and the glaze recipe is 20% kaolin, so 20/258 = 0.078 x $Al_2O_3$ and 0.078 x $2SiO_2$
- $2H_2O$ (ignore the water, which is lost during firing)

So we now know that for every 0.054 molecules of feldspar in the glaze, there are 0.5 molecules of silica, 0.2 molecules of calcia, and 0.078 molecules of kaolin. A 'molecule of feldspar' can be seen from its formula to contain one molecule of potassium oxide, one molecule of aluminium oxide and six molecules of silica.

Some of the glaze materials, kaolin and feldspar, contain two or three different oxides. Add together the total silica, total alumina and total fluxes.

- **Alumina**: Alumina is found in kaolin and feldspar. We know that there is 30/556 (or 0.054) in feldspar and 20/258 (or 0.078) in kaolin. Adding these gives 0.131.
- **Silica**: Silica is found in kaolin, feldspar and quartz. The chemical formula shows that there are six silica molecules within the feldspar molecule, two in the kaolin molecule and one in quartz so the total number of silica molecules relative to other molecules is 6 x 0.054 + 2 x 0.078 + 1 x 0.5 = 0.979.
- **Flux**: Finally we add all the fluxes together, in this case calcium from the whiting plus potassium from the feldspar. The relative amounts are 0.2 CaO + 0.054 $K_2O$ = 0.254 total flux.

## UNITY

By convention, the total flux needs to be one as this is the unity formula. This allows us to see the amounts of alumina and silica relative to one unit of flux.

To do this, we take our total for the fluxes (0.254 in the example) and everything is divided by that total number to make the sum of the fluxes one, so that the number of molecules of silica and alumina are relative to one unit of flux.

Again, the convention is to divide 1 by this total for the fluxes and then to multiply the other glaze ingredients by this number, rather than dividing them by the total for the fluxes (which is effectively the same thing). So, we will divide 1 by the total fluxes and then multiply the total silica, alumina, and fluxes by this number to get the unity molecular formula. This formula gives the relative proportions of each type of oxide in the glaze, relative to one unit of flux.

Add 0.2 CaO and 0.054 K2O = 0.254
Multiply everything by 1/0.254 = 3.94 to make sum of fluxes = 1.0
So now the sum of the fluxes is 3.94 x 0.254 =1.0

Potassia is 3.94 x 0.054 = 0.21
Calcia is 3.94 x 0.2 = 0.79
Alumina is 3.94 x 0.131 = 0.52
Silica is 3.94 x 0.979 = 3.86

| Glaze ingredient | Kaolin | Potash feldspar | Whiting | Quartz | Total | Convert to unity formula |
|---|---|---|---|---|---|---|
| Formula | $Al_2O_3.2SiO_2.2H_2O$ | $K_2O.Al_2O_3.6SiO_2$ | $CaCO_3$ | $SiO_2$ | | |
| Molecular weight | 258 | 556 | 100 | 60 | | |
| Percentage in recipe | 20 | 30 | 20 | 30 | 100 | |
| Molecular equivalent | 0.078 | 0.054 | 0.2 | 0.5 | | |
| Molecules of alumina per molecule of glaze ingredient | 1 | 1 | | | | |
| Amount of alumina from glaze ingredient | 0.078 | 0.054 | | | 0.131 | 0.52 |
| Molecules of silica per molecule of glaze ingredient | 2 | 6 | | 1 | | |
| Amount of silica from glaze ingredient | 0.156 | 0.324 | | 0.5 | 0.979 | 3.86 |
| Molecules of flux per molecule of glaze ingredient | | 1 | 1 | | | |
| Amount of flux from glaze ingredient | | 0.054 | 0.2 | | 0.254 | 1 ($K_2O$: 0.21, CaO: 0.79) |

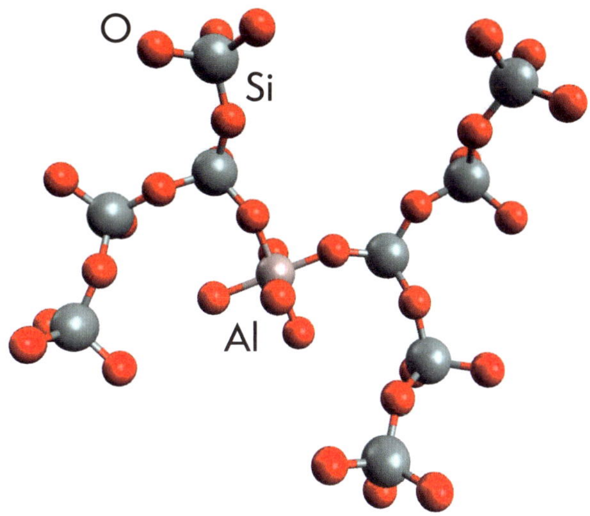


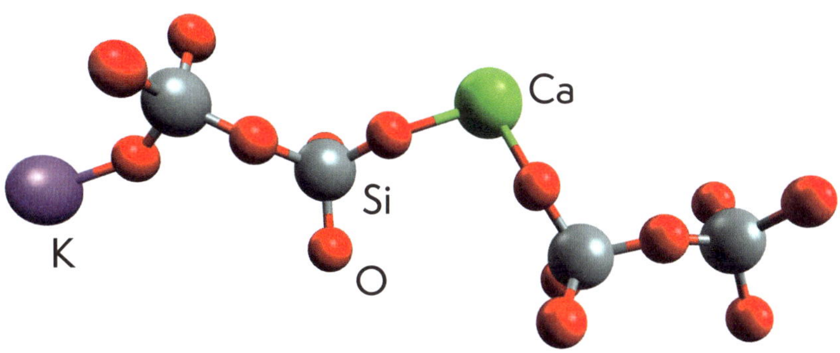


**Above:** Diagrams of molecule. Top diagram shows one molecule of alumina and nine of silica. Bottom diagram shows fluxes potassium, calcium oxide and four silica molecules. Oxygen atoms are shown in red

## MOLECULAR FORMULA

Rearranged into three columns, the unity molecular formula (UMF) is:

| Flux $R_2O/RO$ | Stabilizer $R_2O_3$ | Glass former $RO_2$ |
|---|---|---|
| $K_2O$ 0.21 | $Al_2O_3$ 0.52 | $SiO_2$ 3.86 |
| CaO 0.79 | | |
| Total 1.00 | | |

The molecular formula is arranged into three columns: alkaline fluxes, amphoteric stabilizer alumina and acidic glass former silica. All the fluxes in the first column add up to one. Alumina ($Al_2O_3$) (and boric oxide ($B_2O_3$)) are in the middle. Silica, $SiO_2$ is on the right.

## INTERPRETING THE MOLECULAR FORMULA

We can get two important pieces of information about the glaze from this molecular formula:

### 1. Flux ratio

Looking down the first column of the formula chart, the flux ratio potassium:calcium ($K_2O$:CaO) or in general $R_2O$:RO is 0.21:0.79.

$R_2O$ is the chemical formula for alkali metals, RO is alkaline earths.

This means it is a chemically stable glaze. Any glaze with a ratio 0.3:0.7 ± 0.1 will be stable. A glaze with too much alkali metal, more than 0.4:0.6 will be chemically unstable as there will be alkalis not tied up in the glaze that are free to react with acids in food. To make the glaze food safe, increase the level of alkaline earths relative to alkali metals. This particular glaze has potassium as the alkali metal and calcium as the alkaline earth flux in the ratio 0.21:0.79.

### 2. Alumina to silica ratio

Looking across the first row of the formula, from potassium to alumina and silica, the ratio of alumina to silica ($Al_2O_3$:$SiO_2$) is 0.52:3.86 or 1:7.

This tells us it is a glossy glaze. Glazes with around 8 silica to 1 alumina will be glossy, while glazes with around 4 silica to 1 alumina will be matt.

Now you know how to interpret the molecular formula, you can use glaze calculation software such as glazy.org or Hyperglaze.

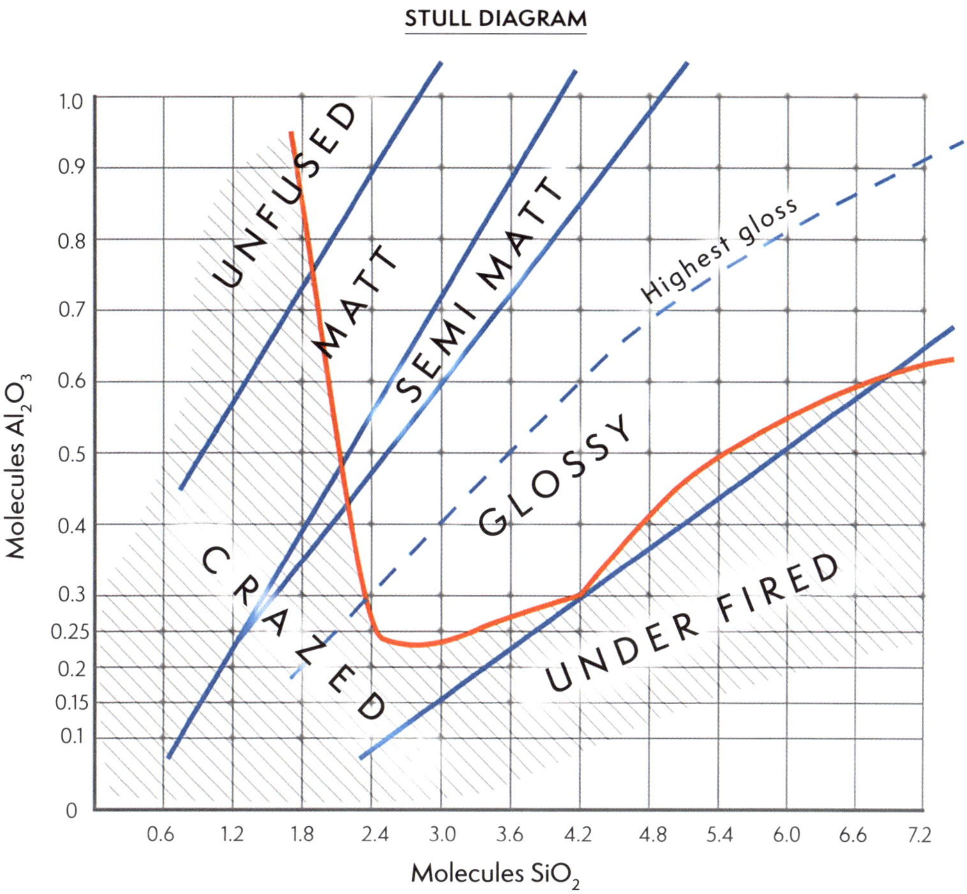


## STULL CHART

The Stull diagram shows the molecular formula on a graph of alumina against silica (always assuming there is one unit of flux). The straight lines on the graph represent the alumina to silica ratios 1:4 for matt glazes, 1:5 for semi-matt and 1:12 on the edge of the glossy region; if there is too much silica, the glaze will not melt. The dashed line shows the experimental results for the glossiest glazes. RT Stull was a ceramic engineer who in 1912 made a series of glaze tests using potassium (0.3 $K_2O$) and calcium (0.7 CaO) on porcelain, fired to cone 11. Each grid intersection on the graph represents a glaze test. The flux is constant and the alumina and silica vary. The most glossy glazes lie along the dashed line around 1:8 alumina to silica. The more silica and alumina, the higher the temperature needed to melt the glaze. The hatched area shows crazed glazes where the glaze contracts more than the clay body owing to the relatively high level of fluxes compared to silica and alumina in the glaze.

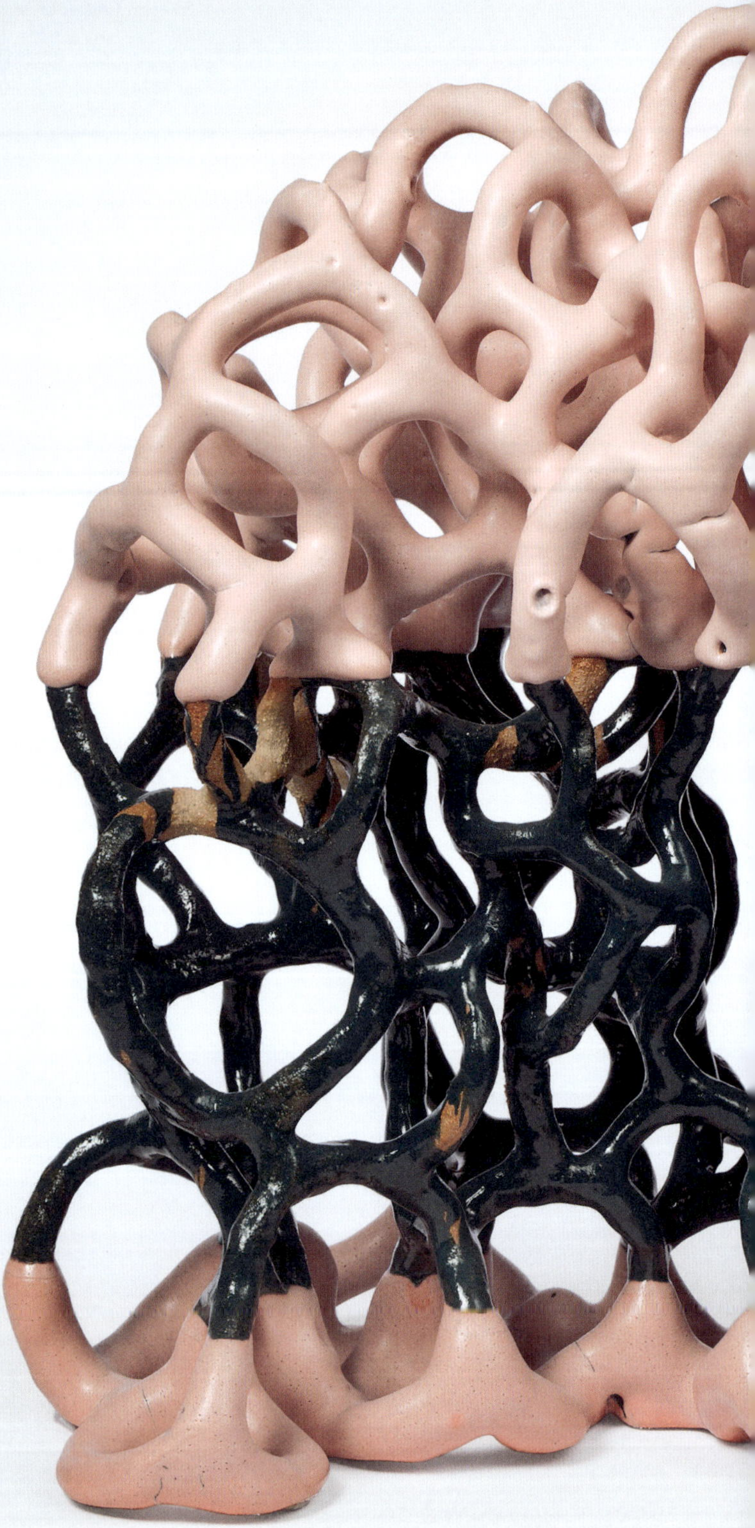

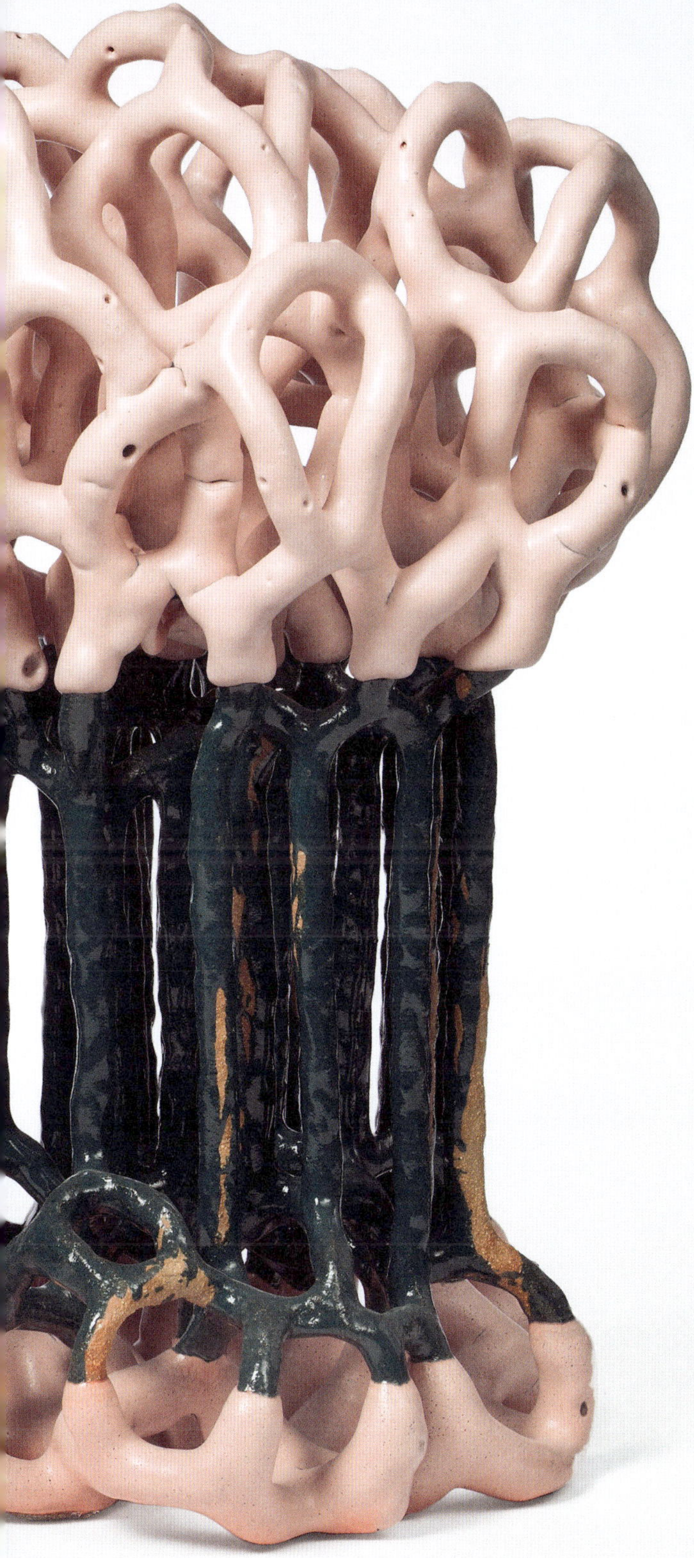

## ADJUSTING GLAZES: GLOSSY OR MATT

If you plot a glaze on the Stull chart, it then becomes clear how to adjust it. To make the glaze more glossy, adjust the level of silica until the alumina:silica ratio is close to 1:8. To make the glaze matt, reduce the silica or add alumina in the form of clay until the ratio is 1:5. To correct crazing, increase both the silica and alumina until the glaze moves out of the crazed region and into the uncrazed region.

**Previous:** Bente Skjøttgaard, *Pink Cloud no 1033*, 2010. Matt and glossy glazes activate and define zones of this sculpture. (*Image credit: Ole Akhø*)

**Opposite:** Linda Swanson, *Spectre 1117204*, 2020. Zinc dominated glazes with cobalt, nickel and titanium over a transparent glaze on porcelain. (*Image credit: Luuk Smits*)

## WHERE DO THE OTHER FLUXES GO IN THE MOLECULAR FORMULA?

We have looked at potassium and calcium. Most other fluxes can be grouped into alkali metals, alkaline earths, and amphoteric oxides, according to their chemical formula. Boron is usually included in the $R_2O_3$ column as it has a similar chemical formula to alumina.

| **Flux** Alkali metal **$R_2O$** | **Stabilizer** Amphoteric **$R_2O_3$** | **Glass former** Acidic **$RO_2$** |
|---|---|---|
| Potassium $K_2O$ | Alumina $Al_2O_3$ | Silica $SiO_2$ |
| Sodium $Na_2O$ | Boron $B_2O_3$ | |
| Lithium $Li_2O$ | | |
| Alkaline earth **RO** | | **Other** |
| Calcium CaO | | Phosphoric oxide $P_2O_5$ |
| Magnesium MgO | | |
| Barium BaO | | |
| Strontium SrO | | |
| Zinc ZnO | | |

## WHERE DO COLOURING OXIDES FIT IN? THE EXTENDED MOLECULAR FORMULA

Traditionally, colourants were not included in the molecular formula. However, these colourants, as well as adding colour to the glaze can also act like fluxes. Research done by Matt and Rose Katz[1] has extended the molecular formula to include many of the colourants, such as copper, cobalt and manganese. Copper oxide acts as a strong alkali metal flux, while cobalt and manganese act as alkaline earth fluxes. These do not fit neatly into $R_2O$, RO or $R_2O_3$ categories corresponding to their chemical formula, but have been categorized according to their fluxing action when fired in a glaze.

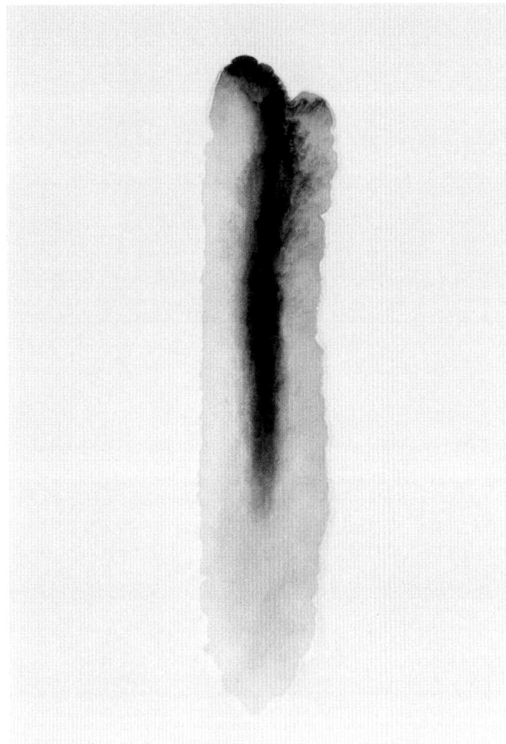

Two of the colouring oxides, titanium and nickel, act as amphoterics or stabilizers, making the glaze more viscous and encouraging crystals to form. While chromium also acts to stiffen the glaze, it is not very soluble in the glaze so is not included in the molecular formula.

| **Alkali metal fluxes $R_2O$** | **Amphoteric $R_2O_3$** | **Acidic $RO_2$** | **Other** |
|---|---|---|---|
| Copper oxide ($CuO$) | Nickel oxide ($Ni_2O_3$) | Zirconium oxide ($ZrO_2$) | Chromium oxide ($Cr_2O_3$) |
| Tin oxide ($SnO_2$) | Titanium dioxide ($TiO_2$) | | |
| **Alkaline earth fluxes RO** | | | |
| Cobalt oxide ($CoO$) | | | |
| Manganese dioxide ($MnO_2$) | | | |
| Iron oxide ($Fe_2O_3$) | | | |

[1] Matt and Rose Katz, Ceramic Materials Workshop NCECA lecture, *Colorforms*, 2019

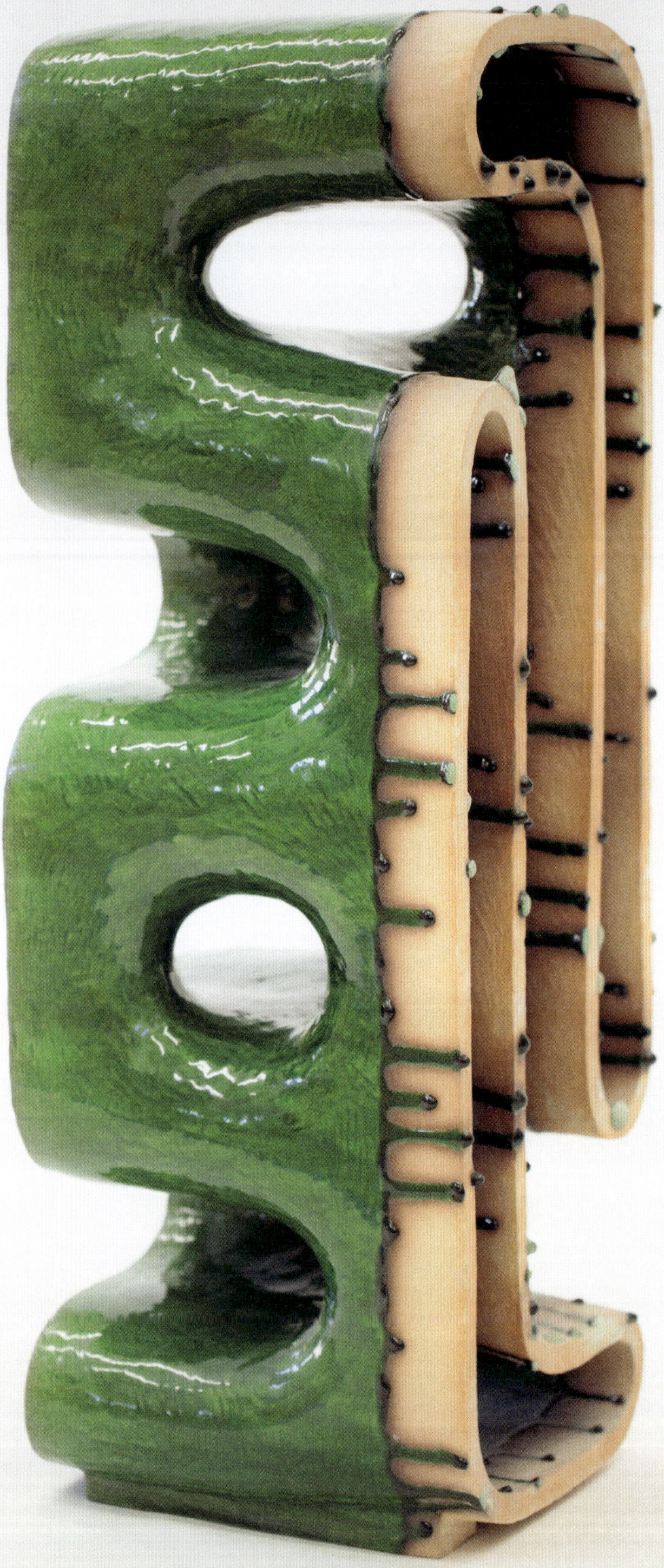

CHAPTER EIGHT

# COLOURANTS AND OPACIFIERS

In this chapter we look at how colourants are added to a glaze recipe, the properties of the colouring oxides and how opacifiers work.

## COLOURANTS AND THEIR EFFECTS, CARBONATES, OXIDES AND SULPHATES

Colouring oxides can be added to slips and glazes to provide colour. They are concentrated mineral pigments and only a small amount (1–3%) is needed in a glaze; larger amounts of 5–10% are used to colour clays, slips and engobes. They are usually considered to be an addition to the base glaze recipe, for example, 1% colouring oxide would be 1 g added to 100 g base glaze (dry weight). Carbonates are less concentrated than oxides and convert to oxides during firing, so more carbonate is needed to get the same intensity of colour as an oxide. The colour of both oxides and carbonates usually changes during firing, depending on the dominant alkaline earth flux; for example, cobalt turns purple in a magnesium glaze. Sulphates are sometimes used to give watercolour effects, but as they are soluble in water they can be more toxic than oxides, which are insoluble.

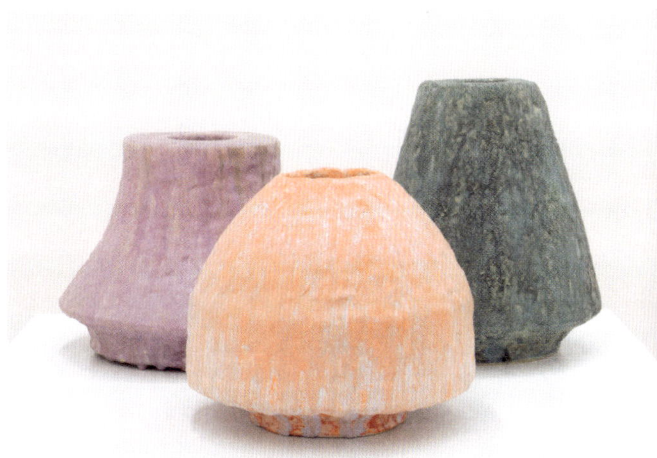

**Opposite:** Del Harrow, *Green Matrix*, 2021. This bright green glaze on stoneware exemplifies the colour that chrome typically brings to a glaze. *(Image credit: Del Harrow)*

**Above:** Anders Herwald Ruhwald, *Glasurstykke #80, #69, #91*, 2022. Glazed stoneware vessels. *(Image credit: Morán Morán Gallery, Mexico Ctiy and Los Angeles)*

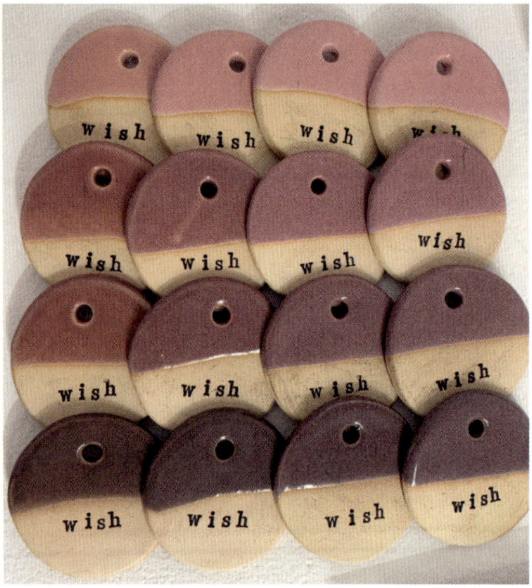


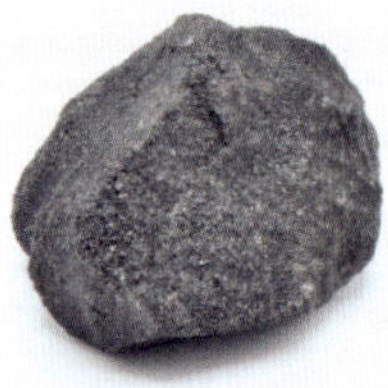

## CHROMIUM

Chromium oxide ($Cr_2O_3$) (toxic)
Iron chromate ($Fe_2Cr_3O_{12}$) (iron + chromium)

Chromium oxide is supplied to potters as a green powder, ($Cr_2O_3$), which is toxic. It is refractory (it withstands high temperatures and stiffens the glaze) and only small amounts (0.2–1.0%) will dissolve in a glaze, while larger amounts cause the glaze to turn a flat, opaque green. Chromium (0.1%) and tin (5%) will produce a pink colour in glazes high in calcium, while chromium and cobalt will give an opaque teal colour. Chromium can also be used by potters in the form of iron chromate ($Fe_2Cr_3O_{12}$), which gives grey or brown colours in glazes.

## COBALT

Cobalt carbonate ($CoCO_3$)
Cobalt oxide ($CoO$, $Co_3O_4$)
Cobalt sulphate ($CoSO_4$) (toxic)

The cobalt oxide supplied to potters is a usually a mixture of oxides $Co_3O_4$ and $CoO$. The monoxide $CoO$ is more toxic than the tetroxide $Co_3O_4$. They are both black powders, which change to blue on firing in a glaze. Cobalt is a very strong colourant and only a small amount is needed (0.1–2.0%). Cobalt carbonate ($CoCO_3$) is a pink powder which is two-thirds the strength of the oxide, so 1.5x more is needed in the glaze recipe. Cobalt chloride, sulphate and nitrate can be used to give watercolour effects but they are soluble in water and therefore more toxic. Cobalt gives a very strong, bright blue in glazes, which can be toned down by adding iron and manganese. It turns purple in magnesium glazes containing dolomite or talc. It acts as a flux and can cause the glaze to run. If too much cobalt (more than 3% oxide) is added to a glaze, it may form a black, metallic sheen on the surface.

**Above top:** Isabella Bursey, *Chrome Tin Pink Wishes*, 2023. Increasing amounts of chrome added to a tin glaze generates a range of pinks to maroon on test tiles with a second life as wish tokens. *(Image credit: Isabella Bursey)*

**Above bottom:** Chrome source: chromite

**Opposite:** Cobalt source: skutterudite (microscopic view)

**Overleaf:** Adam Chau, *Swipe*, 2019. The cobalt used to create the imagery on these porcelain forms was recovered from the lithium ion battery of a smartphone. *(Image credit: Northern Clay Center)*

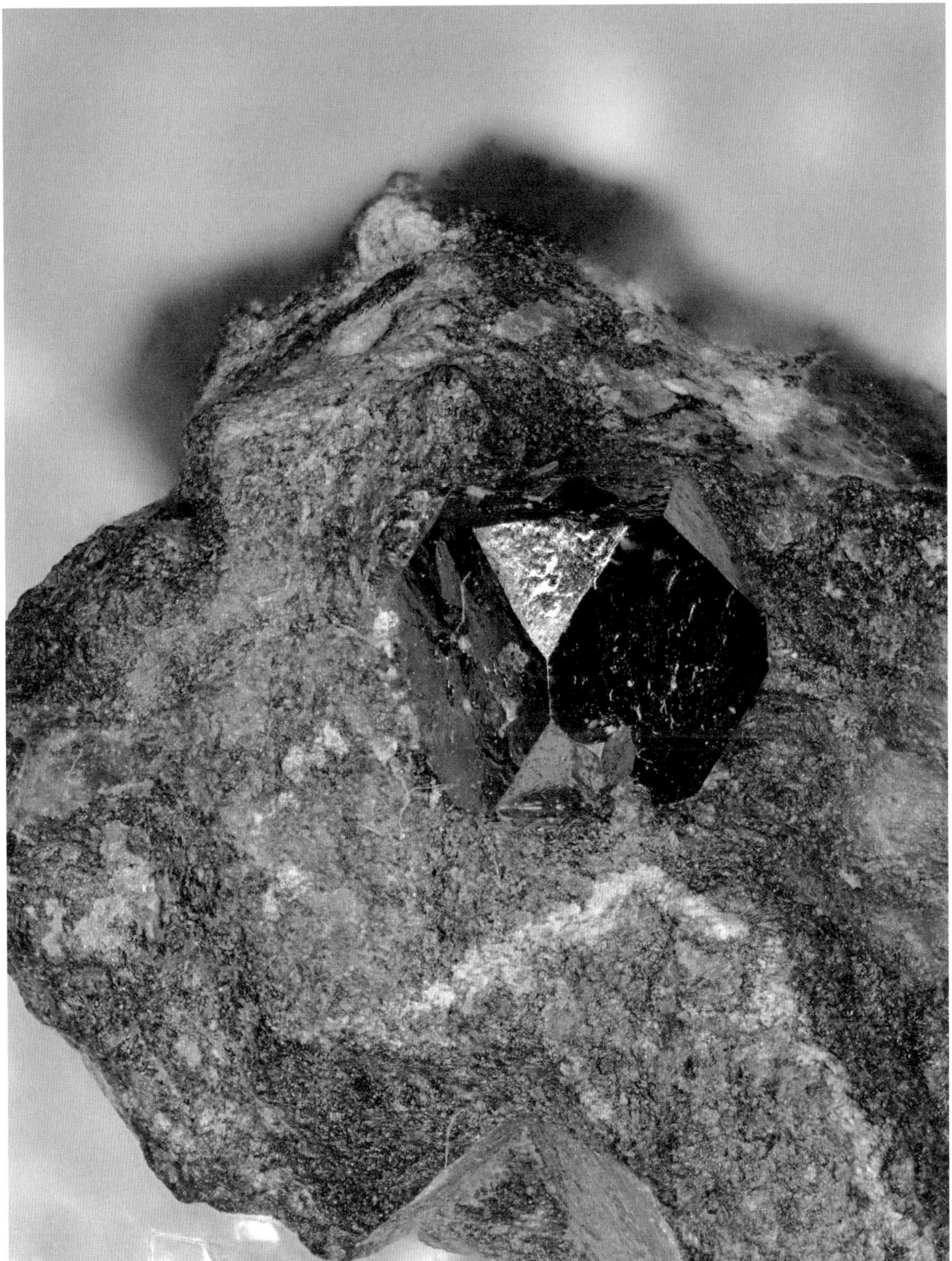

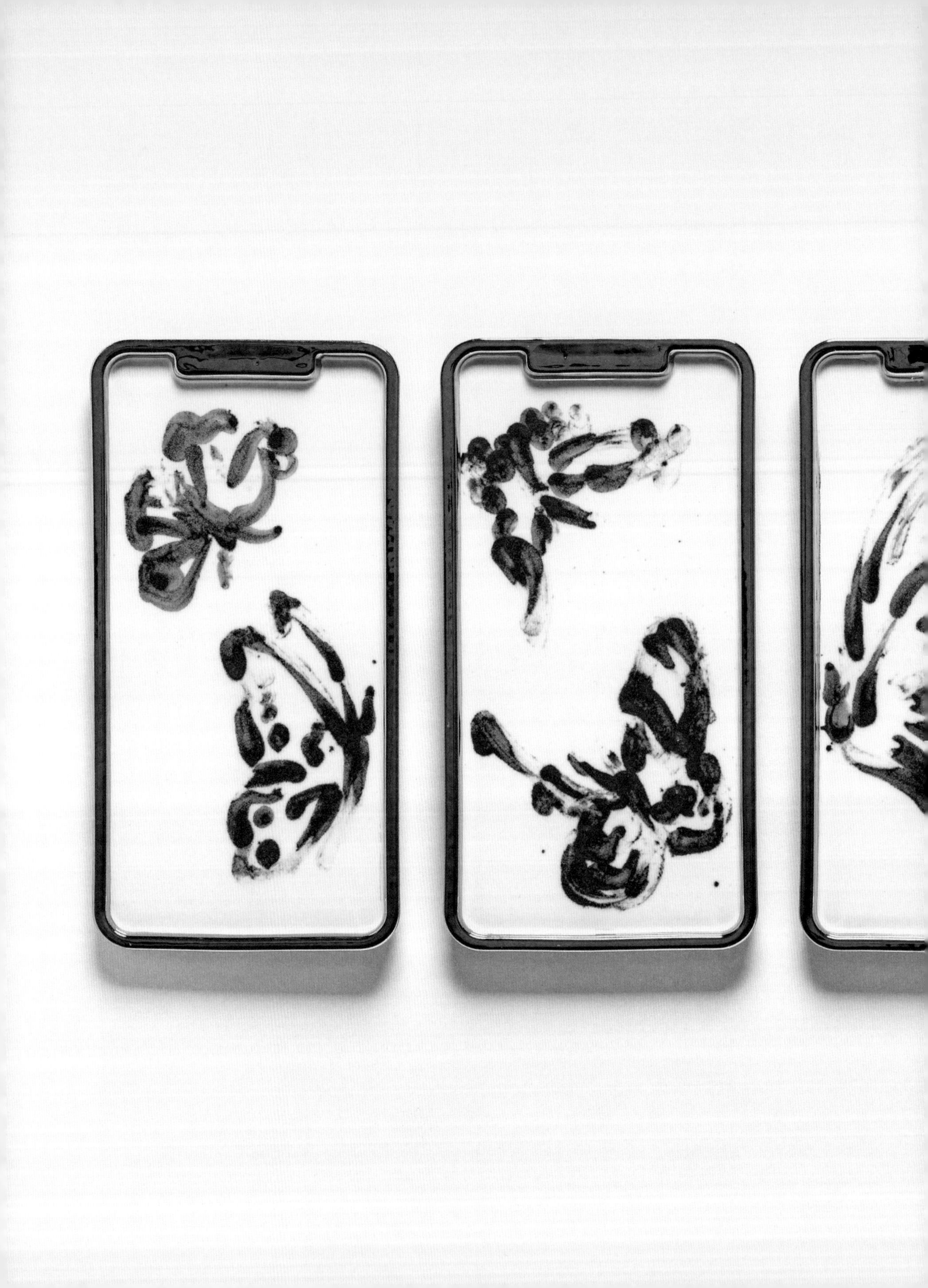

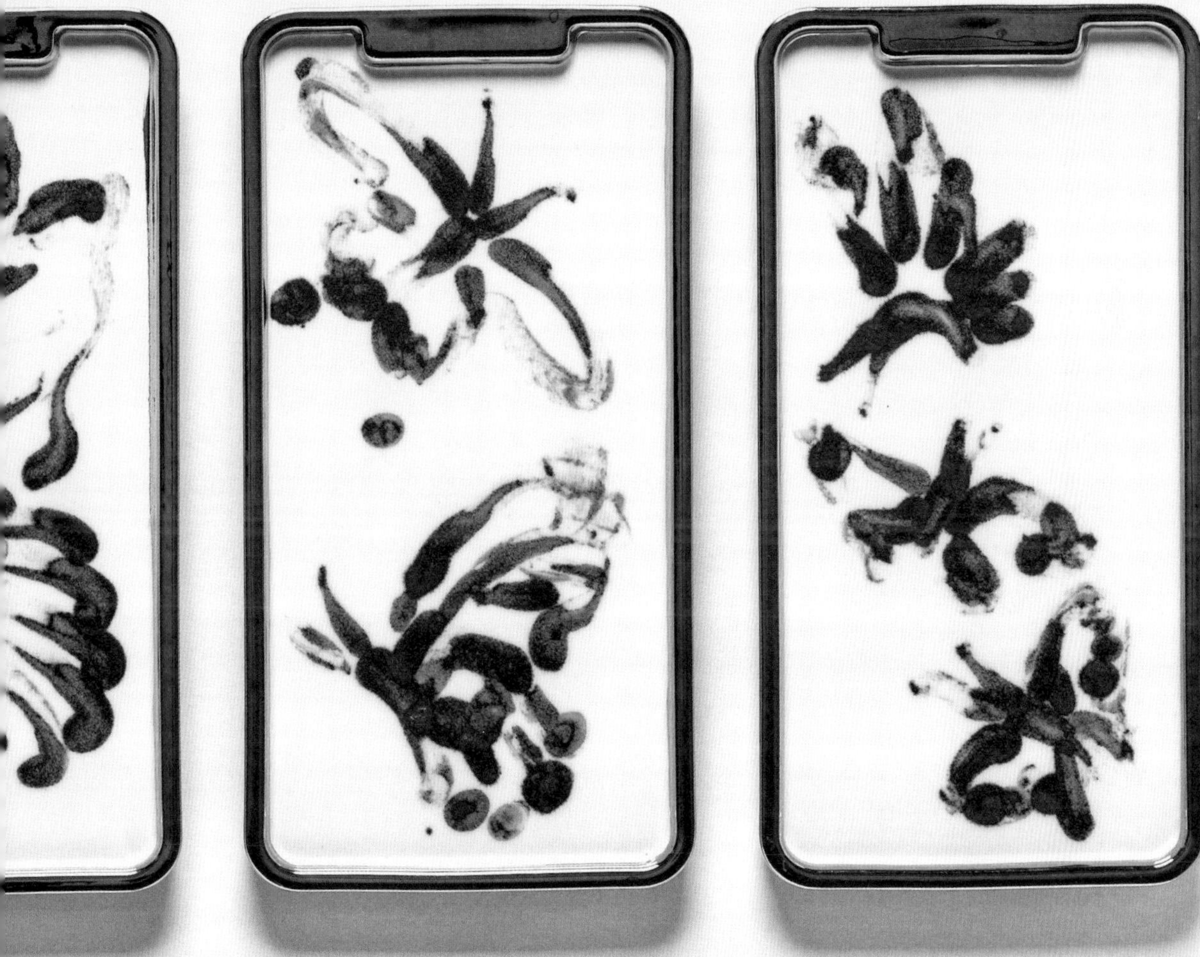

## COPPER

Copper carbonate ($CuCO_3$)
Copper oxide ($CuO$)
Copper sulphate ($CuSO_4$) (toxic)

Copper oxide ($CuO$) is a black powder, while copper carbonate ($CuCO_3$) is verdigris green and is a weaker colourant. To provide the same strength of colour, you would need to add 55% more carbonate than oxide. A small amount of oxide (0.5–3.0%) is used in glazes to give green or turquoise in oxidation and maroon red or pink in reduction. Copper turns turquoise in calcium and barium glazes and green in magnesium and zinc glazes. Copper sulphate is more toxic than the oxide or carbonate as it is water soluble. Copper acts as a strong flux, making the glaze more runny. Excess copper oxide (more than 3%) added to a glaze will precipitate on the surface to form a black, metallic sheen.

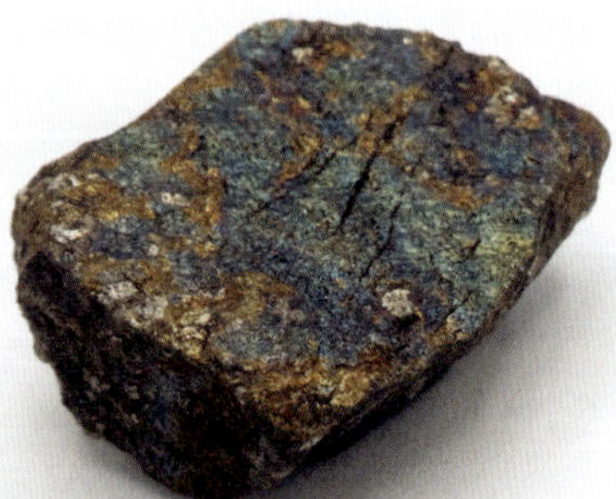

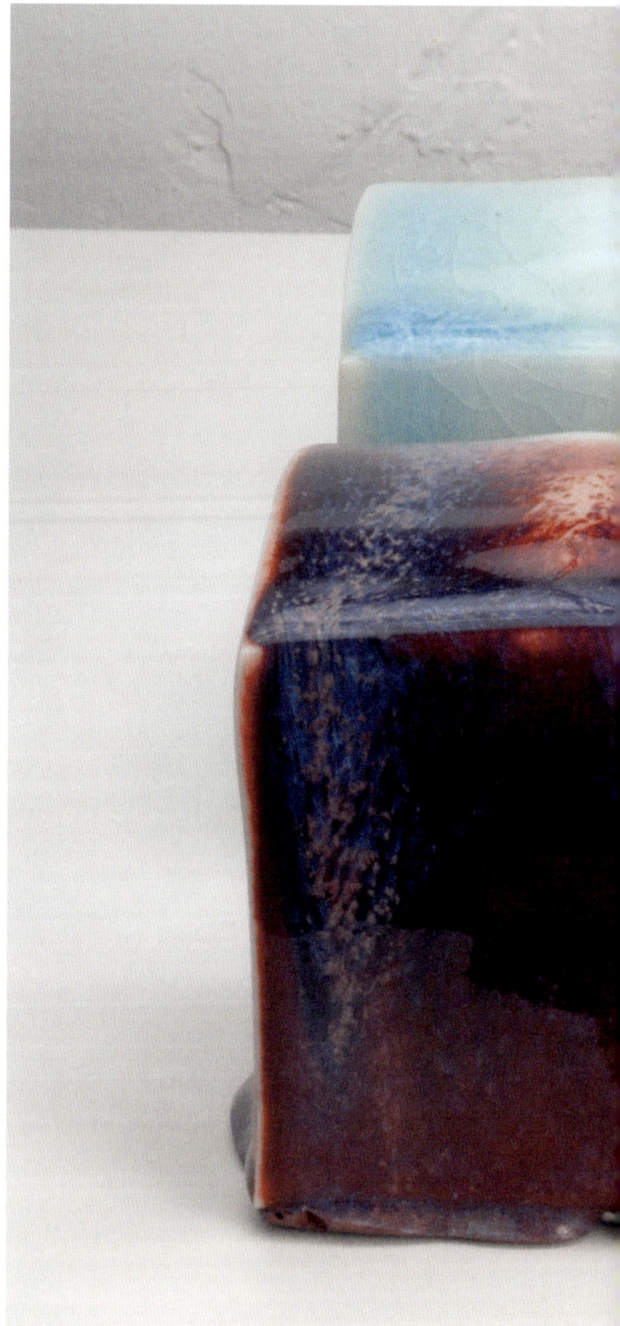

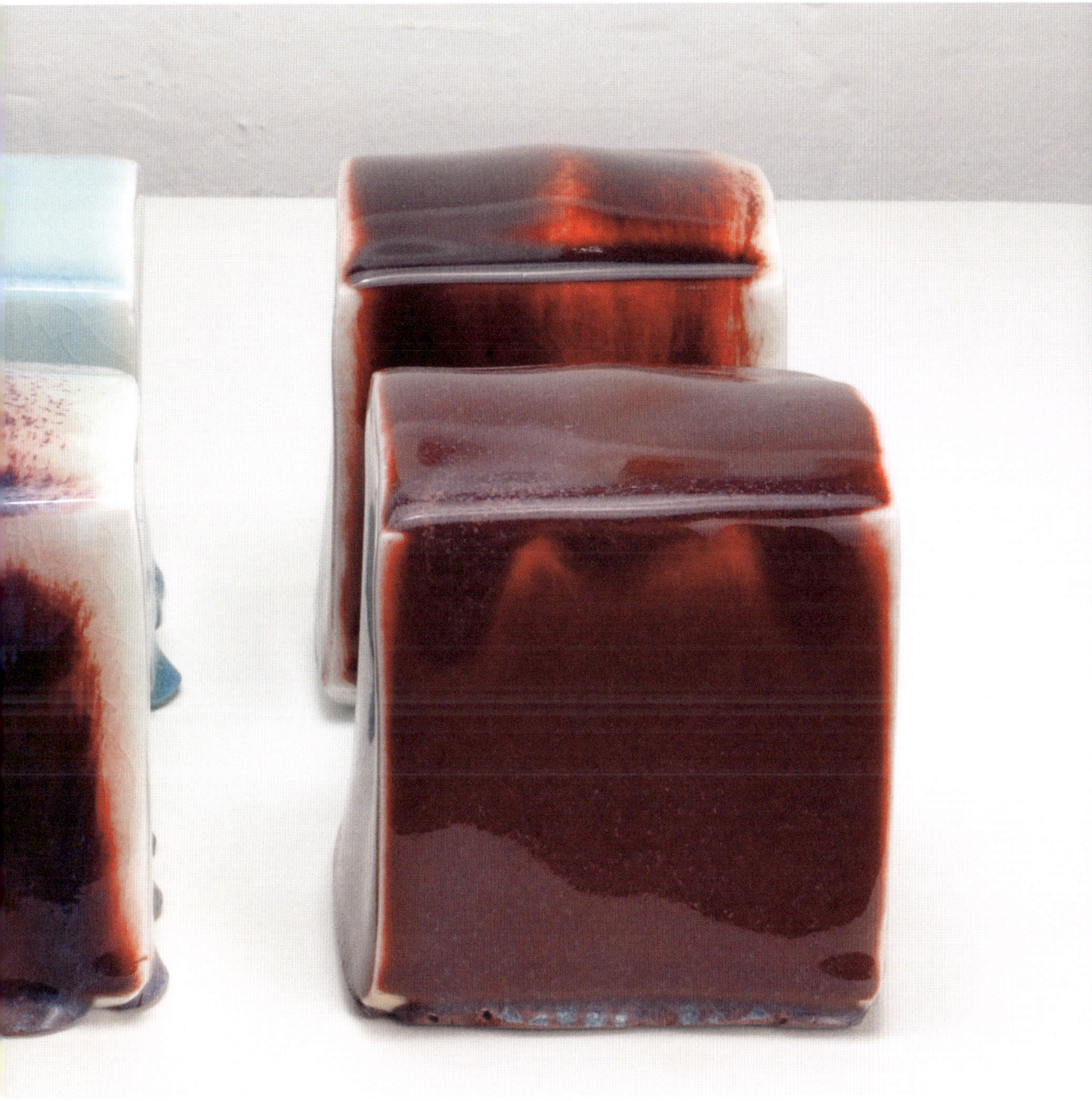

**Opposite:** Copper source: chalcopyrite

**Above:** Emmanuel Boos, *Cobblestones*, 2012. Copper is sensitive to the firing atmosphere, turning turquoise in oxidation and red in reduction. *(Image credit: Emmanuel Boos)*

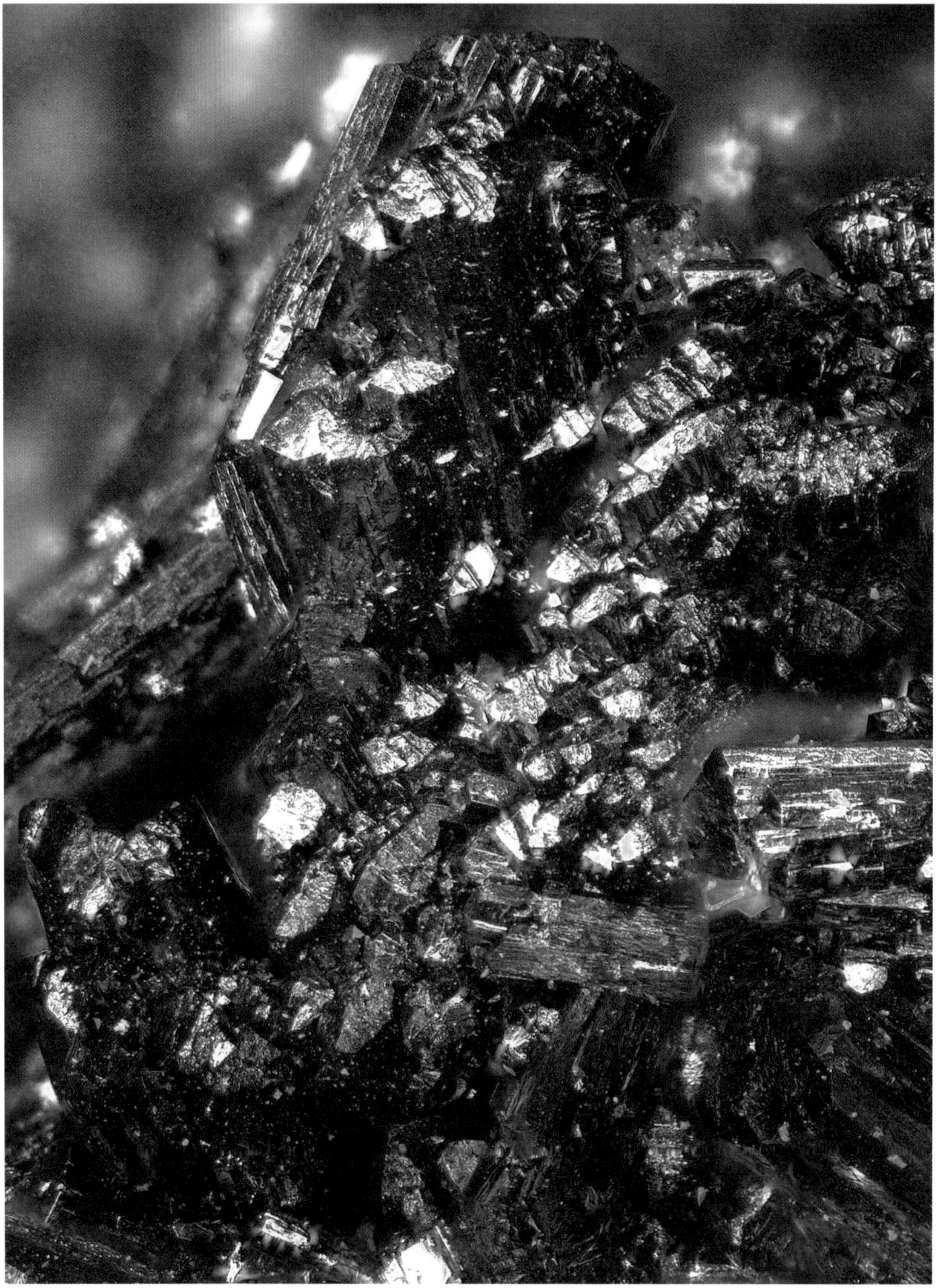

## IRON OXIDE

Red iron oxide ($Fe_2O_3$)
Yellow iron oxide ($Fe_2O_3.H_2O$)
Black iron oxide ($FeO$, $Fe_3O_4$)
Iron sulphate ($FeSO_4$) (toxic)
Rutile, ilmenite (titanium + iron oxide)

Iron oxide is present in many rocks and clays such as red earthenware clay. Yellow ochre (limonite), is a type of clay containing hydrated iron oxide while brown ochre (goethite) is a partially hydrated form. Sienna and umber are clay pigments containing both iron and manganese. Iron oxide is available to potters as hematite or red iron oxide ($Fe_2O_3$), yellow iron oxide ($Fe_2O_3.H_2O$) or black iron oxide ($FeO$). The black form is slightly more concentrated than the red, while the yellow is the least concentrated. Magnetite ($Fe_3O_4$) is used to add speckle to glazes; it is hard to grind and the large particle size causes the speckles. Crocus martis (calcined iron sulphate) is soluble in water and can be used to give watercolour effects. The red and yellow forms of iron oxide are non toxic but the black monoxide and the sulphate are slightly toxic.

Rutile and ilmenite are both impure forms of titanium dioxide that contain varying amounts (10–50%) of iron oxide. Iron oxide is used as a colourant in varying amounts (0.5–12.0%) in many classical Chinese glazes such as celadon, tenmoku and kaki. It fires to tan yellow, brown or orange red in oxidation and celadon blue or black in reduction. With bone ash and magnesium it will produce a kaki red glaze. Red iron oxide thermally decomposes and converts to magnetite above 1210°C (2210°F), giving off oxygen, a property used to good effect in oilspot glazes. Iron oxide acts as a flux in reduction and can cause the glaze to run. In oxidation it is less active and more refractory.

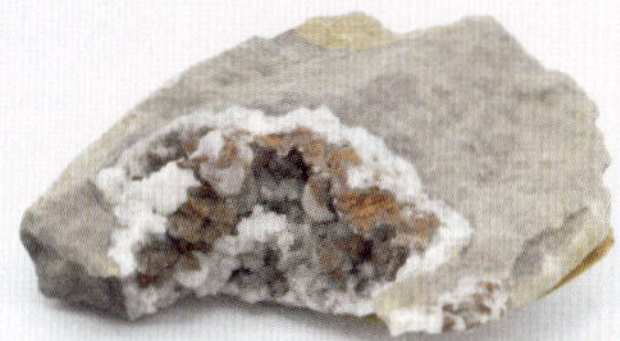

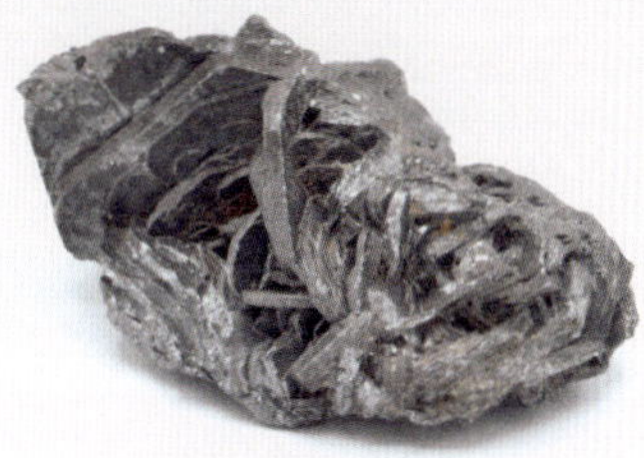

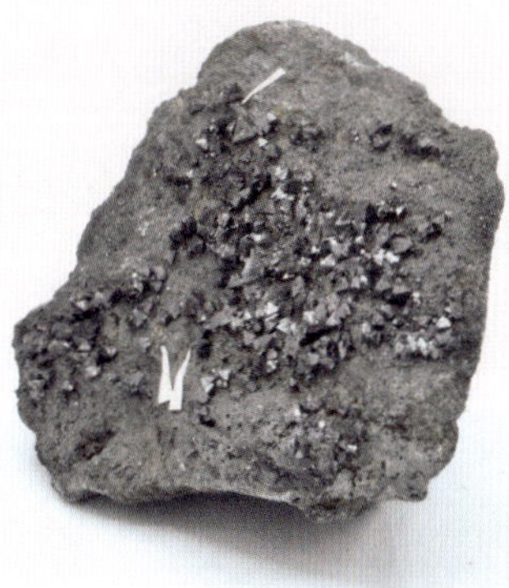

**Opposite:** Iron source: magnetite (microscopic view)
**Above top:** Iron source: limonite
**Above middle:** Iron source: hematite
**Above bottom:** Ilmenite
**Overleaf:** Linda Swanson, *Black Hole*. A combination of oil spot and metallic glazes primarily containing iron and other oxides

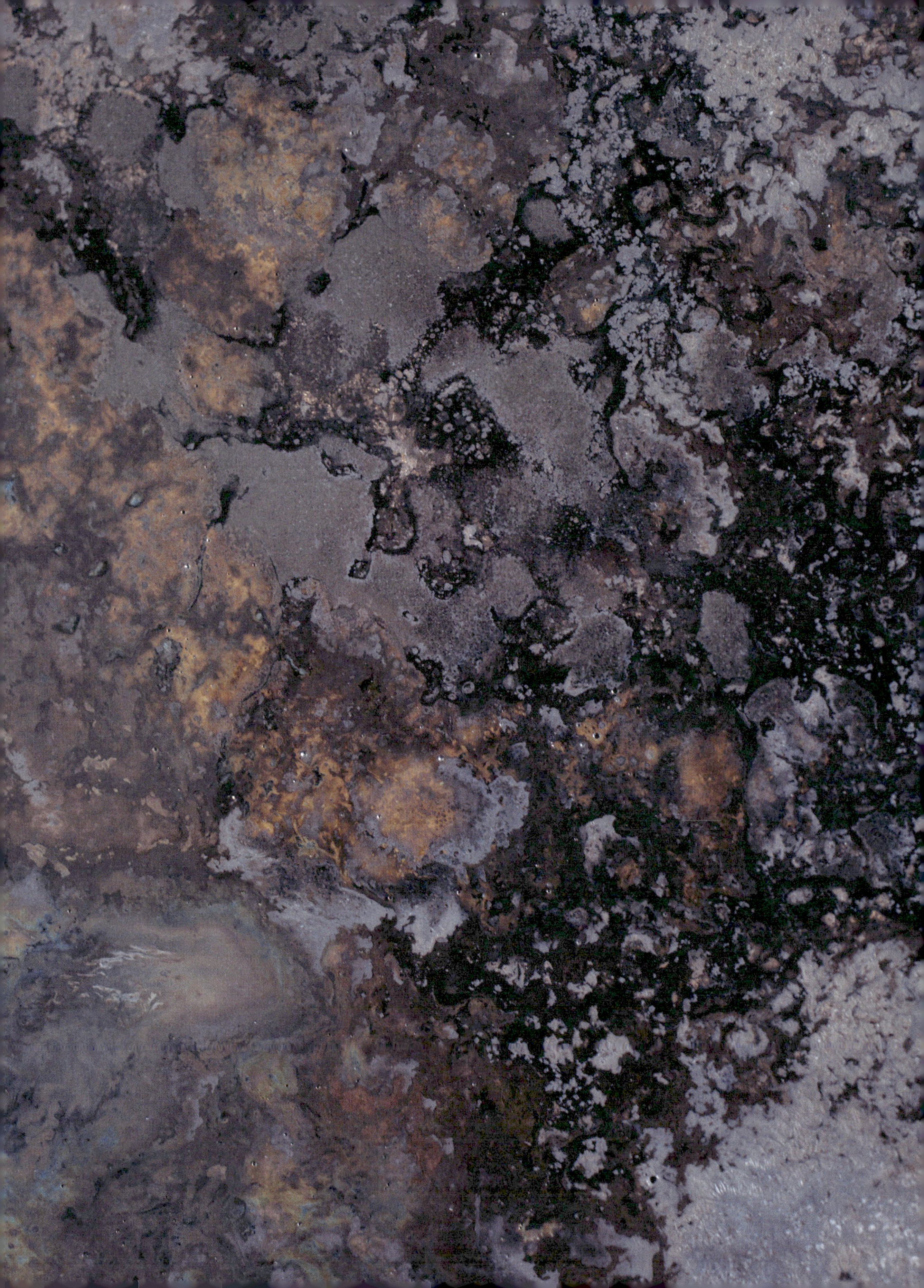

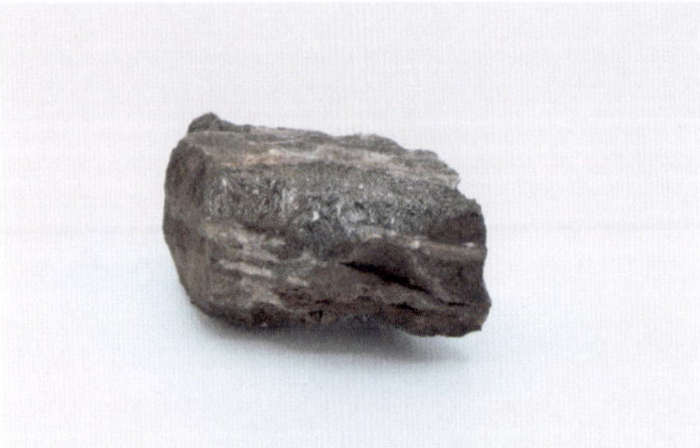

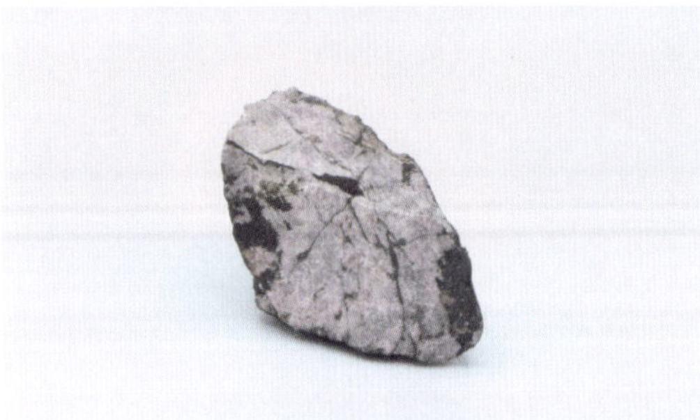

## MANGANESE

Manganese dioxide ($MnO_2$) (toxic)
Manganese carbonate ($MnCO_3$)
Granular manganese ($MnO_2$)

Manganese dioxide ($MnO_2$) is supplied to potters as a dark brown powder. Manganese carbonate ($MnCO_3$) is also available and is a pale brown powder; the carbonate is less toxic to use than the dioxide but 1.3x more is needed to obtain the same intensity of colour. Manganese is used to make brown and black clay bodies and glazes. The granular form is added to clay bodies to make speckles. It is a relatively weak colourant so a large amount is needed in glazes (1–15%), giving a pink to brown colour. Manganese acts as a flux. The firing fumes from manganese are toxic.

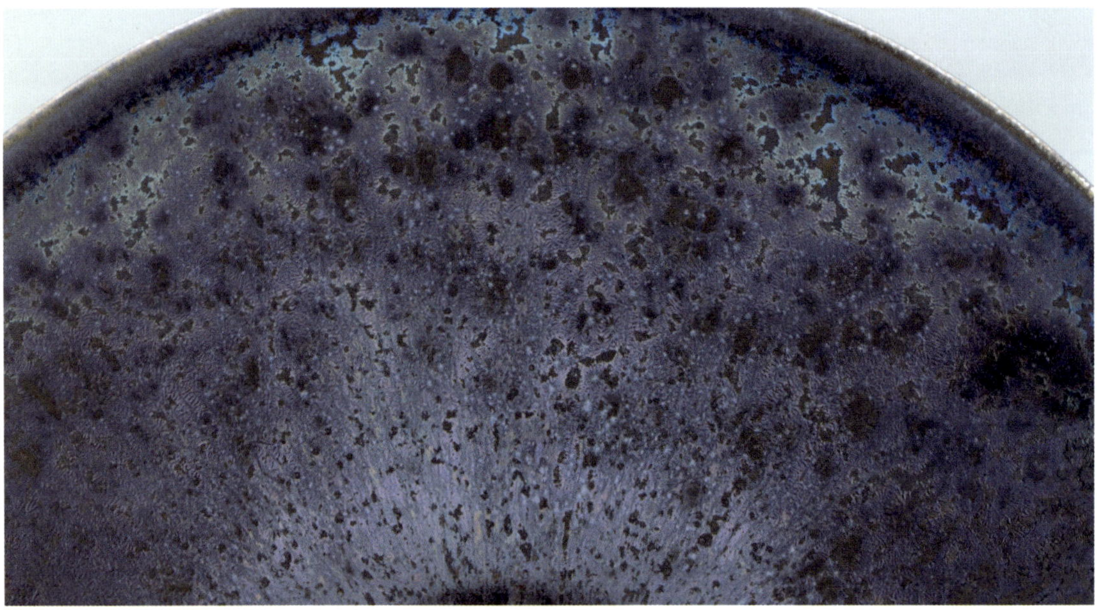

**Above top:** Manganese source: pyrolusite
**Above middle:** Manganese carbonate source: rhodochrosite
**Above bottom:** Matt Fiske manganese purple bowl. The presence of manganese leads to an iridescent surface. *(Image credit: Matt Fiske)*

## NICKEL

Nickel carbonate ($NiCO_3$) (toxic)
Nickel oxide ($NiO$, $Ni_2O_3$) (toxic)

Nickel oxide is available as black nickel oxide ($Ni_2O_3$), green nickel oxide ($NiO$) or green nickel carbonate ($NiCO_3$). The black oxide is 1.4x stronger than the carbonate. It is toxic and only a small amount is required in glazes (0.1–2.0% oxide). Nickel is often brown in glossy glazes and green in magnesium matt glazes. It is used in some glazes to tone down other colours; with cobalt it will make grey, but will also produce pink, steel blue and purple in barium-zinc glazes and green or mustard yellow with titanium in magnesium matt glazes. Nickel is refractory and can cause crystallization and an increase in viscosity and mattness of a glaze.

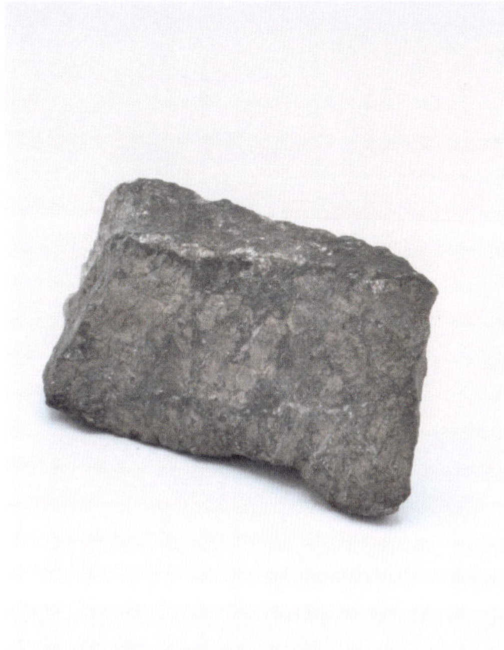

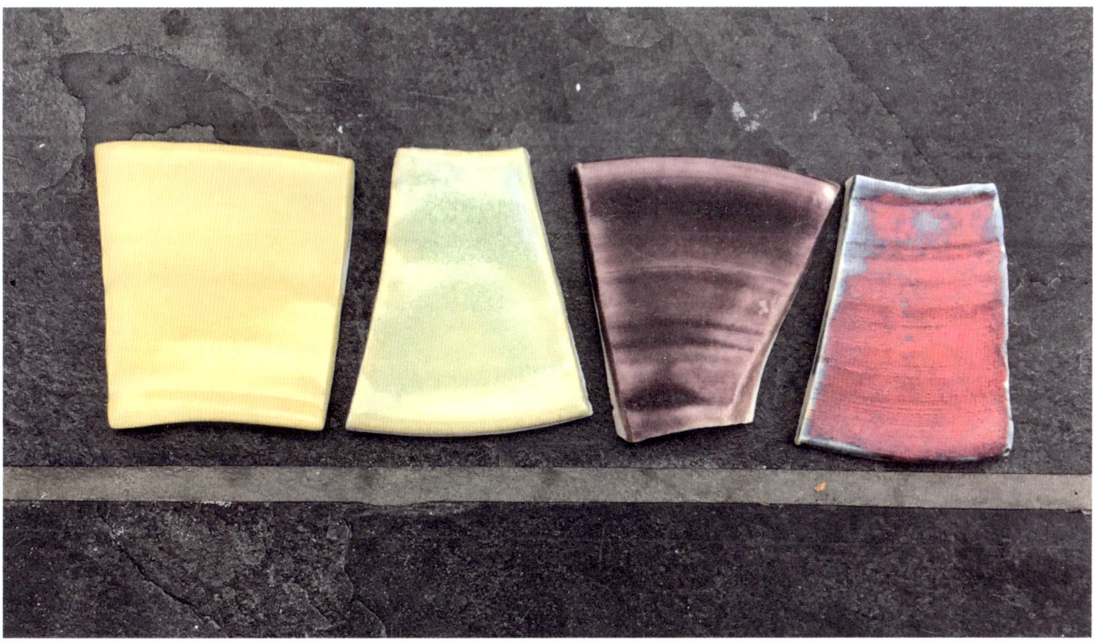

**Above top:** Nickel source: nickel ore containing pentlandite
**Above bottom:** Linda Bloomfield, glaze test tiles with nickel oxide in various glazes containing magnesium, titanium, barium and zinc oxide

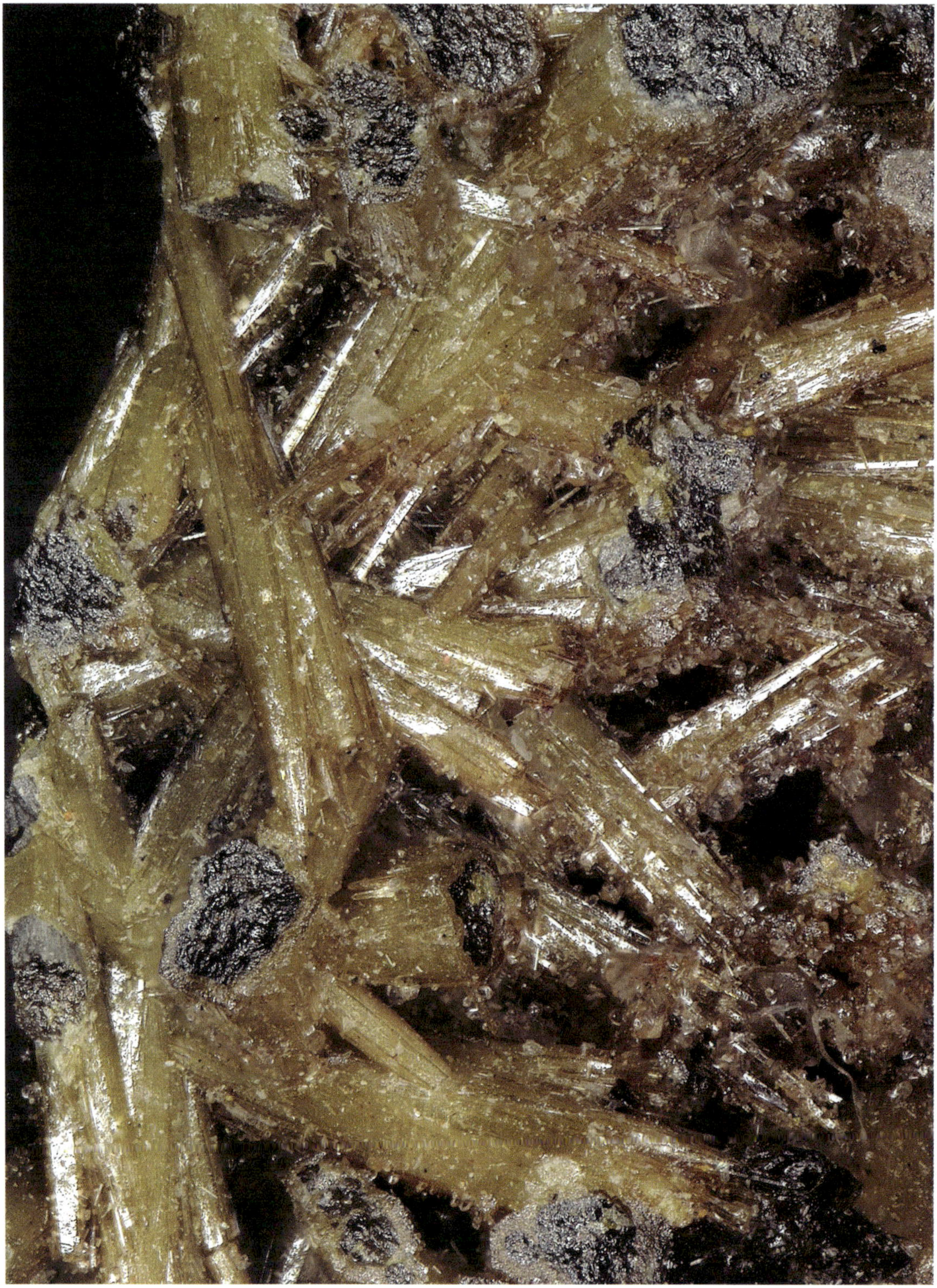

## VANADIUM

Vanadium pentoxide ($V_2O_5$) (toxic)

Vanadium is used together with zirconium silicate in commercial yellow and turquoise stains. It must be made into a stain before adding to the glaze to obtain these colours and is safer to handle in stain form. Vanadium pentoxide ($V_2O_5$) can be added directly to dry matt glazes and 2–8% will give mottled green or yellow. Vanadium acts as a flux and can sometimes cause cloudy mottling and streaking, similar to titanium. It is slightly water soluble and is moderately toxic.

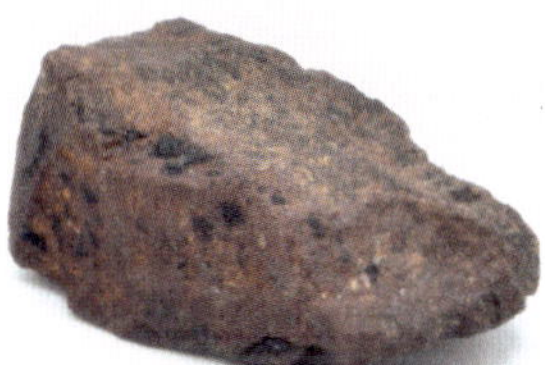

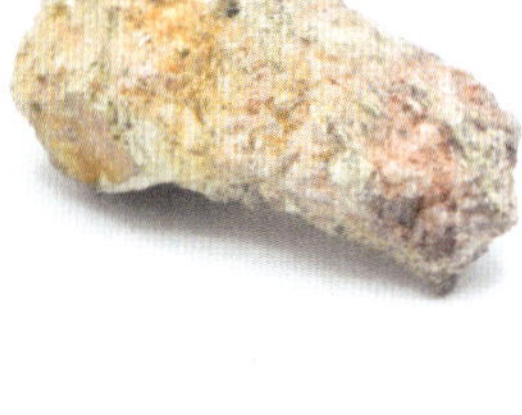

## RARE EARTHS

Cerium oxide ($CeO_2$)
Erbium oxide ($Er_2O_3$)
Holmium oxide ($Ho_2O_3$)
Neodymium oxide ($Nd_2O_3$)
Praseodymium oxide ($PrO_2$)

The rare earth elements cerium (Ce), praseodymium (Pr), neodymium (Nd), holmium (Ho) and erbium (Er) are used to colour glass and, in the case of praseodymium, to make a yellow stain for glazes. The oxides act as fluxes and are weak colourants that can be used in glazes; around 6–8% is needed to colour a glaze. Neodymium give a pale lavender while praseodymium gives a yellow-green colour and erbium gives a pink colour in glazes. Cerium oxide is tan yellow and is combined with titanium to make a yellow-orange coloured glass. Holmium oxide changes colour from tan yellow in daylight to pink under fluorescent light while neodymium changes from lavender to pale blue. Some of the less well-known rare earths (samarium, europium, terbium, dysprosium and thulium) can give bright fluorescent glaze colours but only under ultraviolet light. Though not particularly rare, they are currently only mined in China and have recently become very expensive. They are not toxic to use but the firing fumes should be avoided.

**Left top:** Rare earth source: monazite
**Left bottom:** Rare earth source: bastnäsite
**Opposite:** Vanadium source: vanadinite (microscopic view)

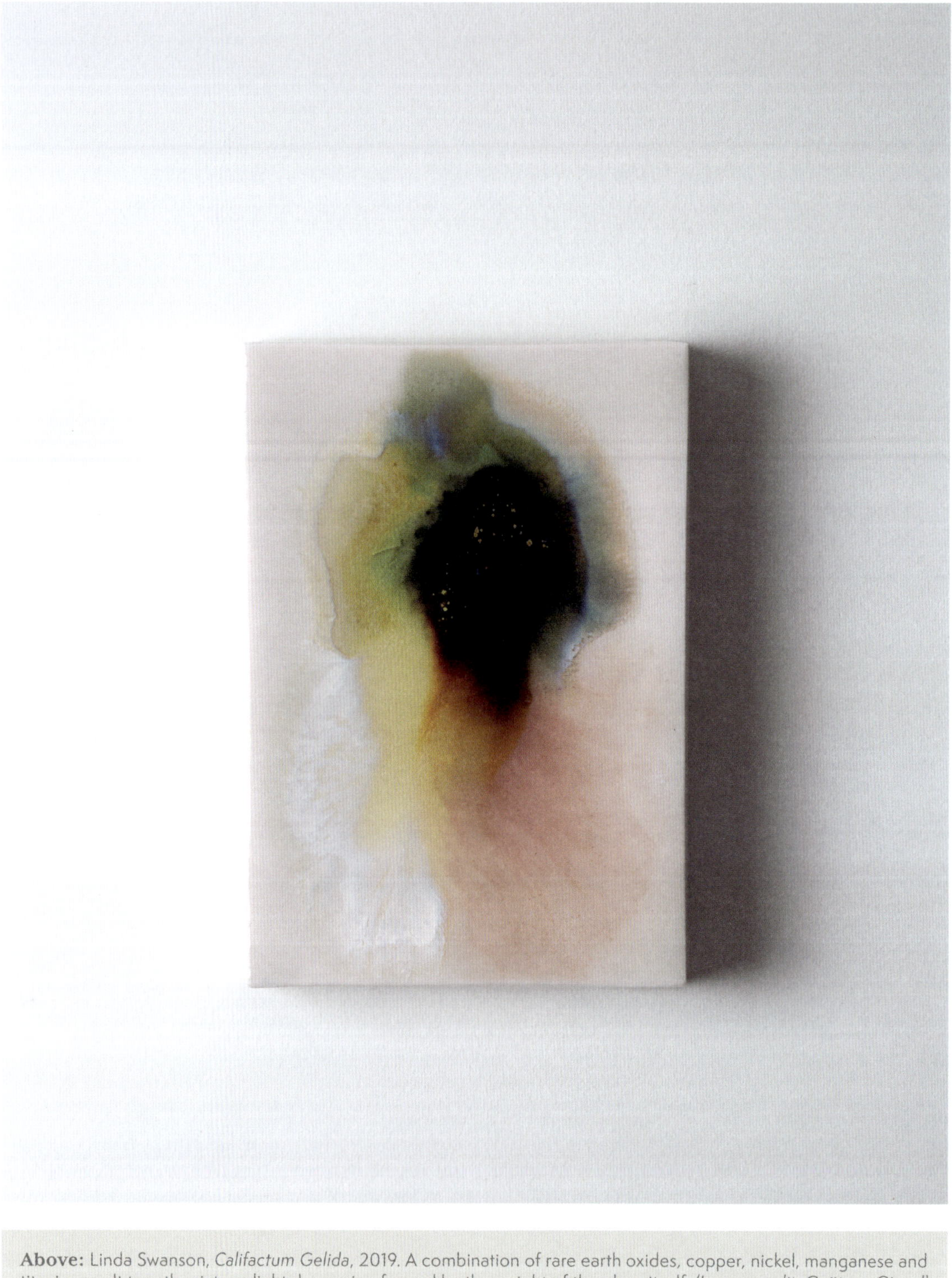

**Above:** Linda Swanson, *Califactum Gelida*, 2019. A combination of rare earth oxides, copper, nickel, manganese and titanium melt together into a slight depression formed by the weight of the glaze itself. *(Image credit: Gaëtane Girard)*

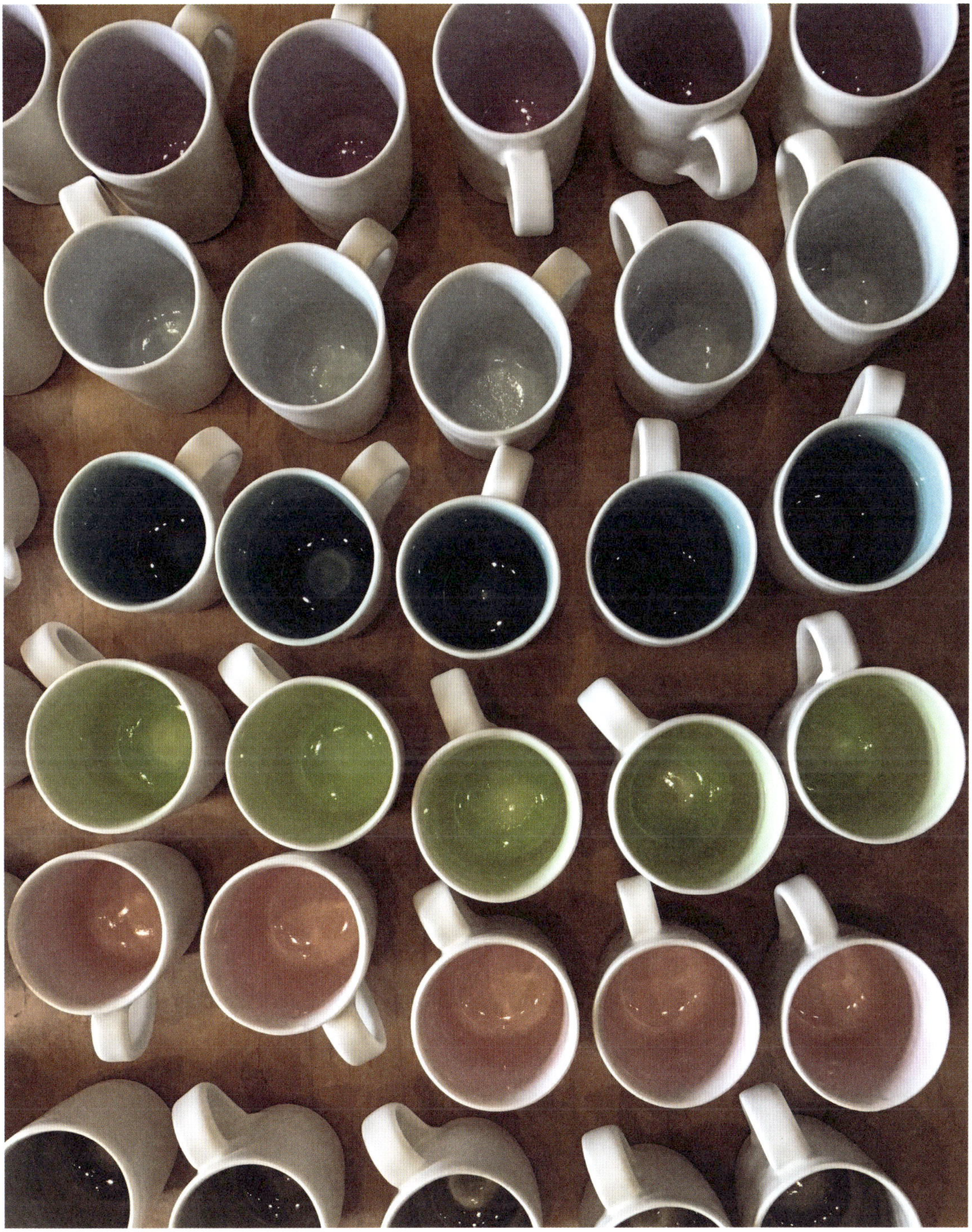

**Above:** Linda Bloomfield, porcelain mugs with glossy glaze with colouring oxides: neodymium, copper and tin, copper oxide, praseodymium, rutile and tin, cobalt and nickel, fired in an electric kiln to cone 8

## OPACIFIERS

Opacifiers make glazes opaque and white. They withstand high temperatures and do not completely melt in the glaze, remaining suspended in the glaze or re-crystallizing on cooling.

### TIN OXIDE ($SnO_2$)

Tin oxide ($SnO_2$) is supplied to potters as a refractory white powder and additions of 5–10% will make glazes opaque. Small additions (1–2%) dissolve in the glaze and can act as a flux while larger amounts remain as unmelted white particles suspended the glaze. Tin oxide is the strongest opacifier and gives soft whites, creamy or bluish milk white. Opacity may be lost in high fire reduction causing grey cloudiness or scumming. Tin may cause glazes to pick up a pink blush from volatile copper (in reduction) or chromium (in oxidation) in the kiln.

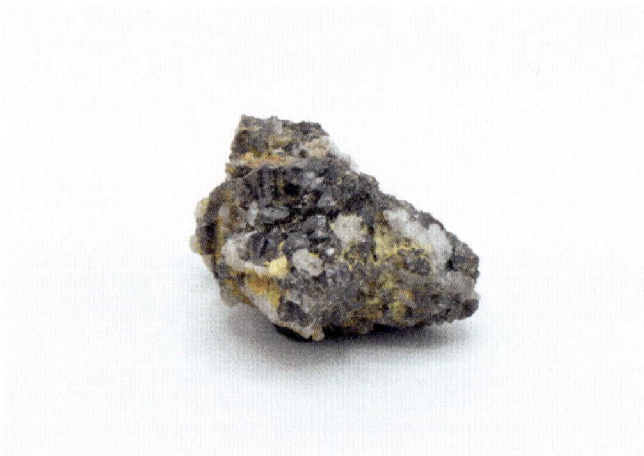

### TITANIUM DIOXIDE ($TiO_2$)

Rutile, ilmenite (titanium + iron oxide)

Titanium dioxide ($TiO_2$) will pick up colour from iron in the clay body so does not produce such white glazes as tin or zirconium. It provides a creamy opacity to glazes; 5–10% gives a crystalline soft matt surface and promotes the formation of crystals. Titanium is found in rutile and ilmenite, both of which also contain iron oxide; rutile can contain up to 15% iron while ilmenite contains 50% iron oxide. Traces of other colouring oxides including vanadium and chromium may be present in rutile causing a pink colour when combined in a glaze with tin oxide.

In small amounts, titanium can intensify colours, particularly chrome pinks; moderate amounts (2–6%) will break up colour by mottling while large amounts (6–10%) will subdue colour and may sparkle. Titanium turns iron yellow in oxidation and green in reduction. It turns cobalt green in high-alumina glazes and slips.

**Above:** Tin source: cassiterite
**Opposite:** Titanium dioxide source: anatase microscopic view
**Overleaf:** Lindsay Montgomery, *Hellmouth Censer 3* (Front/Back), 2023. Long used to disguise earthenware as porcelain, tin glazed earthenware serves as a base for colourful narrative painting. *(Image credit: Lindsay Montgomery)*

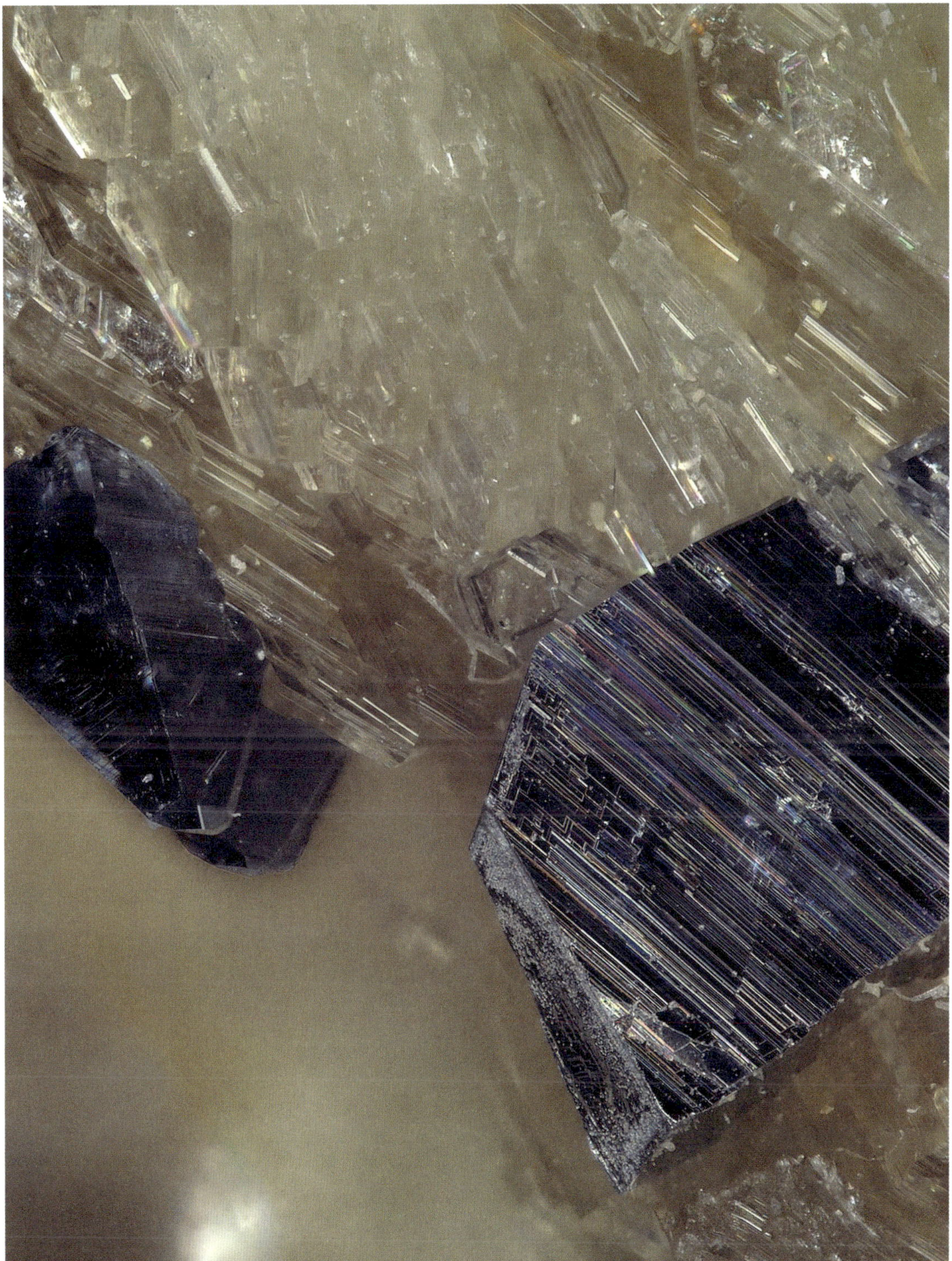

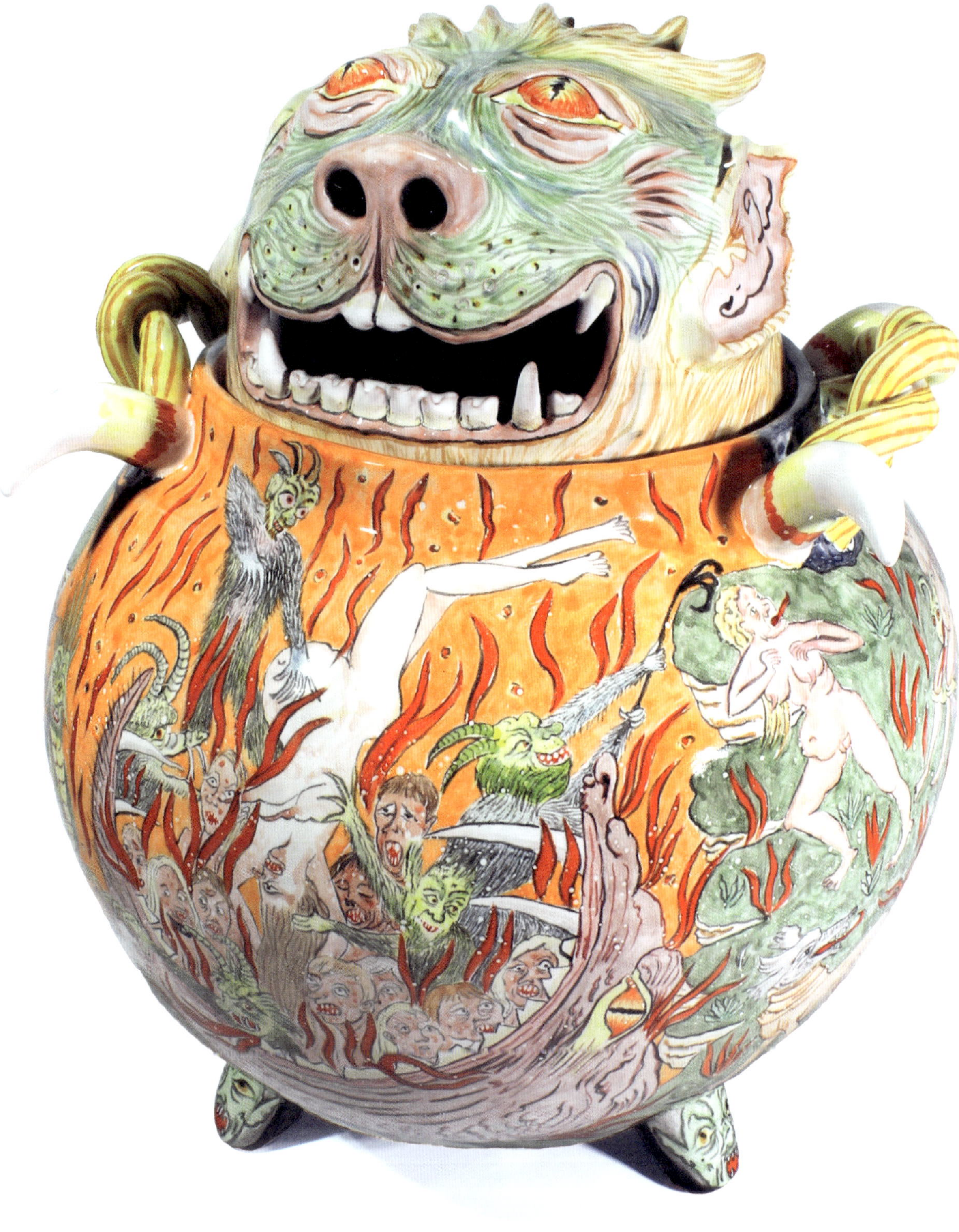

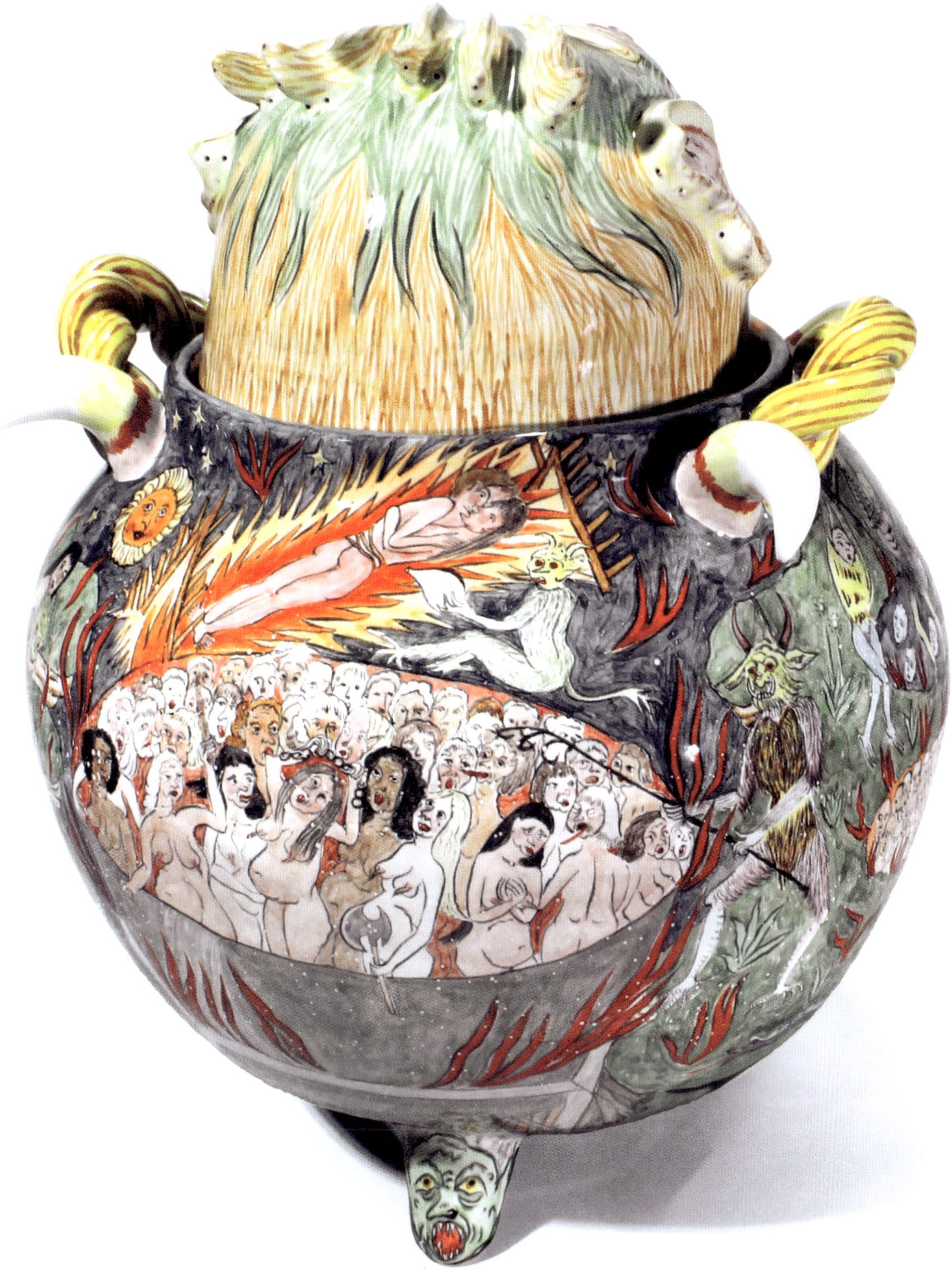

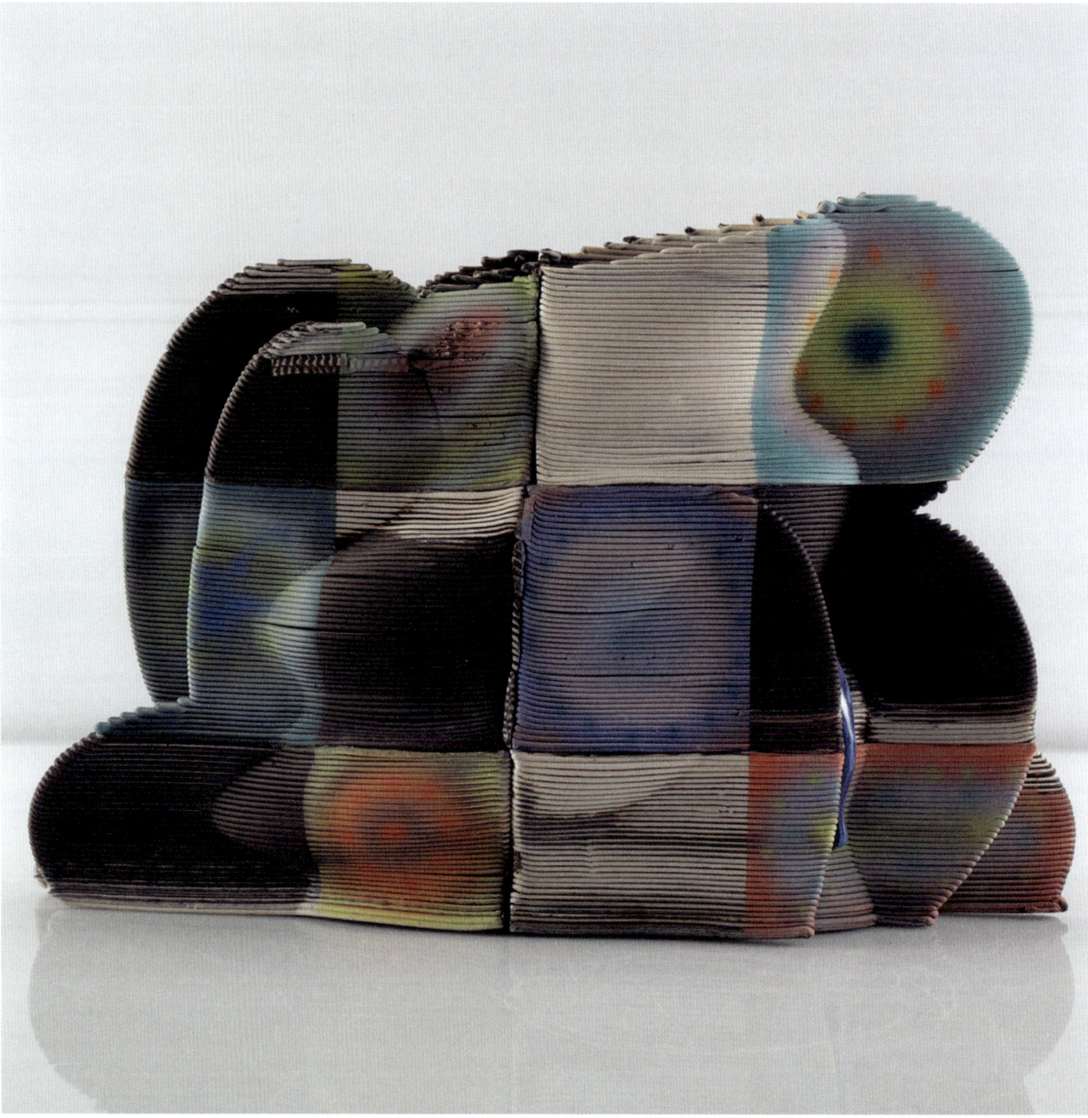

## ZIRCONIUM SILICATE ($ZrSiO_4$)

Zirconium silicate or zircon ($ZrSiO_4$) is less expensive than tin oxide, but a larger quantity of 10–15% is required to fully opacify a glaze. It gives a cold, bright, opaque white and is commonly used in industrial whitewares. It is refractory and stiffens the glaze melt without making it matt; if the glaze looks underfired, reduce the silica or increase the flux in the glaze. It is also used as a stabilizer in commercial stains, particularly in high temperature reds and yellows. Adding an additional 2–3% zirconium silicate can help stabilize commercial red and yellow stains and reduce pinholes. Zirconium also improves glaze durability and hardness but large amounts may cause cutlery marking. In the US it is also known by the brand names Zircopax, Superpax or Ultrox, depending on particle size; a finer particle size gives better opacity.

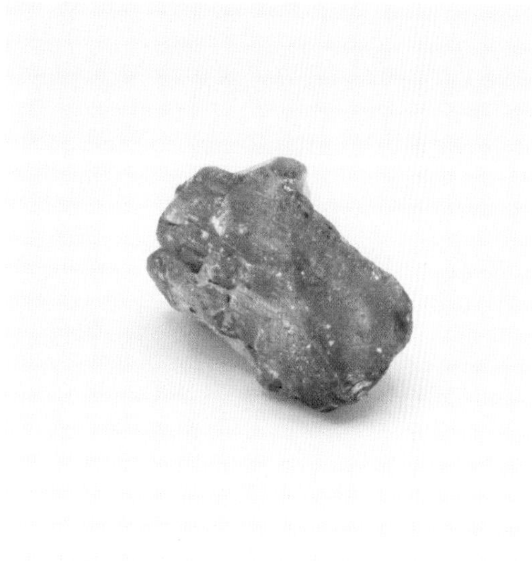

## STAINS

Some colours such as red, yellow and orange are difficult to make using colouring oxides. Glaze stains are made using colouring oxides combined with stabilizers such as zirconium silicate, fired and then ground to a powder. As they have already been fired, they do not change colour during firing in a glaze. Cadmium sulphide and cadmium selenide are used to make yellow, orange and red stains. High temperature reds and yellows are stabilized by encapsulating the pigment in a matrix of zirconium silicate. Another type of yellow stain is made using the rare earth, praseodymium oxide combined with zirconium silicate.

**Above:** Zircon source: zirconium

**Opposite:** Jolie Ngo, *Flatpack Vessel in Space Blanket*, 2023. This 3D printed stoneware clay form is modulated with glaze coloured with ceramic stains. *(Image courtesy of R & Company and the artist)*

CHAPTER NINE

# FIRING

Firing is the heating of clay to high temperatures inside a kiln to change it into ceramic. Once it is fired it can no longer be turned back into clay. There are several stages of firing.

## IMPORTANT STAGES DURING FIRING

Clay goes through several physical changes when fired.

### WATER SMOKING

The first step is the evaporation of water from between the clay particles. Pots must be completely dry before firing otherwise the steam escaping could cause them to explode. Initially, the kiln should be heated very slowly to give time for all the water to evaporate. Drying can be speeded up by pre-heating the ware in the kiln to 80°C (176°F) for a few hours. This part of the firing, up to 100°C (212°F), is called water smoking, which causes the water in the pores between the clay crystals to evaporate.

### CHEMICALLY COMBINED WATER DRIVEN OFF

The next stage is to drive off the water which is chemically combined with the clay (remember the formula for clay is $Al_2O_3.2SiO_2.2H_2O$). This is known as dehydroxylation and occurs up to 550°C (1022°F). Once this chemical change has taken place the process cannot be reversed and the clay cannot be returned to its plastic state.

### QUARTZ INVERSION

At 573°C (1063°F), the crystalline quartz in the clay body increases in volume by 1%, known as the quartz inversion. This may cause cracking if the temperature in the kiln is increased too rapidly.

**Opposite:** Linda Bloomfield, stoneware vase, 2022. Lichen-effect glaze wood fired at Guldagergaard

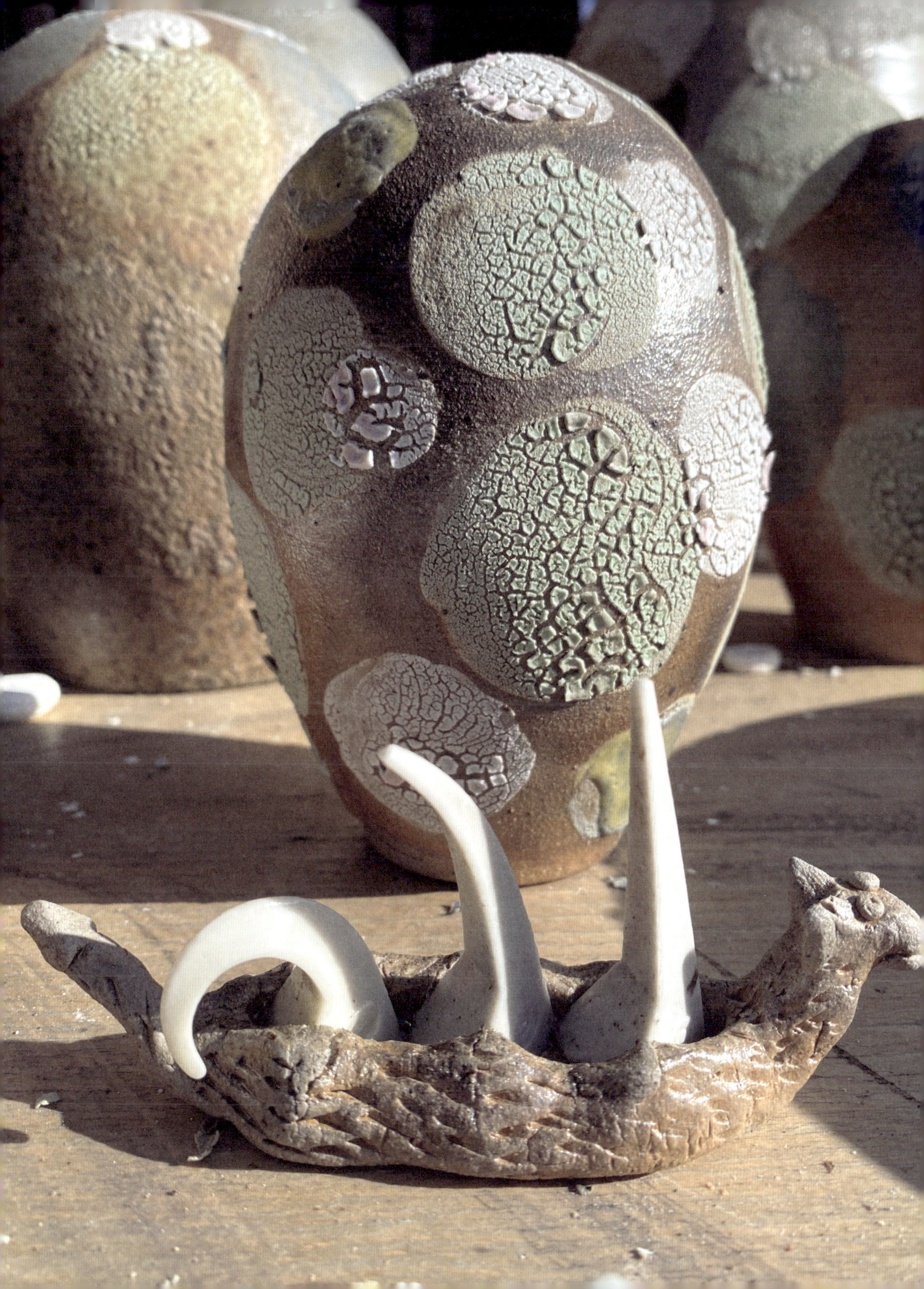

## CARBON BURNED OFF

Organic matter in the clay is burned and oxidised to carbon dioxide, and fluorine and sulphur dioxide from materials in the clay body are driven off at 700–900°C (1292–1652°F). The kiln should be in a well-ventilated room to avoid breathing these fumes. At this point the biscuit firing is completed. The clay particles are sintered or welded together. The ware has undergone little shrinkage since the bone-dry stage, but it is durable enough to withstand handling and glazing.

## VITRIFICATION BEGINS

Above 900°C (1652°F), the clay body begins to shrink and vitrify. The silica starts to melt, filling the spaces between the clay particles and fusing them together. The fired clay is known as metakaolin. At 1000°C (1832°F) the clay crystals begin to break down and melt.

## MULLITE STARTS TO FORM

At 1050°C (1922°F), needle shaped crystals of mullite $3Al_2O_3.2SiO_2$ begin to form, giving the fired clay strength and hardness. When mullite forms from metakaolin $Al_2O_3.2SiO_2$, extra free silica is released.

## CRISTOBALITE FORMS

Above 1100°C (2012°F), any free silica (not chemically combined) in the clay changes to cristobalite, which has a different structure from that of quartz. When the kiln is cooled down, cristobalite contracts suddenly by 3% at 226°C (439°F), known as the cristobalite inversion. This can cause cracking if the kiln is cooled too rapidly by opening too soon, causing some areas to drop in temperature and stressing the ware. It is advisable not to open it until it has cooled down to below 100°C (212°F).

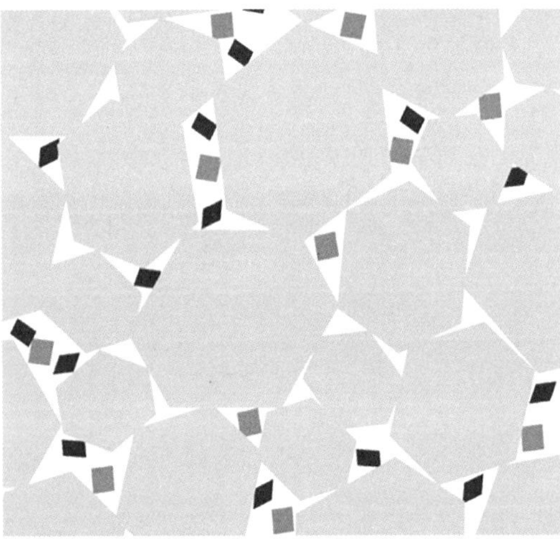

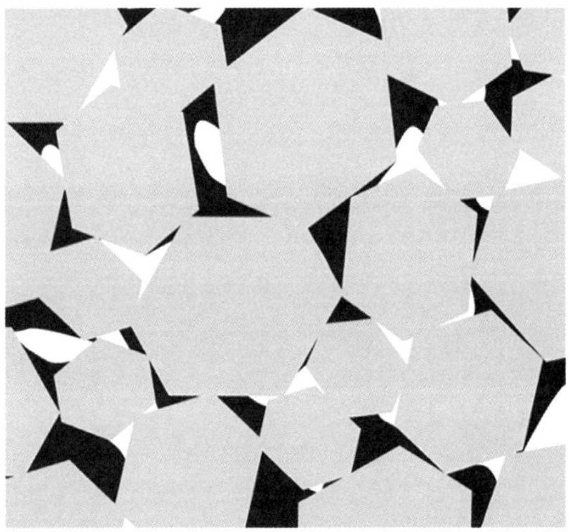

**Above left:** Enlarged view of clay body after biscuit firing , pre vitrification showing hexagonal clay particles, quartz and feldspar
**Above right:** Enlarged view of vitrified stoneware clay body, where the quartz and feldspar have melted into glass and filled the pores

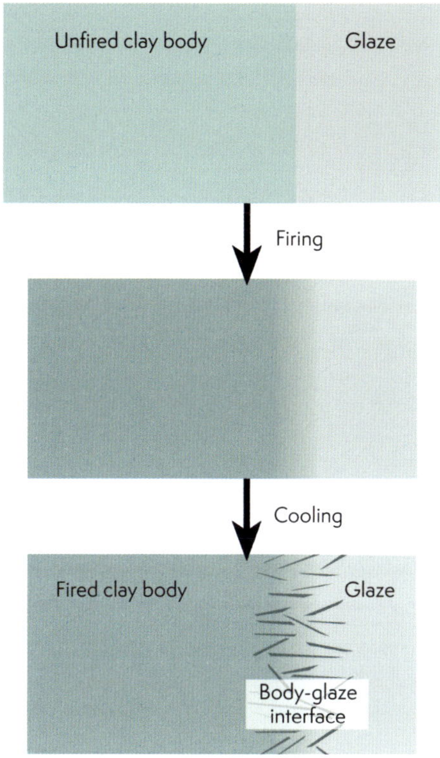


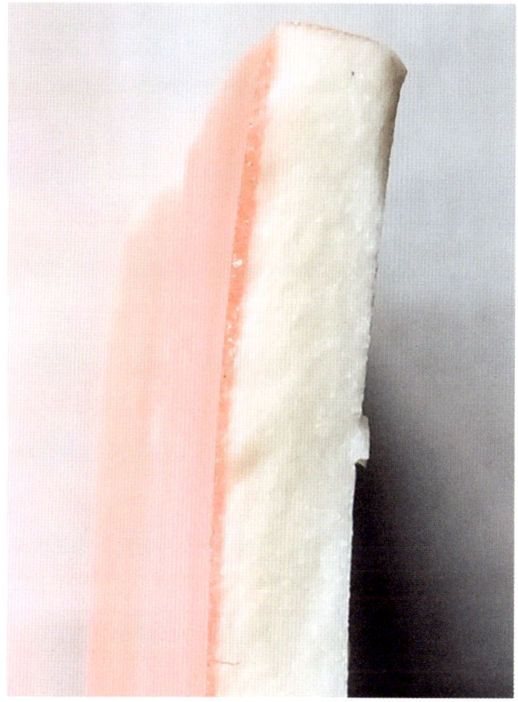

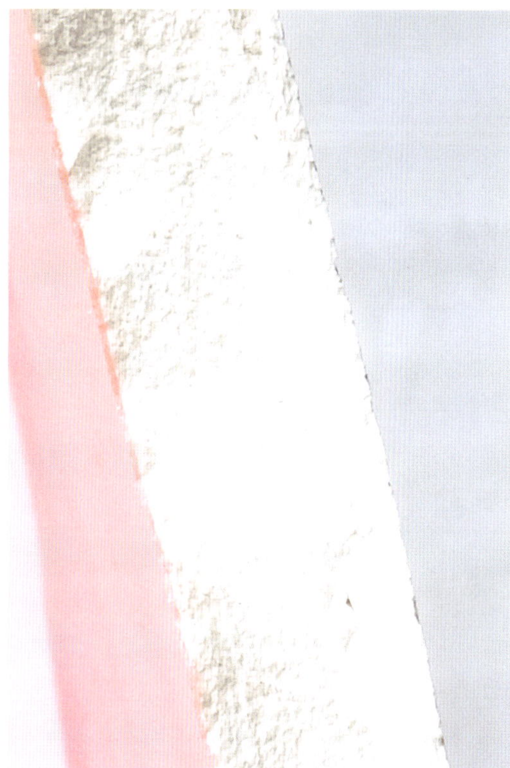

**Above:** Clay-glaze interface. During stoneware firing, needle-shaped crystals of mullite grow at the body-glaze interface, forming a strong bond between the clay and glaze and giving strength to the fired piece

**Above right:** Magnified view of glazed porcelain. The fired clay body is hard and vitrified

**Right:** Magnified view of glazed earthenware. The fired clay body is relatively soft and porous

## MEASURING TEMPERATURE AND HEAT WORK

There are several ways to measure the temperature in the kiln. One way is by looking through welder's goggles at the colour:

Red heat 700–900°C (1292–1652°F)
Orange 1100°C (2012°F)
Yellow 1200°C (2192°F)
White 1300°C (2372°F)

A pyrometer is a device attached to a thermocouple probe made from two metals sheathed in porcelain and inserted into the kiln. The difference in expansion of the metals causes a change in voltage of the thermocouple during heating and gives a measure of the temperature inside the kiln.

A more accurate way to measure the heat inside the kiln is by using pyrometric cones. These are tall, slender pyramids of glaze material that bend after they have undergone a certain amount of 'heat work'. Cones measure not only the temperature, but are also affected by the amount of time at that temperature and it is this combination of temperature and time that gives us a measure of heat work; for example, a fast firing may generate less heat work than a long, slow firing. It is a good idea to use cones in electric kilns to calibrate the reading from the pyrometer. Cones are essential for measuring heat work during firing in fuel-burning kilns. Usually, packs of three cones are placed at various points in the kiln, set at an angle of 8° from the vertical in a pad of clay; a guide cone to let you know you are reaching the peak temperature, a target cone and a guard cone which should still be standing if the kiln has not overfired. At the end of firing, the target cone should be pointing down with its tip touching the base. You can look at the cones through welder's goggles during firing if you set them behind the spyholes in the kiln. See Appendix 12 for cone temperatures.

## OXIDATION AND REDUCTION

In electric kilns, the heat is generated by the resistance to electric current conducted through coiled metal alloy wires known as kiln elements. The neutral atmosphere inside the kiln has plenty of oxygen and the clay and glazes become oxidised during firing; clay turns a buff colour and clear glazes turn pale yellow, owing to the oxidation of small amounts of iron oxide in the clay body.

In fuel-burning kilns (gas, oil and wood), the air intake can be restricted so that the flames need to seek oxygen from the clay and glazes instead; this is known as a reduction atmosphere. The flames grow longer and lick out from any holes in the kiln, searching for oxygen. The clay body turns grey owing to reduction of iron oxide. Celadon glazes turn blue or grey-green while copper glazes can turn red owing to reduction of copper oxide to tiny particles of copper metal suspended in the glaze.

In the current climate of global warming, many reduction potters are turning to wood firing instead of burning fossil fuels. New kiln designs that burn wood more efficiently make it possible to fire without polluting the atmosphere, such as the Girel 3E kiln.

**Opposite top:** Rocket kiln built at 3Ecologies in Québec, Canada with students from Concordia University
**Opposite bottom:** The Girel 3E kiln constructed at Keramis in Belgium

'The three requirements met by this innovative kiln, the "3Es" are: Ecological: release of clean fumes into the atmosphere (post-combustion), low consumption, Economical: 160 kg of dry wood are enough for a firing at 1300°C (2372°F) for a 600 litre load, and Ergonomic: only one person can conduct the firing, without having to bend down or move to feed the fire or make adjustments. The challenge that the Girel 3E kiln takes up finds its place perfectly in today's society and perhaps answers the questions that plague ceramists.' (Jean Girel, The International Academy of Ceramics).

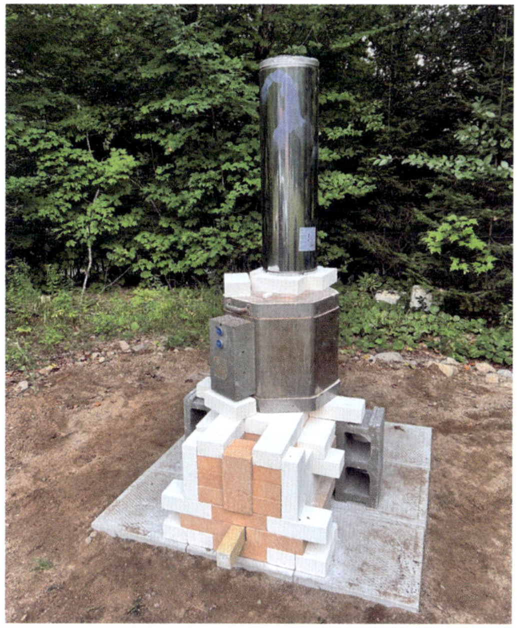

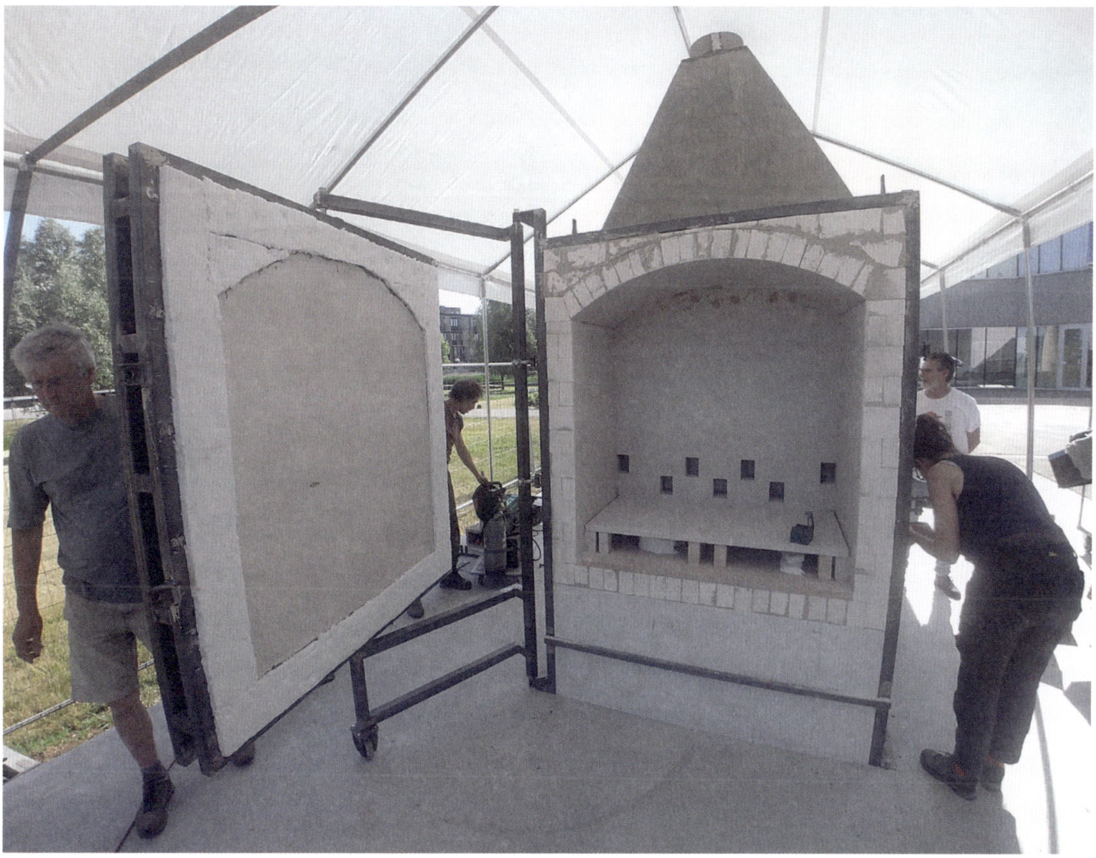

Rocket kilns are another innovative and ecological solution for firing that combines recycling, renewable resources and efficiency. Ceramicists have recently begun adapting the technology developed for rocket stoves into rocket kilns made out of old electric kilns. These designs configure fire bricks into a highly efficient wood combustion chamber that burns wood thoroughly without creating smoke. While most designs are in the experimental phase, there is widespread sharing of ideas, formats and troubleshooting in online forums.

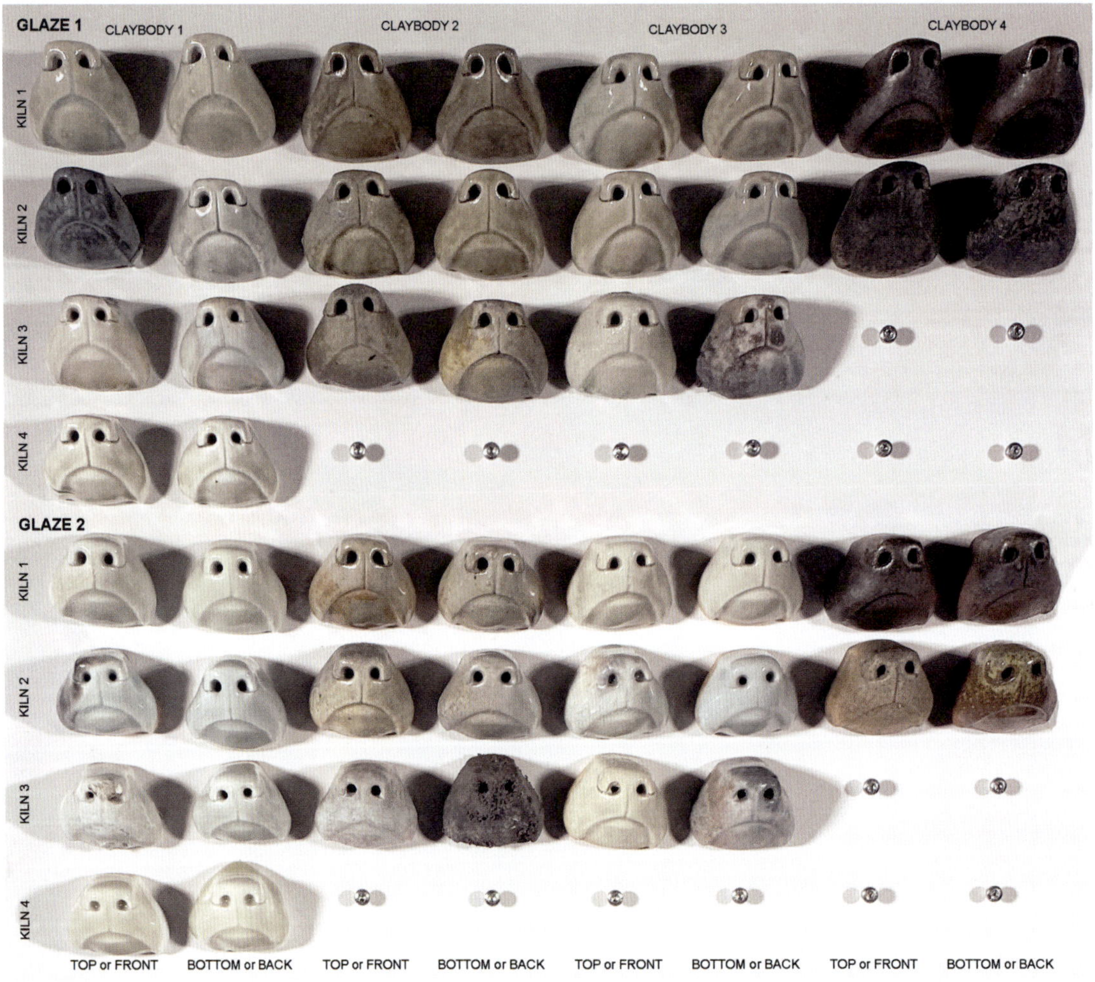


**Above:** *Guldagergaard Dog Nose Test Tiles.* Wide-ranging results in the surface of wood-fired ceramics appear in this test tiles series of two celadon glaze recipes on four different clay bodies in four different kilns at the top and bottom of each kiln. *(Image courtesy of Guldagergaard International Ceramic Research Center)*

**Opposite:** Firing the anagama kiln at Guldagergaard International Ceramic Research Center in Denmark with Sara Mericle and Isabel Anne Fisk Baruque

## WOOD FIRING, SODA FIRING

In wood-burning kilns, soluble salts in the wood are carried on the flames through the kiln. These potassium and sodium salts react with the alumina and iron oxide in the clay to produce a toasted orange colour. Slips containing calcium or titanium added to the clay surface can produce yellows. Many potters leave the clay surface without glaze to encourage marks showing the path of the flames. Glazes will also be affected by reduction and the flame path.

In soda firing, soda ash (sodium carbonate) is introduced into the kiln, either by spraying or in solid form which vaporizes in the heat. The sodium reacts with the alumina and silica in the clay surface and produces a glaze. A thick application of soda can produce a pitted, orange-peel textured surface, while a thin application makes a glossy sheen on the surface, similar to a thin shino glaze. Colour can be introduced using slips coloured with cobalt for blue, titanium dioxide for yellows, iron and manganese for black.

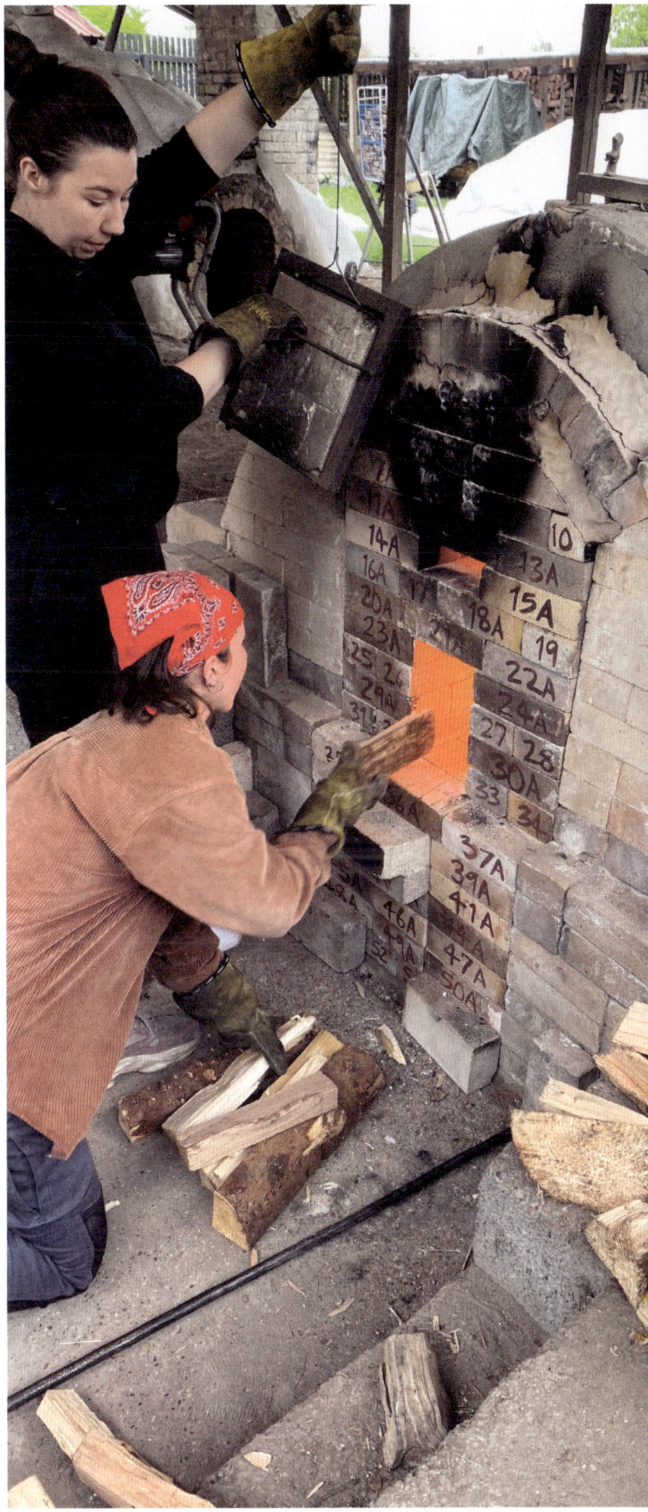


## ADJUSTING FIRING TEMPERATURE

Reducing energy use is a strategy you can use to make your ceramics practice more sustainable. One way to reduce energy use is to lower the firing temperature from cone 10 stoneware to cone 6 mid-range or even cone 06 earthenware. You can adjust glazes and change from cone 10 to cone 6 by adding around 10–20% borax frit (or Ferro frit 3134), while earthenware glazes may need to be composed of 50–90% frit instead of feldspar.

## GLAZE FAULTS

There are many glaze faults that can be corrected when you understand what each material contributes to the glaze (see Chapter 6). A fault is an unwanted glaze effect, some of which can be used as special effect glazes for sculpture or non-utilitarian purposes, such as crackle and crystalline glazes.

### RUNNY GLAZES

Runny glazes are those with too much flux or too little alumina to stiffen the glazes. To make a glaze more viscous, add a few percent china clay (kaolin).

### CRYSTALLINE GLAZES

Glazes that are low in alumina (below 0.3 in the molecular formula) often grow crystals on the surface. If this is unwanted it is called devitrification. The glaze needs to contain enough alumina and silica to make a glossy, clear glaze. Excess flux material such as zinc oxide may form crystals on the glaze surface during cooling. Adding a few percent each of silica and clay can be enough to prevent crystals growing.

### CRAZING

When the glaze has a higher thermal expansion than the clay body, a network of cracks forms during cooling to relieve the stress. This is called crazing or crackle. To correct crazing, the glaze recipe needs to be adjusted by adding a small percentage of a material with low thermal expansion, such as silica or talc (see list of oxides by expansion in Appendix 6). Thinning the glaze with water and applying the glaze more thinly can also help reduce crazing.

**Above:** Linda Swanson, *Doubt Over Certainty*, 2020. Crazing glaze applied alongside a balanced clear glossy glaze

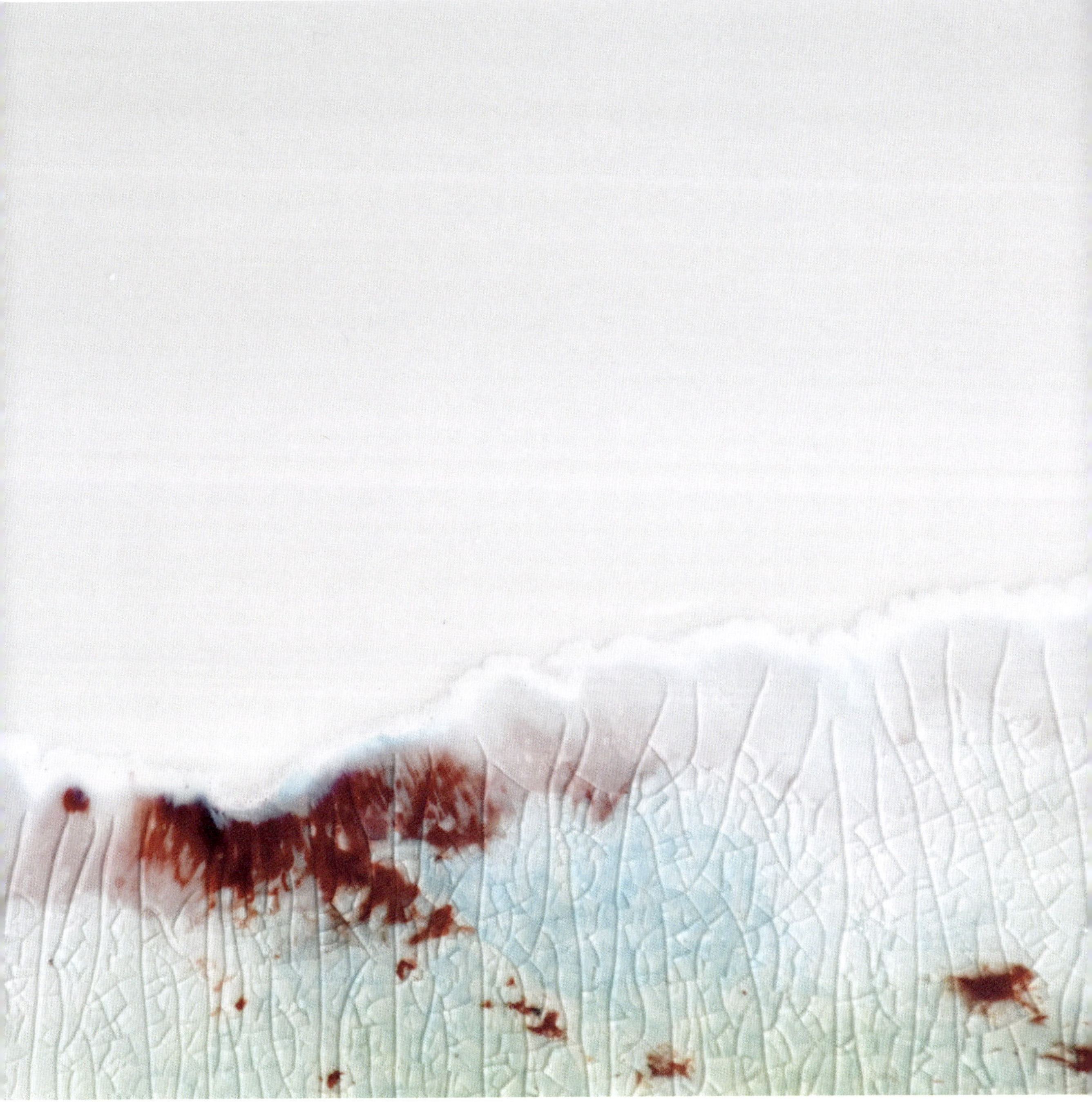

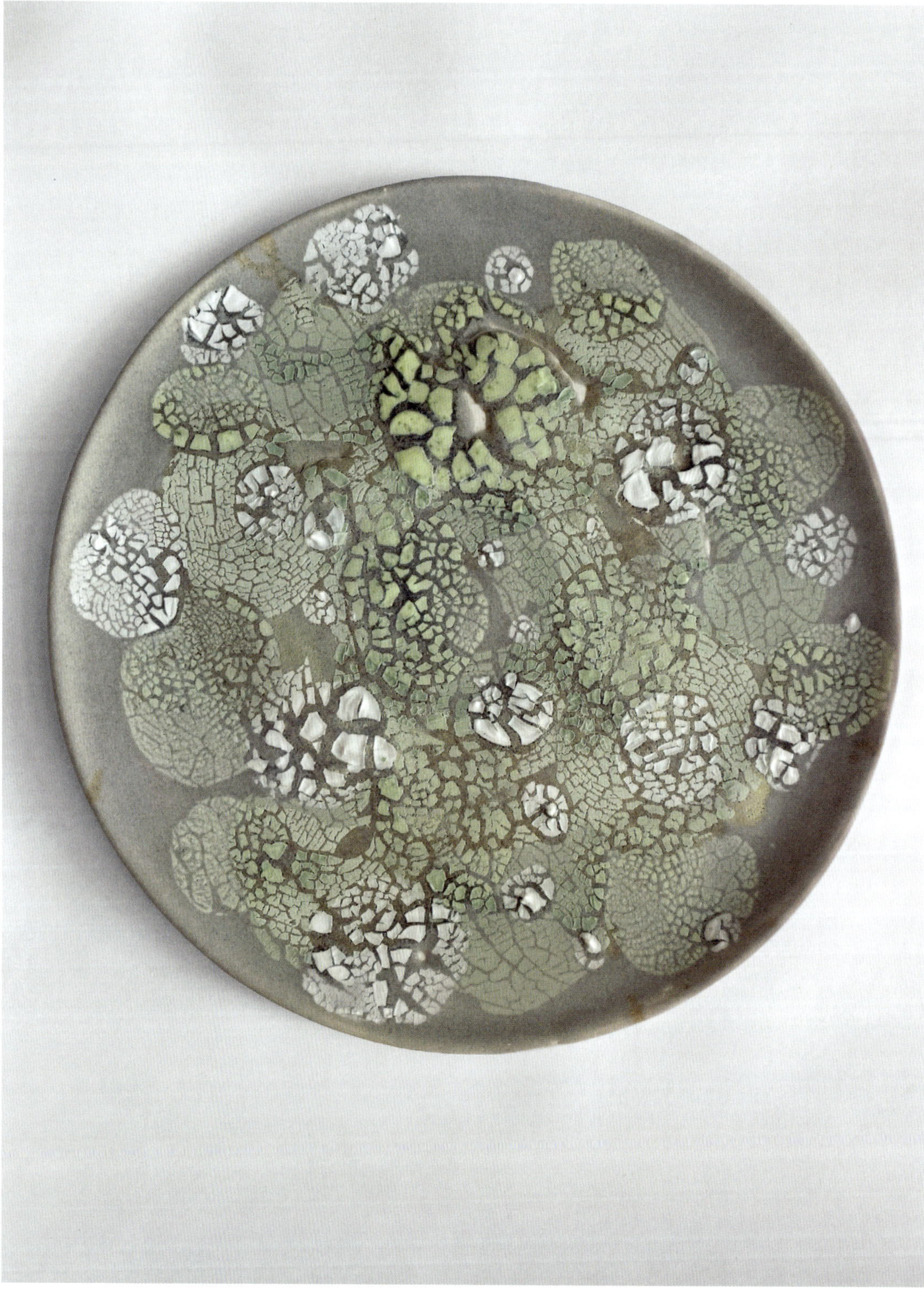

## SHIVERING

When the glaze has a lower thermal expansion than the clay body the clay may crack, either breaking the pot in half or by chipping at the rim, known as shivering or shelling. To correct this fault, a high expansion material such as nepheline syenite or high alkaline frit (Ferro frit 3110) can be added to the glaze to increase the expansion.

## CRAWLING

Some glaze materials shrink a lot during drying and can cause the glaze to crack as it dries. During firing the cracks widen and form bare patches between islands of glaze. These materials include zinc oxide, clay and light magnesium carbonate. The materials can be calcined by biscuit firing before adding to the glaze to reduce shrinkage, or the glaze can be watered down and applied more thinly.

## PINHOLES

Wetting the biscuit ware and glazing before it is completely dry can cause pinholes, as can overfiring the biscuit ware so that it is not porous enough to absorb the water in the glaze. Turning grogged clays so that particles of grog fall out also causes pinholes. Air becomes trapped in these small pits when they are covered by glaze and they form pinholes during firing. Sponging the ware after turning will help smooth over any holes.

**Opposite:** Linda Bloomfield, lichen-effect porcelain plate with crawl glazes

## BLISTERS

Overfiring can sometimes cause blisters, particularly in thickly applied glazes high in fluxes such as borates. Some materials such as bone ash give off a lot of gas during firing. Materials with high surface tension such as zirconium can cause the glaze to trap gas and form blisters. Gases are given off at high temperature as the fluxes start to break down. The blisters need to be ground down before they can be refired, otherwise they will keep expanding. Using cones, check that the kiln is not overfiring. Layering glazes can cause blisters, particularly if the top layer of glaze matures before the glaze underneath, trapping any gas given off.

## IRONING

A metallic sheen on the glaze surface, known in the ceramics industry as ironing, is caused by saturating the glaze with colouring oxides such as cobalt, copper, iron or manganese. These metallic surfaces are not food safe as the colouring oxides are on the surface and are not fully dissolved in the glaze. The total amount of colouring oxides should not exceed around 8% in glazes for use on tableware, particularly the surfaces that come into contact with food.

## SPECIAL EFFECT GLAZES

One potter's glaze defect can be another potter's special effect.

### ASH GLAZE, FAKE ASH

Ash glazes are made from wood ash mixed with feldspar and clay. It is also possible to make a glaze look like an ash glaze by adding excess calcium in the form of whiting. The ash can be made by burning wood, garden waste, vegetable waste, straw or grass. The latter two are high in silica, while wood ash is higher in calcium, depending on the type of tree, the soil where it was growing and the season it was cut. Many potters soak the ash in water for a few days to remove the soluble potassium, which is very caustic. The ash can then be sieved, dried out and used to make a glaze. Reduction firing brings out the blue-green colours from traces of iron oxide in the ash. If fired in an electric kiln, ash glazes usually come out tan yellow to olive brown. A small amount of copper oxide added to the glaze will make it more green in colour.

**Classic ash glaze, cone 10, 1280°C (2336°F) reduction**

| | |
|---|---|
| Wood ash | 40 |
| Potash feldspar | 40 |
| Ball clay | 20 |

**Fake ash glaze, cone 10, 1280°C (2336°F) oxidation**

| | |
|---|---|
| Nepheline syenite | 60 |
| Whiting | 40 |
| + | |
| Copper oxide | 1 |

**Opposite:** Mat Karas, *Warped Ware / Fake Ash*, 2015. This so-called fake ash glaze mimics the effects of ash in wood firing. *(Image credit: Mat Karas)*

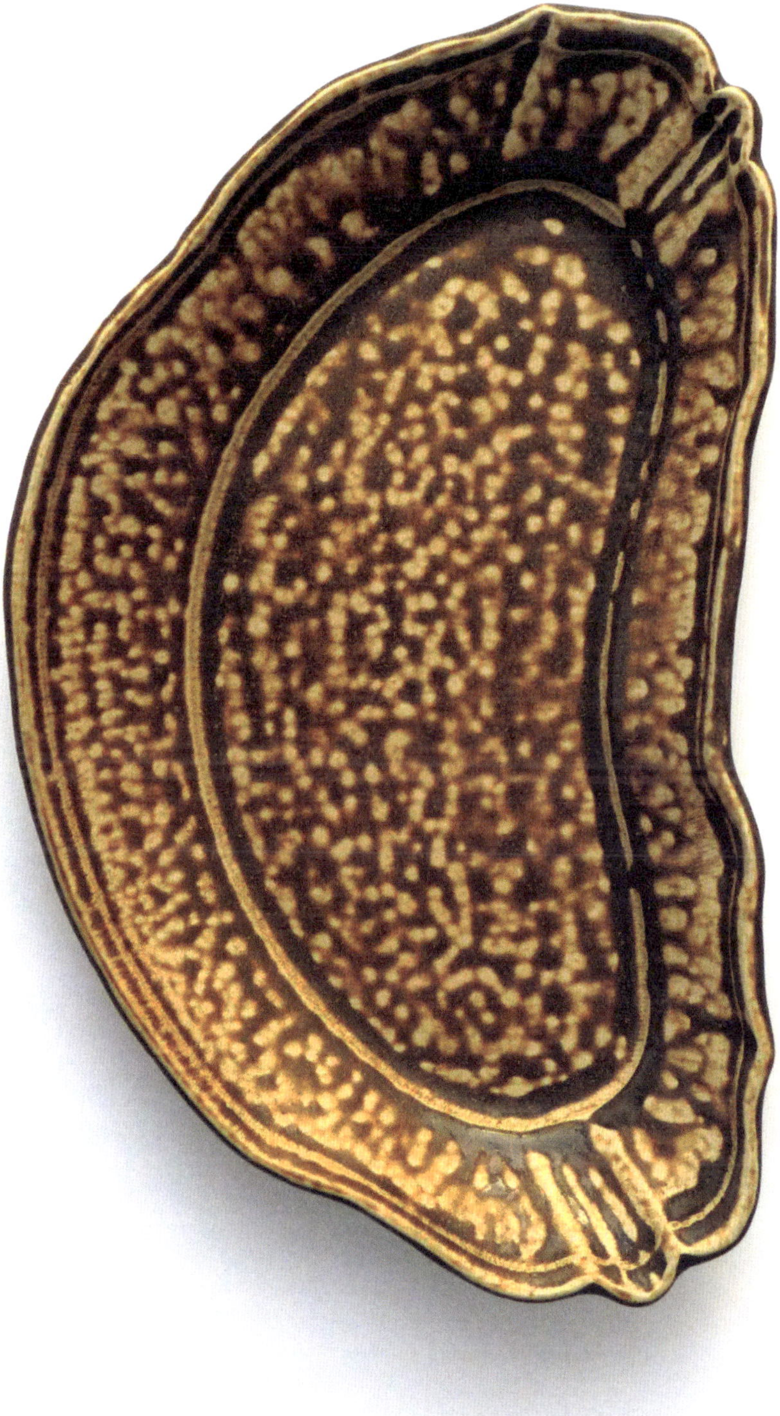

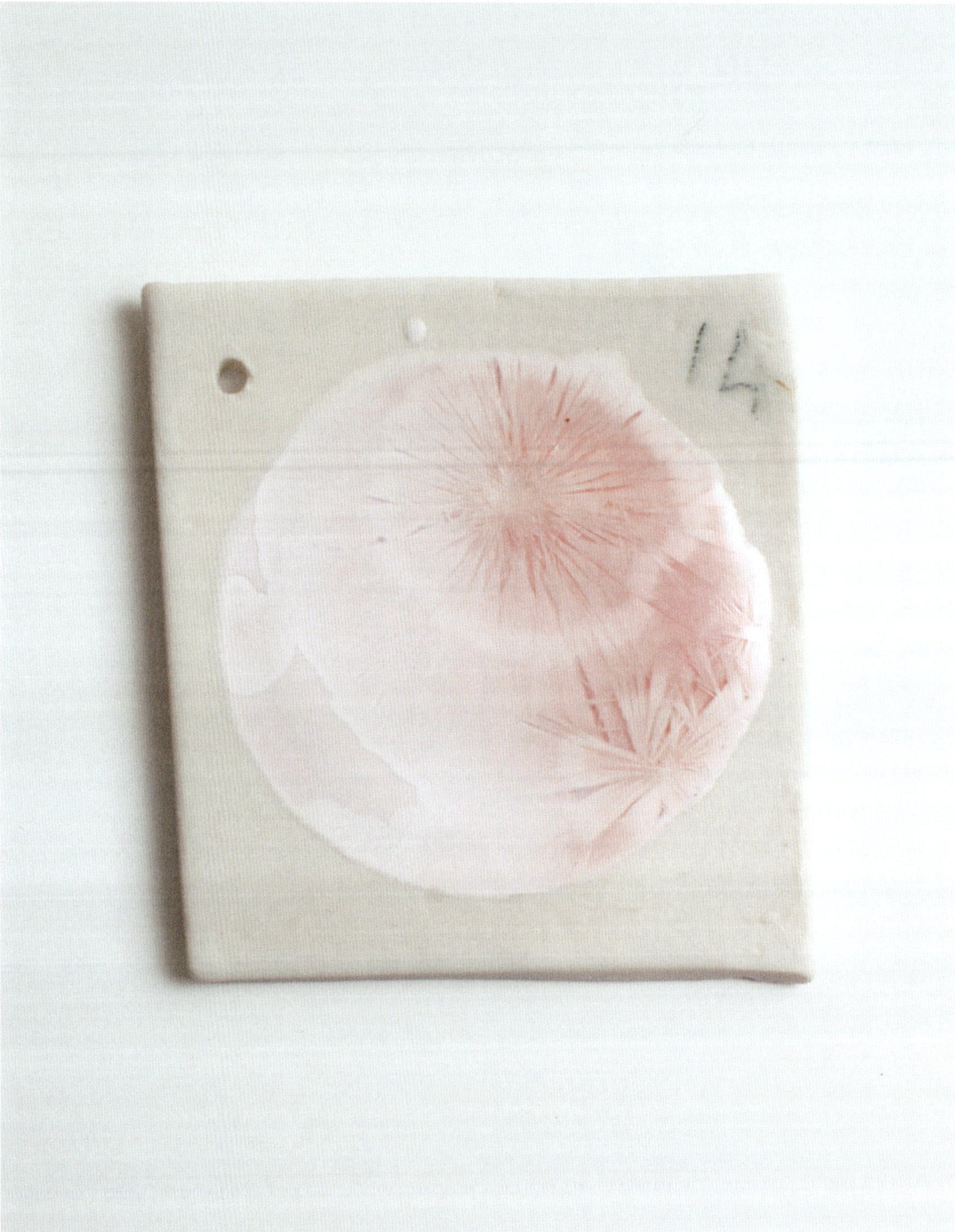


**Above:** Avril Farley, zinc crystalline glaze with erbium

## CRYSTALLINE, MICRO, MACROCRYSTALLINE

Crystalline glazes are low in alumina and high in an alkaline earth flux such as zinc oxide. As the glaze cools, the excess flux combines with silica to form crystals such as zinc silicate. A long, slow cool of several hours below 1100°C (2012°F) can grow larger crystals. Depending on the holding temperature, crystals can be needle shaped or star shaped. Other alkaline earths such as calcium and magnesium grow small crystals which cover the entire surface, making a matt glaze. These are known as microcrystalline glazes as the crystals are very small.

The lack of alumina in the glaze means that it is very fluid in the melt and atoms can move around easily. Only a limited amount of calcia can dissolve in the glaze. The silica in the glaze reacts with the excess calcium to form crystals, which grow in the molten glaze if it is cooled slowly, and this gives time for the atoms to arrange themselves in a crystal structure. These crystals can include wollastonite or calcium silicate ($CaSiO_3$), diopside ($CaMgSi_2O_6$) and enstatite or magnesium silicate ($Mg_2Si_2O_6$). The latter two are types of pyroxene, a chain silicate. The glaze becomes devitrified and is no longer transparent. There can be a few crystals floating in a matrix of glossy glaze, or the crystals can cover the entire surface to form a matt glaze if the kiln is cooled slowly. Barium, strontium and zinc will also form crystals in a similar way.

Crystalline glazes with very large crystals can be made from zinc silicate if held for several hours during cooling at around 1100°C (2012°F) to allow time for the crystals to grow while the glaze is still molten. The zinc silicate mineral willemite ($Zn_2SiO_4$) has crystals which can be coloured with cobalt, copper or nickel. The crystals selectively take up certain colouring oxide in preference to others. For example, cobalt will colour the crystals blue, nickel will give steel blue and manganese will give pink only if there is no cobalt or nickel present. Manganese and copper oxide will usually colour the background if there are other colouring oxides present.

Titanium and rutile help to seed crystals. In a molten glaze, they initially form zinc titanate ($ZnTiO_3$) or calcium titanate ($CaTiO_3$), from which other crystals can grow on cooling. At stoneware temperatures, mullite in the glaze-body interface will act as a seed for crystals, as will the addition of wood ash or bone ash to the glaze.

**Macro-crystalline glaze, cone 6–7, 1240°C (2264°F)**

This glaze has large crystals if held at 1070°C (1958°F) for several hours

| | |
|---|---|
| Ferro frit 3110 | 45 |
| Zinc oxide | 21 |
| Silica | 20 |
| Kaolin | 9 |
| Gerstley borate | 4 |
| Titanium dioxide | 1 |

**Frosty crystal glaze, cone 6, 1220°C (2228°F)**

| | |
|---|---|
| Nepheline syenite | 40 |
| Silica | 29 |
| Zinc oxide | 17 |
| Whiting | 10 |
| Kaolin | 5 |
| Bentonite | 2 |

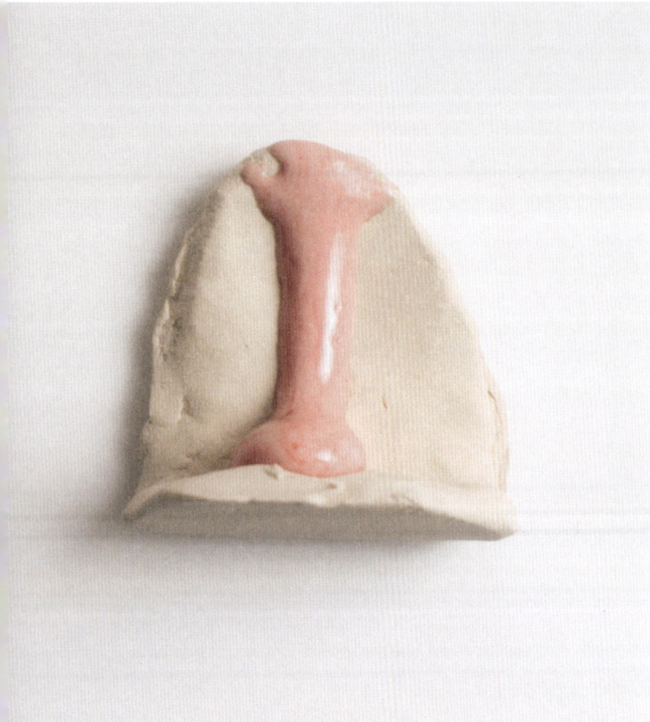

## GLOOP

In between clay and glaze, gloop is a thickly applied, viscous glaze. It is high in clay and silica and low in fluxes, so it is glossy but not very fluid unless applied very thickly. Gloop is usually applied as a slip-trailed drip or solid ball of clay consistency. It can be made more runny by adding more frit in 5% increments

**Medium gloop**
**(from Derek Au, glazy.org) cone 6, 1220°C (2228°F)**

| | |
|---|---|
| Silica | 34 |
| Kaolin | 26 |
| Soda feldspar (Minspar) | 21 |
| Whiting | 10 |
| Ferro frit 3134 | 9 |

Add 40% water, blend and dry to a clay-like consistency.

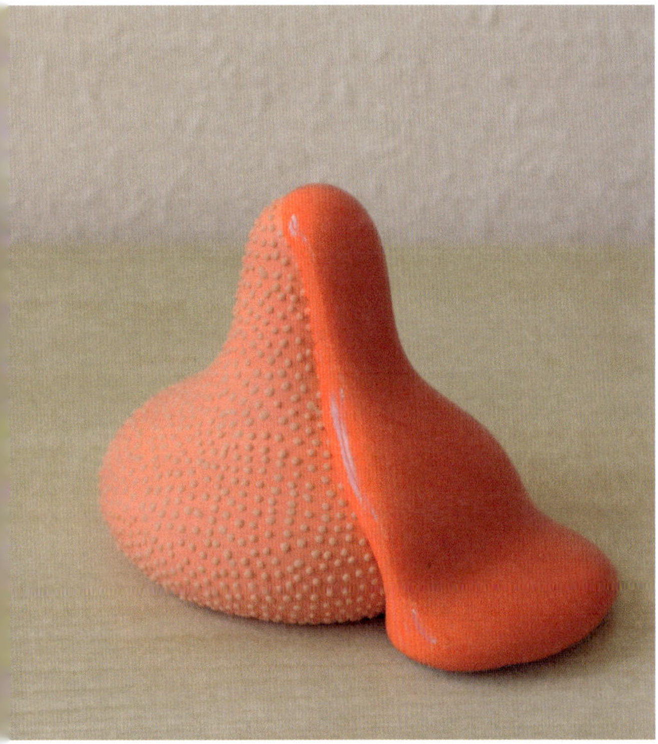

**Above left:** Test tile gloop. *(Image Credit: Henry Bloomfield)*

**Left:** Jihyun Kim, *Salty Fairy Gnome Red*, 2023. High-fired porcelain with gloop glaze and coloured stain, 1200°C (2192°F), oxidation. *(Image credit: Jihyun Kim)*

## CRATER, LAVA, FOAM

Crater glazes can be made by adding silicon carbide to a viscous matt glaze. The oxides in the glaze react with the silicon carbide to produce carbon dioxide gas. Colouring oxides with high oxidation states produce the most gas: these include vanadium pentoxide, titanium dioxide, manganese dioxide and red iron oxide. The firing temperature, thickness of glaze application and mesh size of the silicon carbide affect the number and size of the craters; coarse grades (60–120) produce numerous pinholes, while ultra-fine grades (800–1200) produce frothy bubbles. Many potters use a medium-fine grade such as 220–500 mesh to make crater glazes.

**Satin white volcanic (Marilee lava), cone 6, 1220°C (2228°F)**

| | |
|---|---|
| Potash feldspar | 50 |
| Whiting | 24 |
| China clay | 13 |
| Silica | 13 |
| + | |
| Titanium dioxide | 11 |
| Silicon carbide | 0.3 |

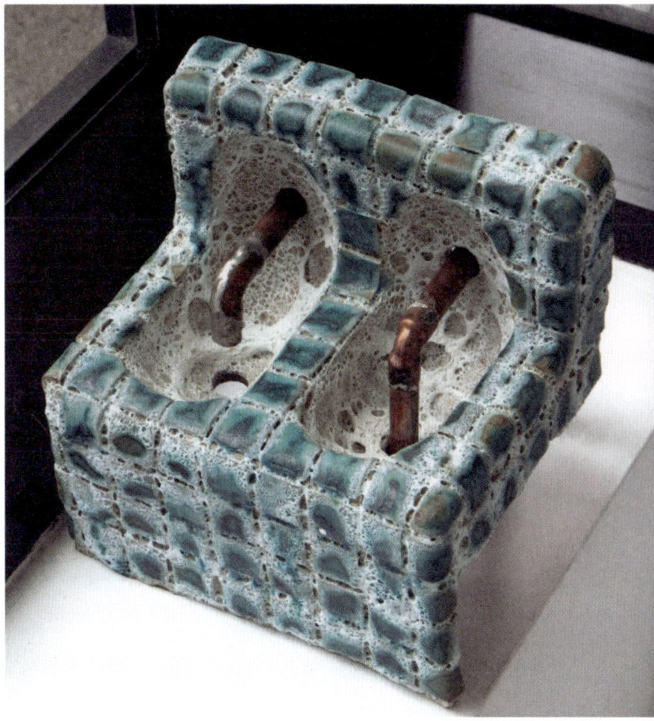

**Above:** Rebecca Ramsey, *Closed Circuit*, 2023. Various glazes including crater glazes enliven this stoneware sculpture with copper elements. *(Image credit: Brandon Brookbank)*

**Below:** Crater glaze, base glaze and with added cobalt 0,5, copper 2, chromium 0.5, tin 6 and manganese 4

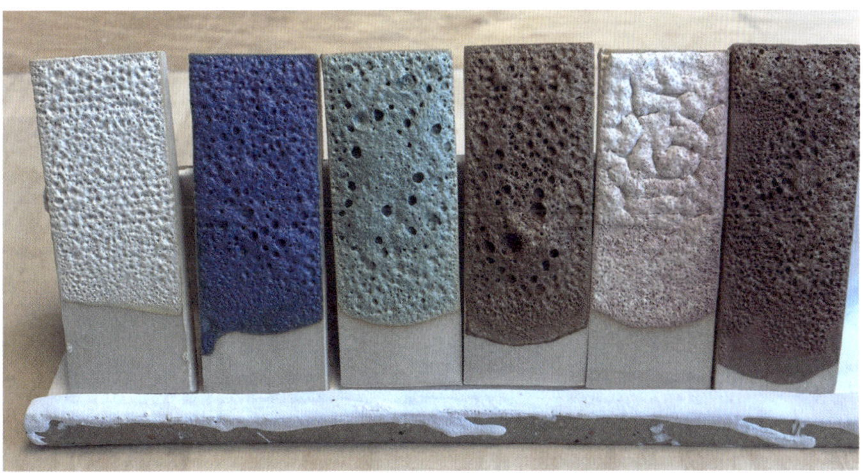

## CRAWL, LICHEN

Crawl glazes are made by using one or more materials that have a high drying shrinkage. These materials are zinc oxide, clay and light magnesium carbonate. Zinc oxide will melt to form beads of glaze, while light magnesium carbonate can produce a dry, lichen-effect glaze. Glazes that are high in clay, such as shino, can also crawl, as can those high in bone ash.

**Lichen glaze, cone 8, 1250°C (2282°F)**

| | |
|---|---|
| Nepheline syenite | 50 |
| Light magnesium carbonate | 40 |
| Standard borax frit | 10 |

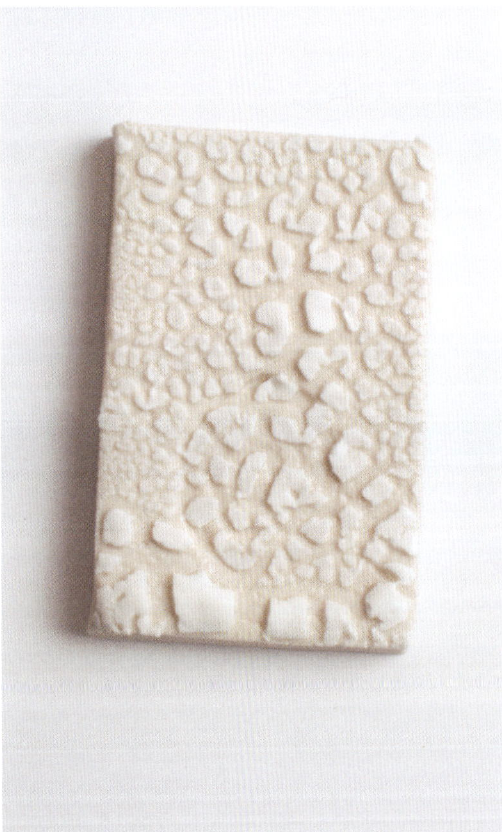

## SHINO

Shino is a Japanese white-to-orange glaze made from feldspar and clay. It is usually fired in reduction, which brings out the orange colour from the iron oxide in the clay, forming red iron aluminate on the glaze surface[1]. The glaze is white where thick and orange where thin, the clay body having more effect in the thin areas. Japanese shinos often have pinholes and crawling where applied thickly. American shinos include soluble soda ash which migrates to the surface during drying and traps black carbon in the glaze during firing. The contrast of black carbon against orange shino is very effective.

**Bray shino, cone 10 reduction, 1280°C (2336°F)**

| | |
|---|---|
| Nepheline syenite | 46 |
| Soda feldspar | 18 |
| Ball clay | 17 |
| Spodumene | 15 |
| Soda ash | 4 |

[1] Daniel de Montmollin, *The Practice of Stoneware Glazes: Minerals, Rocks, Ashes*; 2005

**Left:** Test tile lichen crawl

## COPPER RED

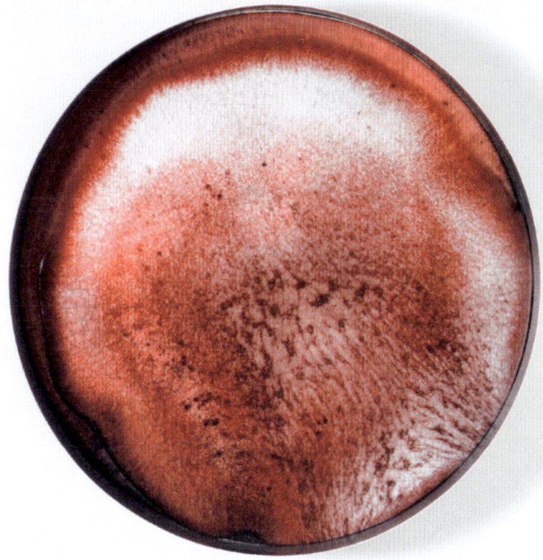

Copper red glazes are glossy, fluid glazes that contain a small amount of copper oxide or carbonate. When fired in reduction, the copper is reduced to tiny colloidal particles of copper metal which appear red when suspended in the glaze. Tin oxide helps to stabilize the red colour. Ultra fine silicon carbide (800–1200 mesh) can be added to achieve reduction in an electric kiln.

**Clotted blood red, cone 6–8, electric kiln (Emmanuel Cooper)**
**1240–1260°C (2264–2300°F)**

| | |
|---|---|
| Soda feldspar | 35 |
| Calcium borate frit | 10 |
| Whiting | 15 |
| China clay | 5 |
| Flint | 35 |
| + | |
| Tin oxide | 1 |
| Copper carbonate | 0.5 |
| Silicon carbide | 0.5 |

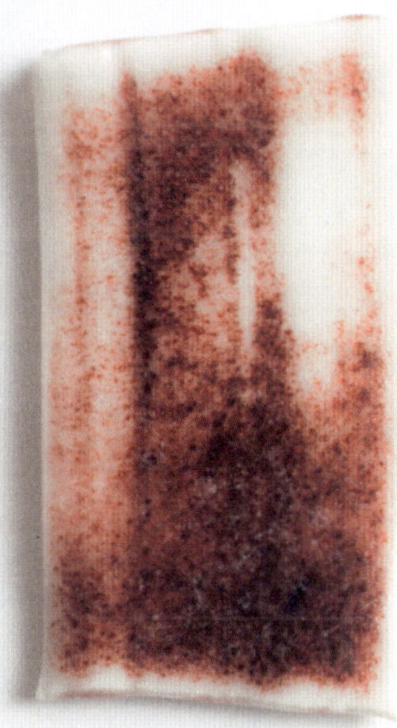

**Above right:** Linda Swanson, *Red View*, 2008. Slip cast porcelain with copper red glaze. *(Image credit: Gaëtane Girard)*

**Right:** Test tile copper red

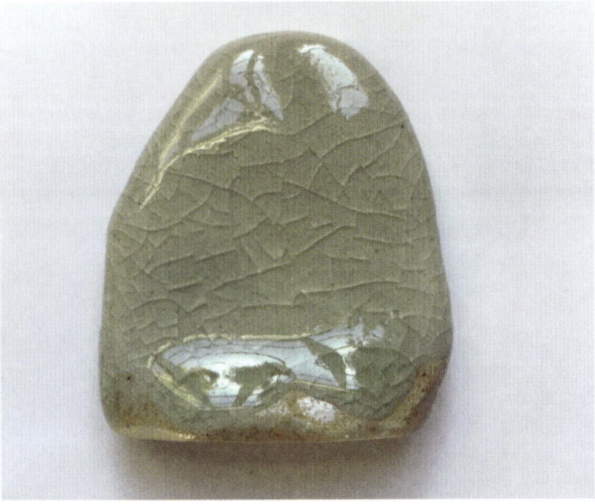

## CELADON

Celadons are pale grey-green to pale blue, glossy glazes containing a small amount of iron oxide, less than 1%. The iron oxide can also come from impure materials such as ball clay. To obtain a pale blue celadon, the glaze materials, particularly the kaolin and the clay body used must be very pure; impurities such as titanium dioxide in the clay will make the celadon turn green. Celadons must be fired in reduction. Alternatively, it is possible to make a celadon in an electric kiln by adding 0.2% ultra fine silicon carbide to cause reduction in the glaze. This works best in celadons that contain zinc oxide, which helps clear any residual black specks of silicon carbide.

**Elisa's celadon, cone 9–10 reduction, 1280–1300°C (2336–2372°F)**

| | |
|---|---|
| Potash feldspar | 36 |
| Silica | 27 |
| Wollastonite | 18 |
| Kaolin | 13.5 |
| Talc | 4.5 |
| Zinc oxide | 4.5 |
| + | |
| Red iron oxide | 0.45 |

**Above:** Test tile, Elisa's celadon on stoneware reduced in a wood firing

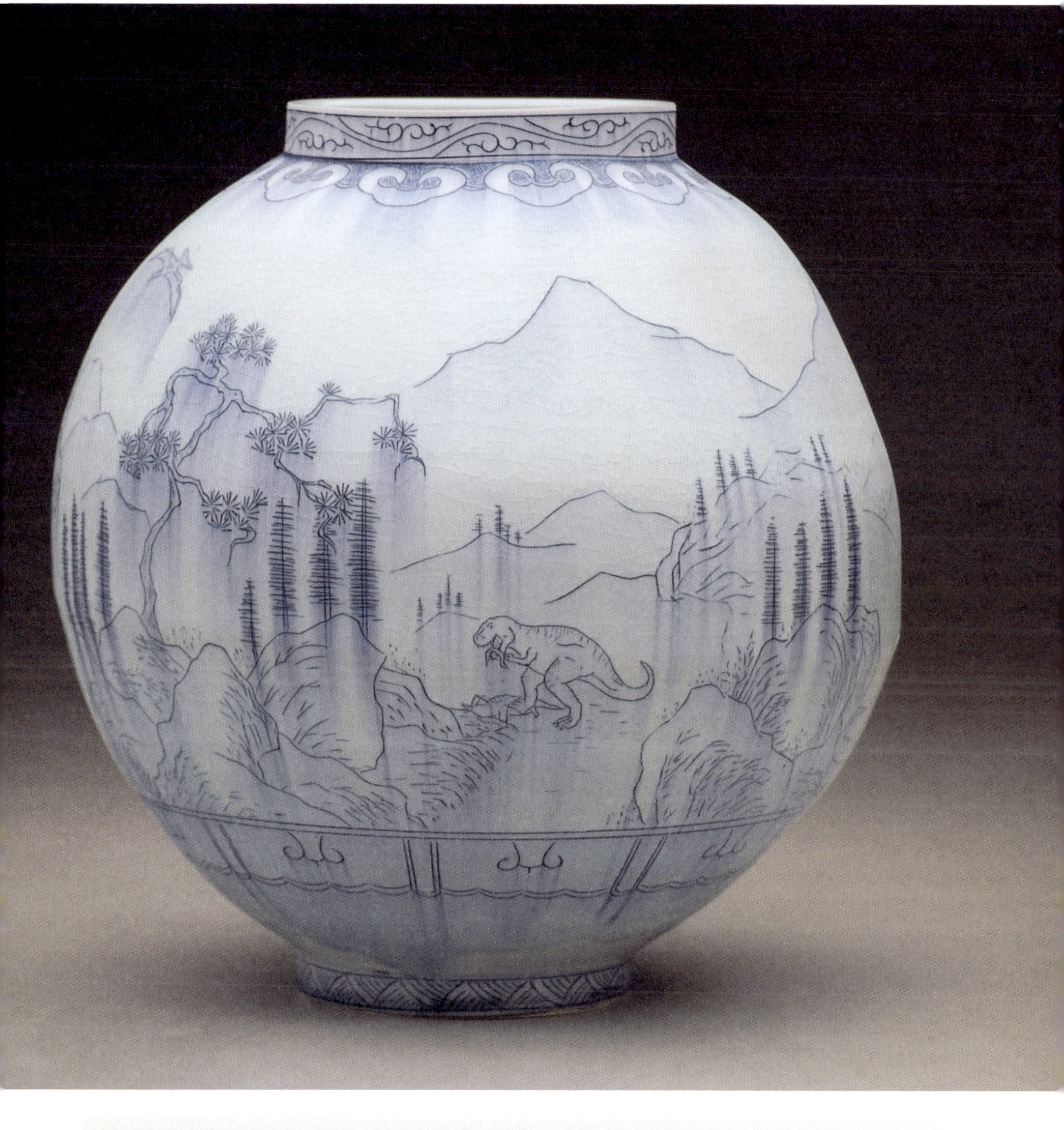

**Above:** Steven Young Lee, *Another Time and Place*, 2010. Atmosphere is created in this scene when cobalt inlay streaks into the celadon glaze. *(Image credit: Steven Young Lee)*

## CHUN

Chun or jun glazes are pale blue opalescent Chinese glazes originally made in northern China in the Song Dynasty (960–1279 CE). The blue colour is an optical effect similar to the effect that makes the sky look blue. Tiny calcium-rich bubbles in the glaze cause blue light to be reflected more strongly than other colours of light, which makes the glaze appear opalescent. The colloidal particles need to be smaller than the wavelength of light for opalescence to occur. It used to be thought that phosphorus from bone ash (calcium phosphate) in the glaze caused this effect, but from scientific analysis we now know it is liquid bubbles of a calcium-rich phase separating out from the silica-rich glaze. Adding a small amount (1–2%) of bone ash does enhance the effect but causes the glaze to become opaque, eventually turning it white with larger amounts up to 10%. The glaze needs to be high in silica (quartz or flint) and relatively low in clay, which makes it quite runny (the molecular ratio of alumina to silica is ideally around 1:12). Either soda feldspar or potash feldspar can be used. Iron oxide can be added in the form of red iron oxide or yellow iron oxide. Calcium is an important ingredient of chun glazes and is added in the form of whiting (calcium carbonate) or wollastonite (calcium silicate). Chun glazes can be used on stoneware or porcelain, and are sometimes applied over a dark, iron-rich glaze such as tenmoku which gives more contrast and enhances the blue colour.

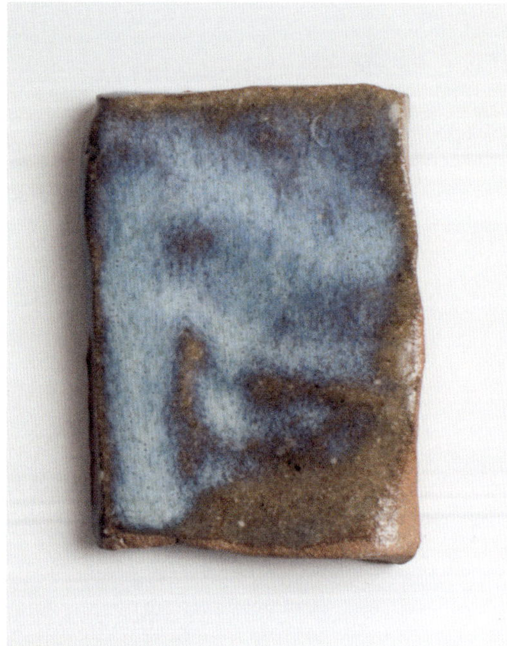

**David Tsabar SiC chun glaze cone 6–8 oxidation, 1240–1260°C (2264–2300°F)**

| | |
|---|---|
| Soda feldspar | 33 |
| Quartz | 30 |
| Standard borax frit | 20 |
| Whiting | 13 |
| China clay | 4 |
| Bentonite | 3 |
| Zinc oxide | 3 |
| + | |
| Bone ash | 1 |
| Silicon carbide 1200 mesh | 1 |
| Tin oxide | 1 |
| Yellow iron oxide | 1 |

**Above:** Test tile chun

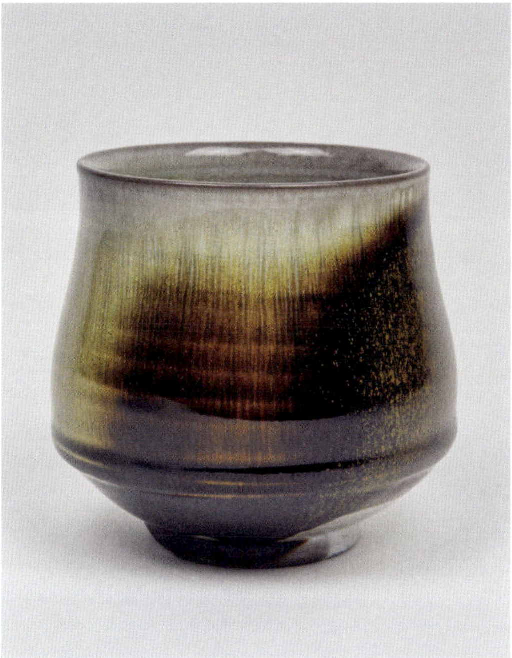

## TENMOKU AND TESSHA

Tenmokus are glossy, brown-black glazes containing around 8% iron oxide, traditionally fired in reduction. They are made by adding iron oxide to a clear glaze. They often break to red-brown on rims and edges. Tessha glazes, also known as tea dust, are higher in iron than tenmoku, and usually form yellow-green crystals on the surface. They contain magnesium, which helps the yellow iron pyroxene crystals to form.

**Tenmoku, cone 9, 1280°C (2336°F)**

| | |
|---|---|
| Potash feldspar | 50 |
| Whiting | 5 |
| Dolomite | 15 |
| Silica | 30 |
| Kaolin | 10 |
| + | |
| Red iron oxide | 8 |

**Tessha iron saturate, cone 10 reduction, 1300°C (2372°F)**

| | |
|---|---|
| Potash feldspar | 52 |
| Whiting | 9 |
| Silica | 19 |
| Kaolin | 20 |
| Magnesium carbonate | 0.13 |
| + | |
| Red iron oxide | 11.5 |

**Above:** Marie-Pier Laverdière, 2019. Cup with tessha iron saturate glaze, wood fired at Guldagergaard. *(Image credit: Laurence Poirier)*

## OIL SPOT, HARE'S FUR

Oil spot glazes are traditionally dark brown or black glazes high in iron oxide, usually containing around 6% red iron oxide. The red iron oxide hematite ($Fe_2O_3$) changes from a trigonal crystal structure to cubic magnetite ($Fe_3O_4$) at 1210–1232°C (cone 6–7), releasing bubbles of oxygen gas which drag to the surface iron spots of darker magnetite. Oil spot glazes are applied thickly and fired in an oxidation atmosphere, traditionally at cone 9–10, but they are also possible at cone 6–8. Adding magnesium in the form of talc or dolomite increases the viscosity of the molten glaze, allowing the spots to form. The spotted effect can be enhanced by applying a layer of opaque white glaze on top of the dark iron glaze. If the glaze is slightly more fluid, the spots can run on vertical surfaces to form a streaked effect known as hare's fur. These glazes can be fired in an electric kiln.

**Mike Bailey oil spot, cone 6 oxidation, 1220°C (2228°F)**

| | |
|---|---|
| Soda feldspar | 36 |
| Potash feldspar | 26 |
| Kaolin | 15 |
| Borax frit | 13 |
| Dolomite | 10 |
| Silica | 8 |
| + | |
| Red iron oxide | 6 |

**Opposite top:** Matt Fiske, lava oil spot yunomi, 2016. *(Image credit: Matt Fiske)*

**Opposite bottom:** Paul Briggs, *Reflection Knots*, 2024. Metallic glaze cone 04 containing manganese dioxide, copper oxide, black stain and zircopax. *(Image credit: Paul Briggs)*

## METALLIC, BRONZE

Metallic bronze glazes are made using colouring oxides manganese and copper as the flux materials, in the ratio of 7 manganese to 1 copper oxide, mixed with clay and sometimes feldspar. The addition of cobalt is expensive and optional. Apply thickly for a gold metallic effect, more thinly for brown-black. Without copper, the manganese and clay make a pewter glaze. This glaze is not suitable for surfaces intended for food.

**Metallic bronze glaze cone 8–10, 1260–1280°C (2300–2336°F) oxidation**

| | |
|---|---|
| Ball clay | 40 |
| Potash feldspar | 30 |
| Manganese dioxide | 26 |
| Copper oxide | 4 |
| Red iron oxide | 4 |

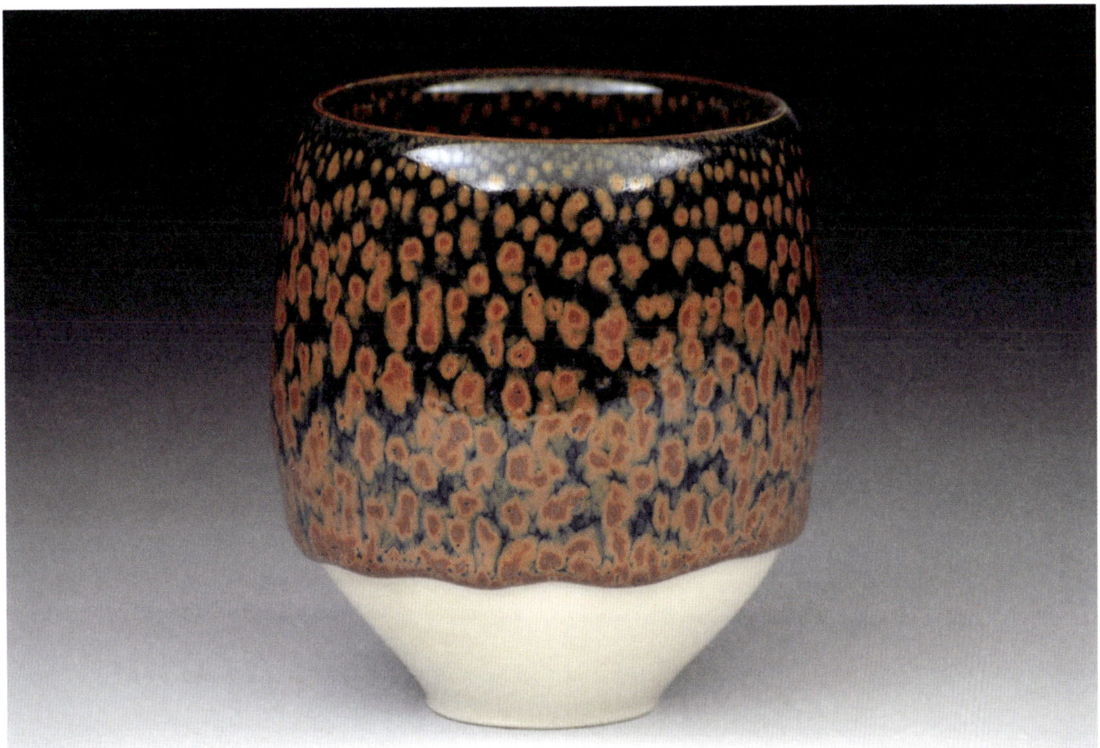

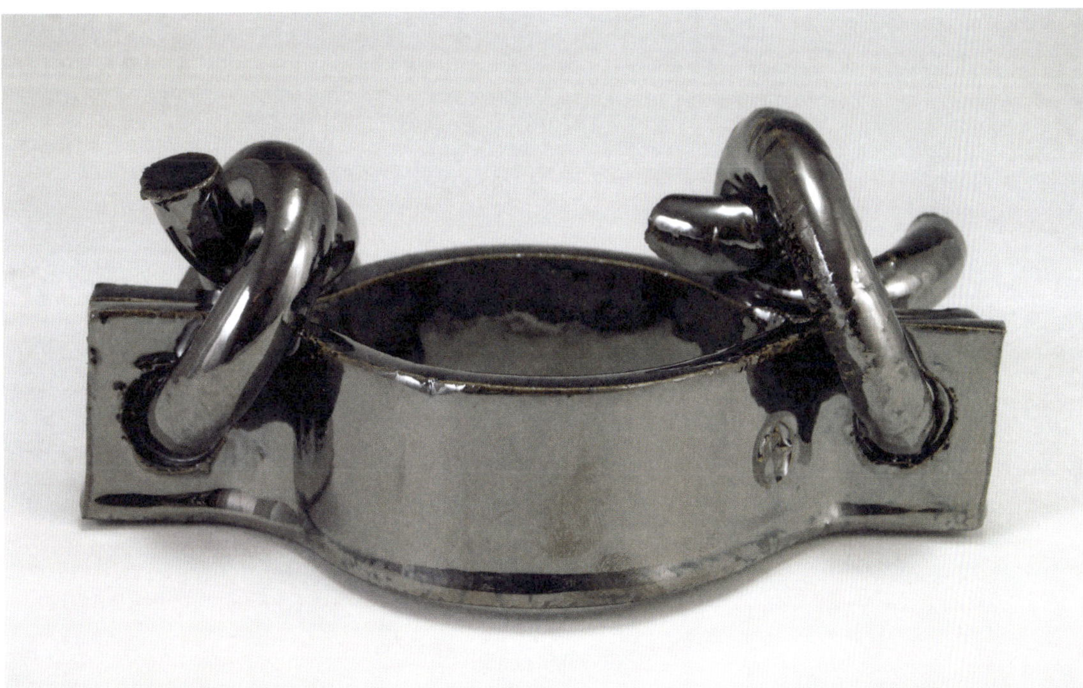

## LUSTRE

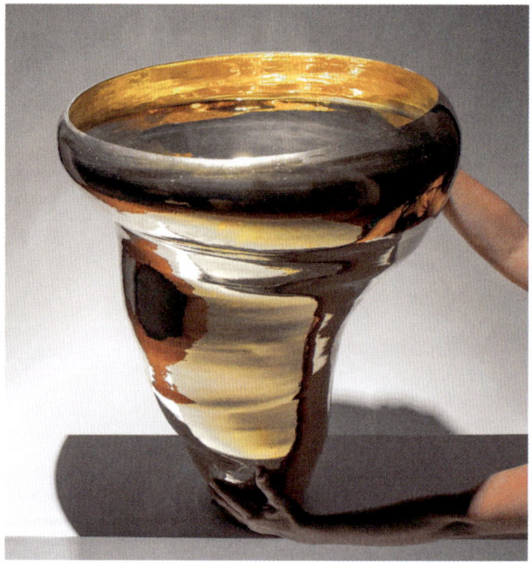

Traditional lustres made from metal salts mixed with clay were reduction fired, although modern resin-based lustres can be fired in oxidation. These lustres contain metal salts such as gold chloride mixed with pine resin, which burns out during firing and reduces the gold to metal. This type of lustre is applied and fired in a second glaze firing to a low temperature, around 700–800°C (1292–1472°F), to soften the existing glaze so that the metal sticks to it permanently. Various metals are available, including gold, platinum and bismuth, which makes an iridescent mother-of-pearl effect.

In raku, post-firing reduction is carried out when red-hot pots are removed from the kiln using tongs and covered in sawdust. Any metal oxides such as copper oxide or metal salts such as silver nitrate and tin chloride will reduce to metal during post-firing reduction in the burning sawdust. The combustion of the sawdust requires oxygen, which is drawn from the glazes, causing reduction from oxide to metal. Metal oxides used by potters are insoluble in water, while metal salts including nitrates, chlorides and sulphates are water soluble. This makes them more likely to form a vapour and travel around in the kiln during firing. However, their solubility makes them more toxic, so care must be taken when making and applying these glazes.

**Above:** Elysia Athanatos, *Gold and Alchemy, 2018.* Stoneware with gold lustre *(Image credit: Elysia Athanatos)*
**Opposite:** Claudia Clare, pomegranate bowl with underglaze, oxides and lustres

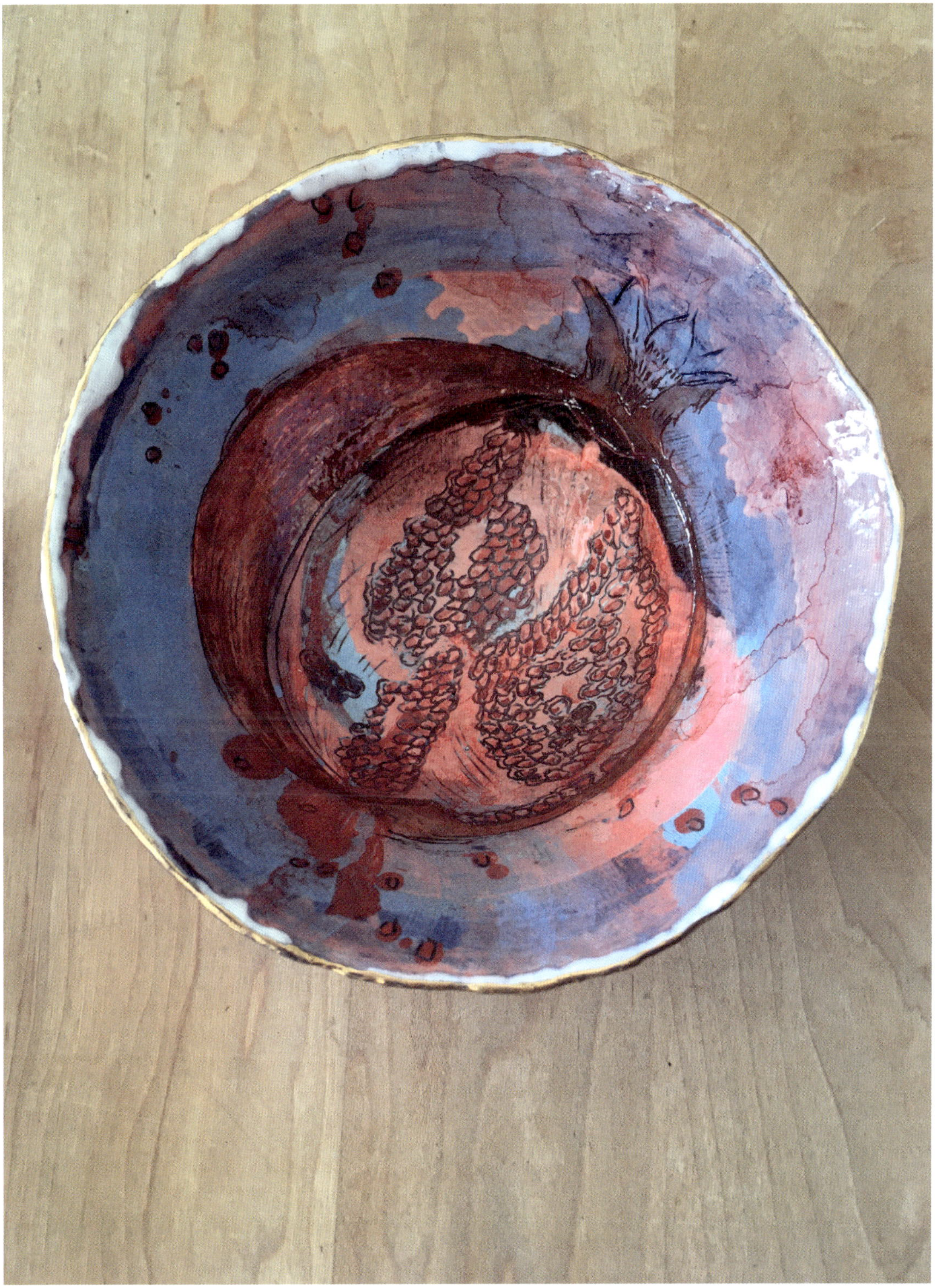

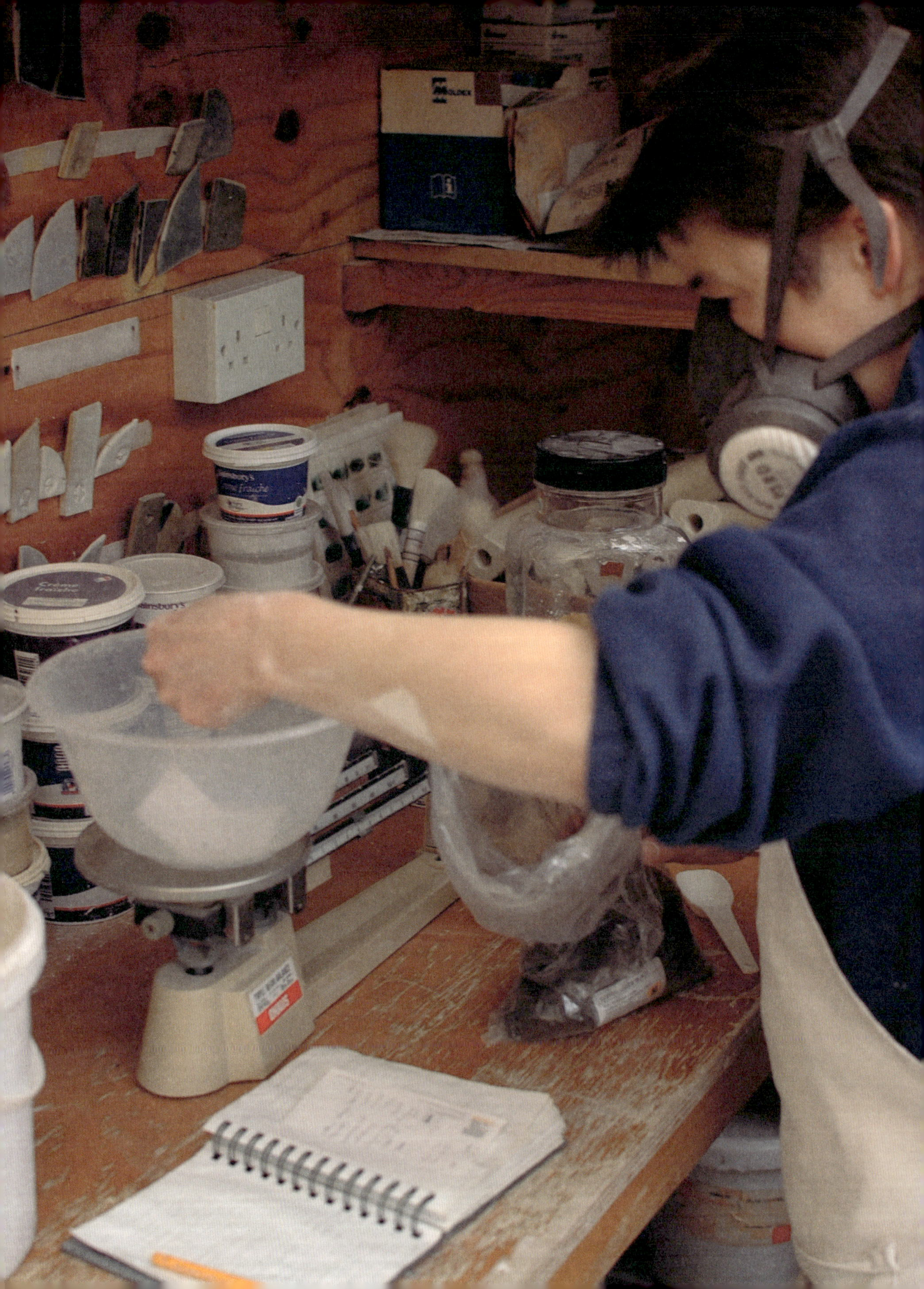

CHAPTER TEN

# HEALTH AND SAFETY

It is important to look after your health in the ceramic studio. Silica dust is harmful to the lungs and can cause silicosis after many years of exposure. It is vitally important to wear a respirator when making glazes and to prevent dust accumulating by cleaning the studio often. Silica is present not only in quartz and flint used for making glazes but also in clay, feldspar, frits, talc, wollastonite and other silicate materials, which include ground up igneous rocks such as granite and pumice (see Chapter 1).

## SOLUBLE MATERIALS

Glazes are normally made of materials that are not soluble in water, otherwise there would be problems with the glaze consistency in the bucket and absorption of soluble salts into the biscuit ware. If an alkaline material dissolves in water, it can cause the glaze to settle in a hard layer at the bottom, referred to as hard panning. This sometimes happens with slightly soluble materials such as nepheline syenite, a rock related to feldspar. This can be reconstituted by digging out the hard layer, re-mixing with the water and adding a small amount of Epsom salts (magnesium sulphate) dissolved in water.

If acid materials dissolve in water, they can cause the glaze to become thick and gel-like, which can happen with Gerstley borate or bone ash glazes. If this happens, it is best to add a deflocculant such as sodium silicate to counteract the acidity. Because most glaze materials are insoluble, they will not be absorbed by the skin, so most glazes are safe to stir by hand, with the exception of ash glazes and American shinos, which contain soluble alkalis, are caustic, and require rubber gloves to be worn.

## TOXIC MATERIALS

Most clays and glaze materials are not toxic, with some exceptions (see full list below). By law glazes must not contain soluble, leachable lead or cadmium. It is best to avoid the use of these materials in functional tableware. However,

**Opposite:** Wearing a respirator mask

many of the colouring oxides and carbonates are toxic, particularly nickel, chromium, vanadium and manganese. They do not dissolve in water so are not absorbed into the skin but should not be ingested or inhaled.

Firing fumes from these metal oxides can also be toxic, so make sure the kiln room is well ventilated and fire at night when you are not in the studio. Chromium oxide is particularly volatile and forms a vapour that wafts around the kiln. Soluble forms of these colouring oxides are very toxic as they can be absorbed easily; these are nitrates, chlorides and sulphates sometimes used in pottery to obtain watercolour effects.

Once the glaze has been fired, the metal oxides will be locked up in the glaze structure, provided it is properly formulated with the correct proportions of silica, alumina, alkali metal (sodium, potassium, lithium) and alkaline earth (calcium, magnesium, strontium) oxides (see Chapter 7).

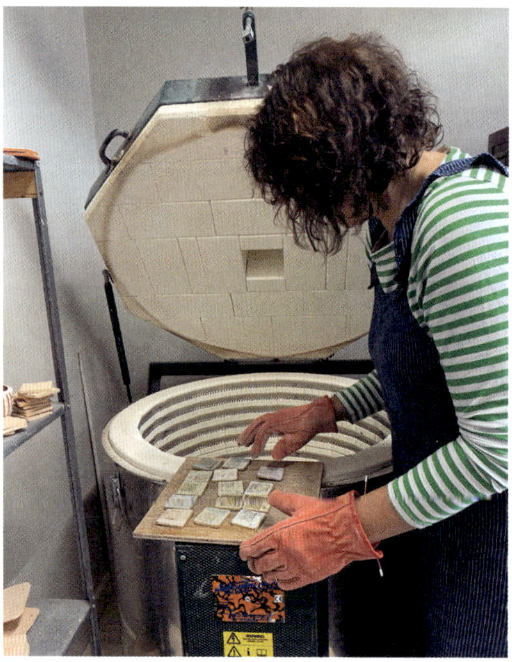

There can be a problem with adding too much copper oxide to a glaze as it acts as an alkali metal flux like sodium and can unbalance the glaze, causing alkali to leach out of the glaze when in contact with acid foods. To avoid this happening, do not add more than 3% copper oxide (or 4.5% copper carbonate) to a glaze; it is preferable to add only 1–2%. Many other colouring oxides (cobalt, manganese, iron) act as alkaline earth fluxes, which are more stabilizing additions to glazes.

Stains and underglazes are made using colouring oxides that have been heated together with silica and ground to a powder, rendering them non-toxic. The disadvantage of stains and underglazes is that they do not dissolve in the glaze and so often lack depth. The advantage is that they are safer to use, particularly in schools and community studios.

## STUDIO PRACTICE

It is most important to keep down clay dust in the ceramic studio. Make sure you clean the studio regularly using a wet sponge, mop or a vacuum cleaner with a dust filter. Clean work surfaces, boards, batts and floors and wash towels, overalls and aprons frequently. Food and drink should not be consumed in the studio as dust could fall into it. If you have a home studio, change your shoes and overalls at the door so that you do not bring clay dust into the house.

Wear a respirator mask when weighing dry glaze materials. Use ventilation or open windows. Wear rubber gloves when using wet glazes, particularly ash glazes and shino. Do not wash glaze colourants down the sink. Dry out and dispose of them in landfill or mix your waste glazes together to use as an unrepeatable mystery glaze.

## MATERIALS CONTAINING SILICA

Clay
Feldspar
Frit
Talc
Wollastonite
Zirconium silicate
Stains

### TOXIC GLAZE MATERIALS

Lead
Cadmium
Barium carbonate
Vanadium pentoxide
Nickel oxide and carbonate
Chromium oxide
Manganese dioxide

### SLIGHTLY TOXIC MATERIALS

Cobalt oxide/carbonate and chloride/nitrate
Copper oxide/carbonate and sulphate
Zinc oxide
Lithium carbonate
Wood ash
Borax

## FIRING SAFELY

When firing, wear welders' goggles to protect your eyes when looking into the kiln. Wear heat resistant gloves for unloading work from the hot kiln.

During wood firing, wear clothes made of natural materials, a face shield and bandana to cover your hair, as well as a leather apron and gauntlets while stoking. During reduction, make sure you do not breathe in any black smoke or carbon monoxide.

When cleaning kiln shelves wear a respirator mask, eye protection and gloves to protect your hands from sharp shards.

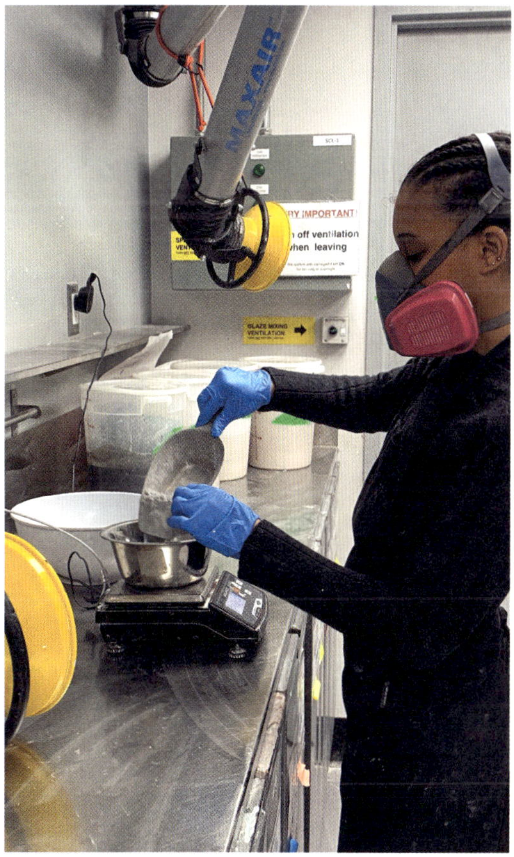


**Opposite:** Unloading warm work from a kiln with gloves

**Right:** Student Camille David-Boudreault mixing a glaze in the ventilated glaze lab at Concordia University wearing protective gear including a respirator mask and gloves

# CONCLUSION

This book has been designed to guide the reader in developing a deeper understanding of the materials central to the practice of ceramics with an eye to the importance of finding more sustainable methodologies. To work sustainably in ceramics it is essential that we not only choose our materials thoughtfully, but also that we utilize them with care. Recycling and reducing waste as much as possible are essential. It is also important to find out all that we can about our materials and their sources so we may make choices that better safeguard the environment.

Seeking out more local materials, minimizing the use of industrially refined materials, and incorporating materials from the waste stream are all ways to mitigate the environmental impacts of ceramics. Firing at lower temperatures will help save energy, while firing in electric kilns supplied with electricity from renewable energy sources, or fast firing with more efficient wood kilns that use waste wood will also help to lessen the environmental impact of ceramic production. As we have stressed throughout, such practices should not be seen as limiting but rather may enhance the variety of clay bodies and glazes that are possible.

Overall, the best way to learn about materials is by working with them and systematically testing them to their limits. Keeping good notes will help you achieve reproducible results. As you gain a more thorough understanding of your materials, you will be capable of targeting specific outcomes and cultivating new directions. We hope that this book will expand your discovery of ceramics and we encourage everyone to be generous and to continue sharing knowledge with each other. We are all standing on the shoulders of those who came before us and it is only by working together that we can find new ways forward.

**Opposite and overleaf:** Laura Pehkonen, *Domino*, 2022. This collage of ceramic elements resulted from the testing of different ceramic techniques and materials. *(Image credit: Paula Virta / EMMA)*

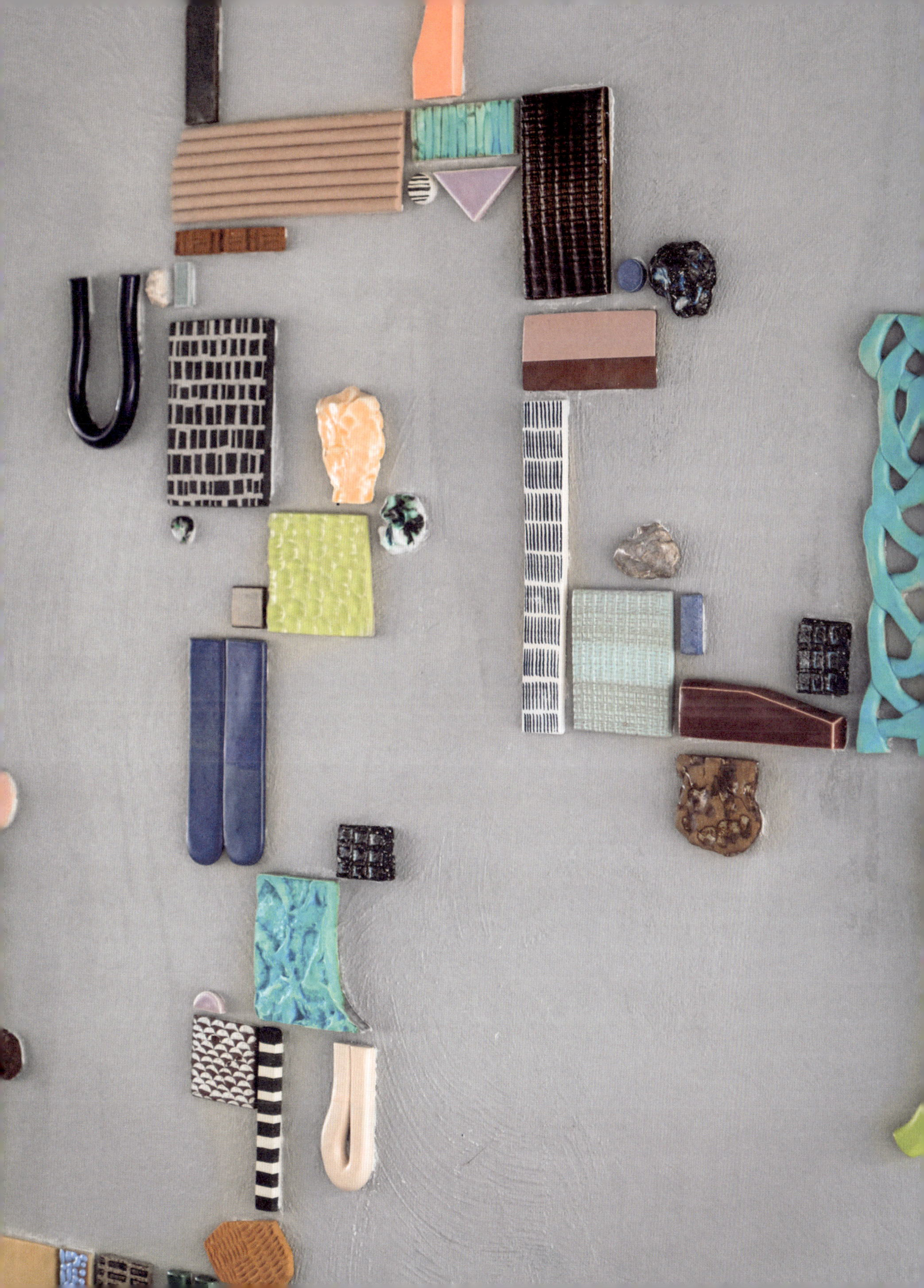

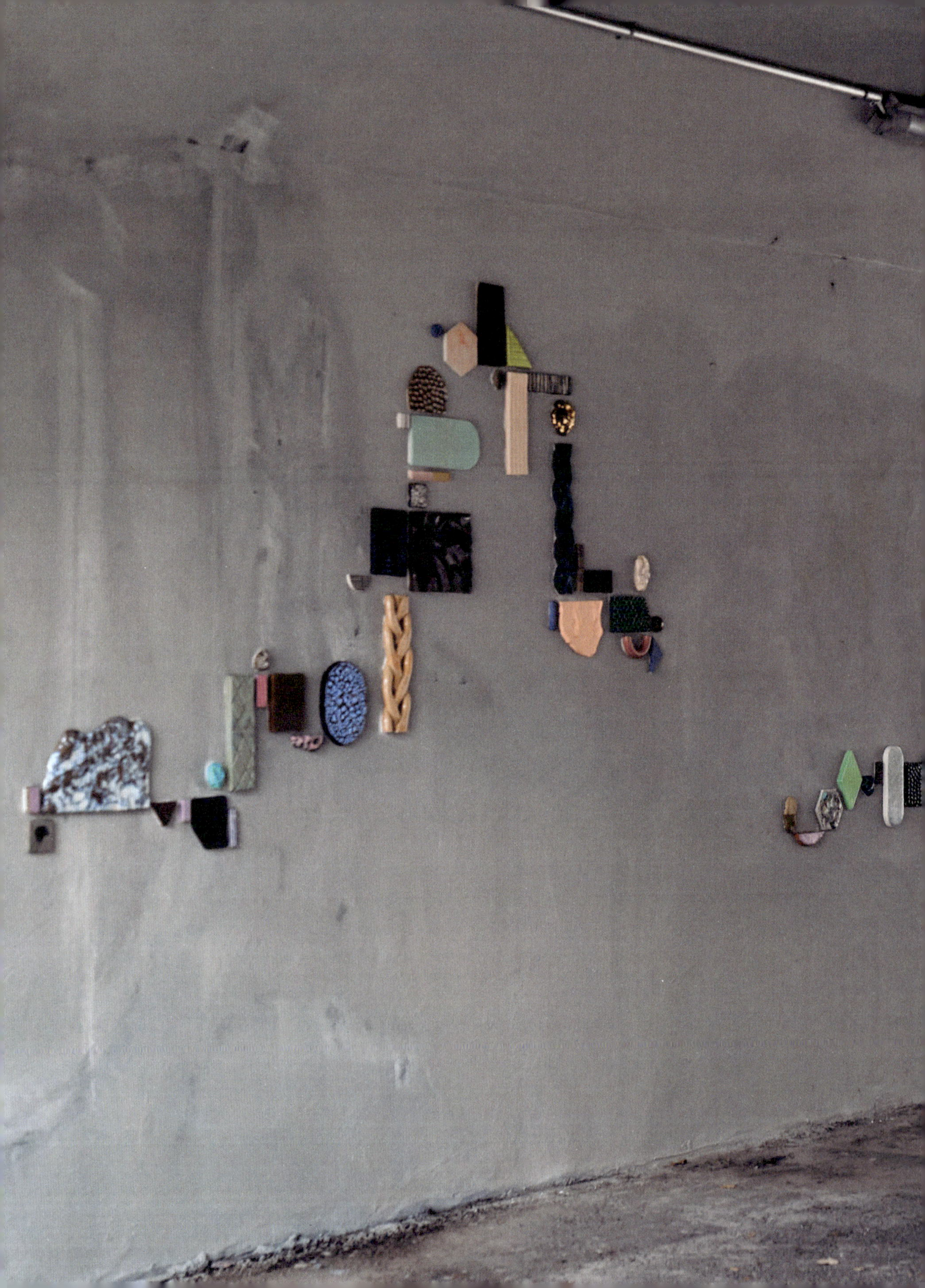

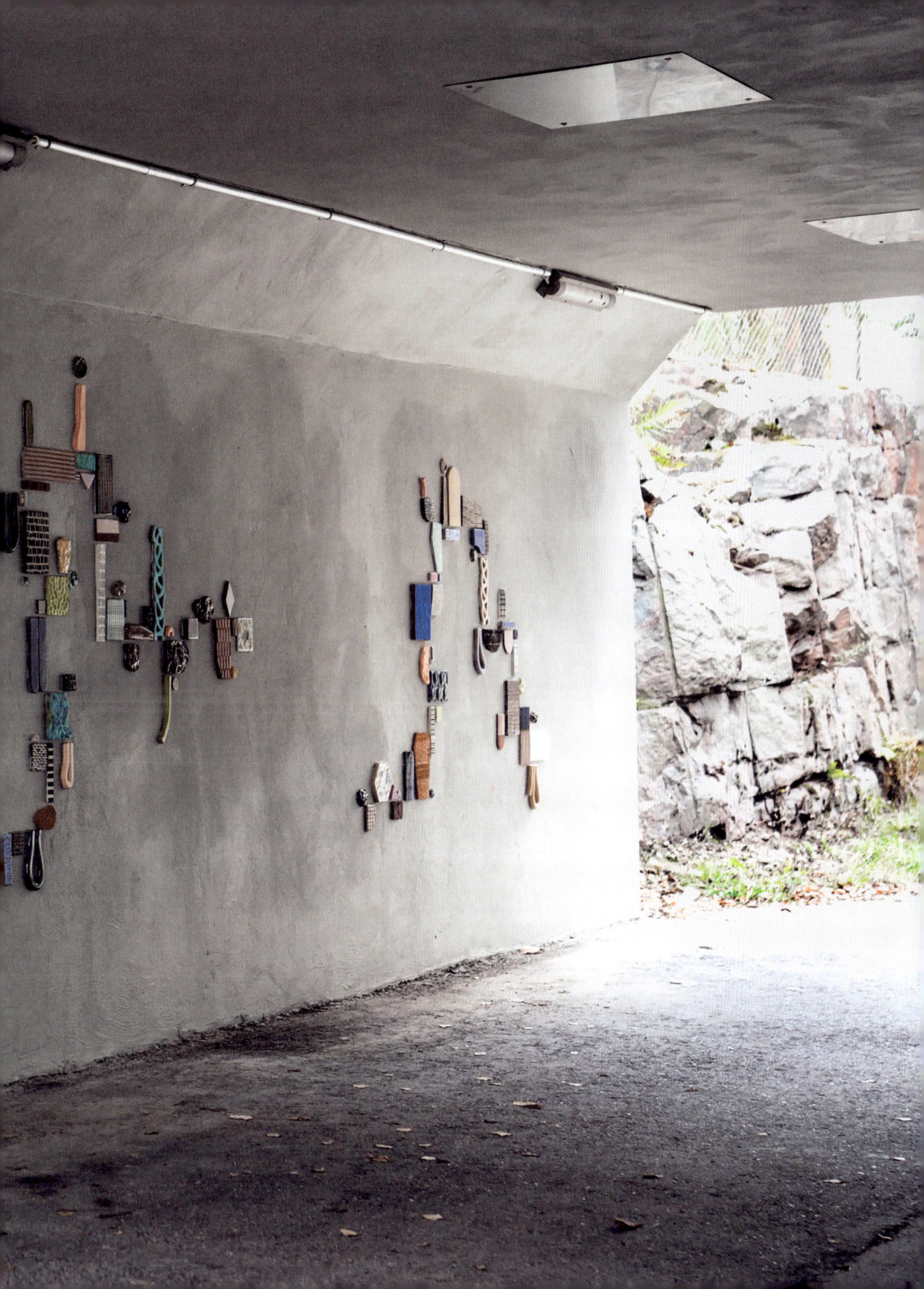

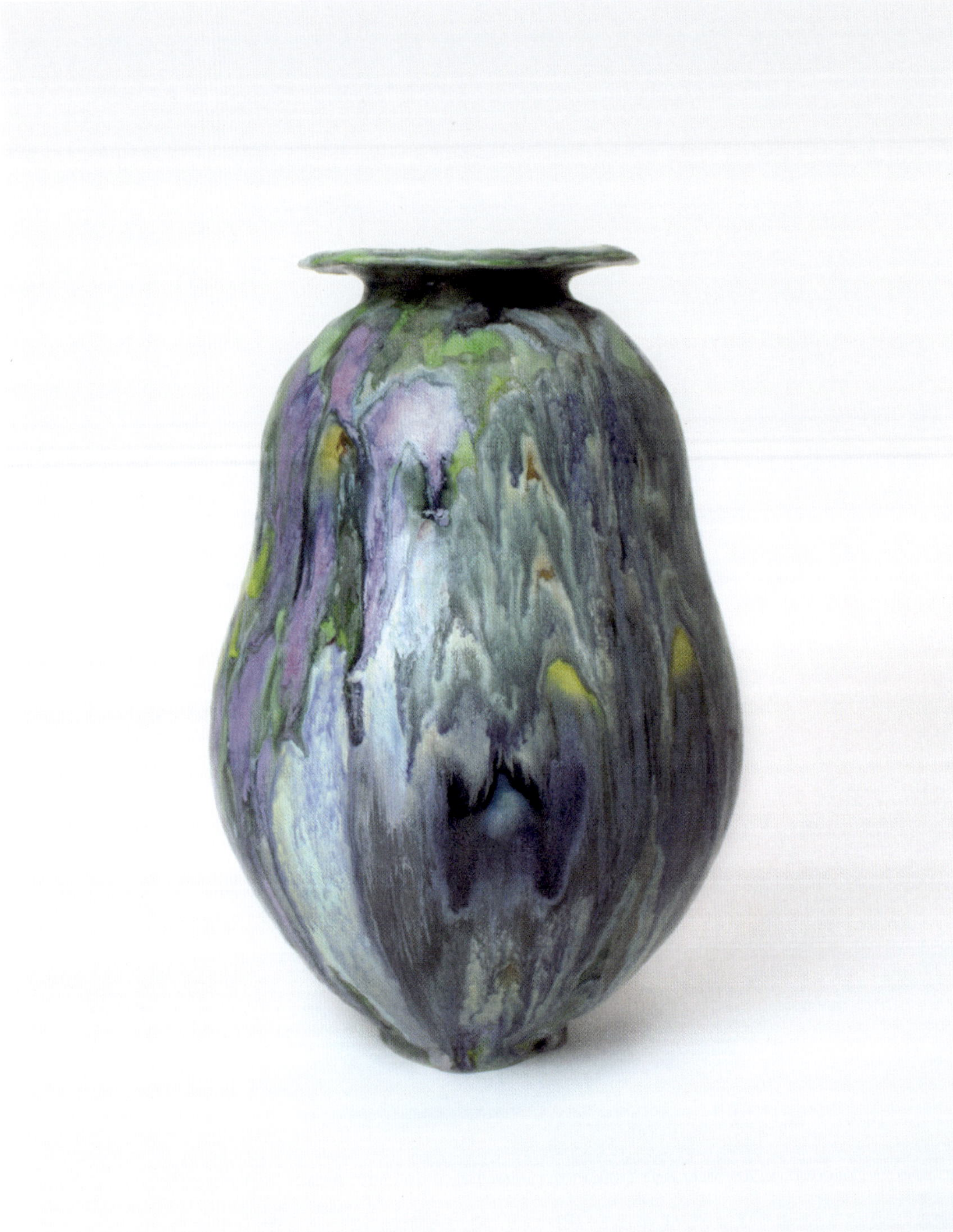

**Above:** Mel Arsenault, *Waterfall*, 2025. Hand-built vessel with glaze exploring connections between urban flora and well-being. *(Image credit: Galerie Florian Daguet-Bresson)*

## REFERENCES


CONRAD, John W, *Advanced Ceramic Manual, Technical Data for the Studio Potter*, Falcon, 1987.

CURRIE, Ian, *Revealing Glazes: Using the Grid Method*, Bootstrap Press, 2004.

DEBOOS, Janet, HARRISON, Stephen, SMITH Leonard, *Handbook for Australian Potters*, Methuen, 1984.

HAMER, Frank and Janet, *The Potter's Dictionary*, Bloomsbury, 6th edition 2015.

HAMM, Nicole, *Raw Material Routes*, Studio Potter Magazine, https://studiopotter.org/raw-material-routes (accessed August 1, 2024).

HOPPER, Robin, *The Ceramic Spectrum: A Simplified Approach To Glaze And Color Development*, Krause Publications, 1984.

HUTCHINSON CUFF, Yvonne, *Ceramic Technology for Potters and Sculptors*, University of Pennsylvania Press, 1996.

MASON, Ralph, *Native Clays and Glazes for North American Potters*, Timber Press, 1981.

MEMMOTT, Harry, *An Artist's Guide To The Use Of Ceramic Oxides: With Examples Of Their Use In Glaze*, Melbourne, Australia: Deakin University Press, 1986.

RHODES, Daniel, *Clay and Glazes for the Potter*, Krause Publications, 1957.

TICHANE, Robert, *Clay Bodies*, New York: The New York Glaze Institute, 1990.


## RESOURCES ON SUSTAINABILITY


BLAKELY, Matthew, *Rock Glazes Unearthed*, self-published, 2021.

FORREST, Miranda, *Natural Glazes: Collecting and Making*, Bloomsbury, 2013.

HARRISON, Robert, *Sustainable Ceramics: A Practical Approach*, New York: Bloomsbury, 2013.

HOWARD, Sara, *Circular Ceramics*, 2023.

LEVY, Matt, SHIBATA, Takuro and SHIBATA, Hitomi, *Wild Clay: Creating Ceramics and Glazes from Natural and Found Resources*, Herbert Press, 2022.

MAKLIUK, Yulia, *Potters Save the World: Learn To Make Sustainable Ceramics And Help Protect The Earth*, self-published, 2023.

ROGERS, Phil, *Ash Glazes: Techniques and Glazing from Natural Sources*, Herbert Press, 2003.

# GLOSSARY

**Acid** a non-metal oxide which, when dissolved in water, releases hydrogen ions $H^+$.

**Alkali** the oxide of an alkali metal or alkaline earth metal which, when dissolved in water, releases hydroxide ions $OH^-$.

**Alkali metals** metals in the first column of the periodic table: lithium, sodium and potassium.

**Alkaline earths** metals in the second column of the periodic table: magnesium, calcium, strontium and barium.

**Alkaline frit** a frit containing sodium, potassium and silica.

**Alkaline glaze** a glaze which is high in sodium, lithium, potassium or barium.

**Amphoteric** able to act either as an acid or as a base/alkali.

**Andesite** a fine-grained rock of intermediate composition between rhyolite and basalt.

**Atom** the smallest unit of an element.

**Ball clay** a fine-grained, plastic secondary clay.

**Ball mill** a rotating drum containing ceramic pebbles for grinding colouring oxides in glazes.

**Base** a metal oxide which does not dissolve in water but will react with an acid.

**Basalt** a fine-grained, dark volcanic rock containing iron and magnesium.

**Bentonite** a very plastic clay, used to add plasticity to clay bodies or to suspend glazes.

**Biscuit firing (or bisque)** a first firing, usually to around 1000°C (1832°F), done before applying a glaze.

**Blistering** blisters in a fired glaze, caused by gas escaping during firing.

**Bloating** blisters in the clay body, caused by trapped bubbles of gas.

**Borax frit** a frit containing sodium, calcium, boron and silica.

**Calcine** to heat materials in a kiln to drive off volatile compounds.

**Casting slip** a clay slip made fluid with a deflocculant and used in the casting process.

**Celadon** a type of pale, grey-green or blue-green glaze containing iron oxide.

**Ceramic change** occurs when the chemically combined water is removed from clay.

**Chemically combined water** is water that is combined within the clay molecule.

**China clay** kaolin, pure white clay.

**Chun** a pale blue, opalescent glaze.

**Clay body** a combination of clay with other ceramic materials suited for a purpose.

**CMC** carboxymethyl cellulose gum, used to suspend glaze ingredients in water.

**Cone** a small, slender pyramid of ceramic material, set in the kiln, which softens and bends when fired to a specific temperature at a specified rate.

**Cornwall stone** a type of feldspar found in Cornwall.

**Crawling** a defect where glaze pulls away from the body, leaving bare patches.

**Crazing** a network of fine cracks caused by a higher coefficient of expansion in the glaze than in the clay body.

**Cristobalite** a crystalline form of silica, which forms in the clay body above 1100°C (2012°F) and contracts suddenly on cooling at 226°C (439°F).

**Crystalline** a glaze in which crystals have grown during cooling.

**Deflocculate** to disperse clay particles in order to make a slip more fluid.

**Devitrify** the growth of crystals from the molten glaze, which is then no longer in the glassy state.

**Diorite** coarse-grained, dark-coloured igneous rock with moderate silica.

**Dunting** cracking caused by cold air entering the kiln during cooling.

**Earthenware** a clay body fired at low temperatures, below 1150°C (2102°F), that remains porous.

**Egyptian paste** a low temperature clay body that contains soluble glaze materials that migrate to the surface and form a glaze during firing.

**Element** a single type of atom.

**Engobe** a clay slip formulated with some flux and applied on unfired clay.

**Eutectic** the lowest melting combination of two or more materials.

**Feldspar** an aluminosilicate mineral containing potassium, sodium and calcium.

**Filler** inert material used in clay bodies to increase workability and reduce shrinkage.

**Fire clay** a high-alumina refractory clay used in stoneware clay bodies and to make kiln furniture.

**Flashing** colour resulting from volatile compounds released during firing.

**Flocculate** to cause glaze particles to clump together and thicken a glaze or slip.

**Flux** a material that lowers the melting temperature of a glaze.

**Frit** glaze fluxes melted with silica and ground to a powder.

**Glass former** a substance that does not regain its crystal structure upon cooling such as silica, boron oxide and phosphorus.

**Glaze** a glassy coating applied to clay forms composed of glass former, flux and stabilizer.

**Granite** a medium-grained, igneous rock composed of feldspar, quartz and mica.

**Grog** crushed, fired clay added to clay bodies to reduce shrinkage.

**High fire** fired above 1200°C (2192°F).

**Igneous** rocks formed from cooled magma.

**Illite** a clay mineral derived from mica.

**Low fire** fired below 1150°C (2102°F).

**Kaki** a red-brown, high-iron glaze. Kaki is the Japanese word for persimmon.

**Kaolin** pure, white clay.

**Kaolinite** pure, white clay mineral.

**Kiln** a structure for firing ceramics made of refractory materials.

**Lime** calcium oxide, CaO.

**Loam** clay soil, a mixture of clay, silt and sand.

**Local clay** clay found in the ground.

**Lustre** metallic decoration obtained by reducing metal salts.

**Macaloid** a suspension agent twice as powerful as bentonite.

**Magma** molten rock.

**Metamorphic** rocks formed from other rocks by pressure or heat.

**Mica** a lustrous, platy mineral.

**Mineral** an inorganic substance with an ordered atomic structure e.g., quartz.

**Molecule** two or more atoms bonded together with no overall charge.

**Molochite** finely-ground, calcined china clay.

**Montmorillonite** a group of clay minerals including bentonite, also known as smectite.

**Opacifier** a material that does not dissolve in a glaze, causing it to become opaque.

**Oxidation** firing in the presence of sufficient oxygen, usually in an electric kiln.

**Oxide** an element bound to oxygen.

**Paperclay** a clay body containing paper fibres.

**Pinholes** small holes in a glaze, caused by burst bubbles that have not healed over during firing.

**Plasticity** the ability of a material to form and hold a shape.

**Porcelain** a white clay body fired to 1250–1400°C (2282–2552°F), highly vitrified and frequently translucent.

**Pyrometer** a device used to measure the temperature inside a kiln.

**Raku** low temperature firing where red hot pots are removed from the kiln.

**Reduction** the act of limiting the air available during firing, so that oxygen is drawn from the clay and glaze.

**Refractory** able to withstand high temperatures.

**Residual clay** clay formed in situ through decomposition and disintegration of rocks.

**Rhyolite** a fine-grained, light-coloured volcanic rock high in silica.

**Rock** an aggregate of minerals e.g., granite.

**Salt firing** introducing salt into the kiln at high temperature, to combine with silica in the clay and produce a glaze on the surface.

**Sedimentary** rocks formed from deposited layers of sediment.

**Shale** clay rock deposited and subjected to pressure.

**Shino** a white Japanese glaze made from feldspar and clay.

**Shivering** flaking glaze caused by poor glaze fit, with the glaze contracting less than the clay body often occurring on edges.

**Silica** silicon dioxide, found in quartz, flint or sand.

**Silt** granular material with low plasticity and a particle size between sand and clay.

**Slip** clay mixed with water.

**Smectite** highly expansive clay mineral (montmorillonite).

**Soak** to maintain the top firing temperature to mature the glaze evenly throughout the kiln.

**Soda firing** introducing soda (sodium carbonate) into the kiln at high temperature to combine with silica in the clay to produce a glaze on the surface.

**Soluble** capable of being dissolved in water.

**Spalling** flaking of clay after firing.

**Specific gravity** the comparison of the weights of a liquid slip or glaze to water.

**Stabilizer** ingredient that has refractory properties and increases viscosity in a glaze.

**Stain** industrially-produced ceramic pigment containing colouring oxides and opacifiers.

**Stoneware** a clay body fired to a high temperature, above 1200°C (2192°F), so that the clay body is vitrified and non-porous.

**Stoneware clay** is a naturally occurring high fire clay.

**Tenmoku** a brown-black, high-iron glaze.

**Terracotta** unglazed, low-fire red clay.

**Terra sigillata** a slip made of the finest particles of clay.

**UMF or unity molecular formula** a formula calculated to compare the proportion of flux, stabilizer and glass former in a glaze.

**Vitrify** to fire to a glassy state.

**Water of plasticity** water added to clay to make it wet.

**Wild clay** a term synonymous with local or foraged clay.

## APPENDIX 1: PERIODIC TABLE

| | Group 1 | Group 2 | | | | | | | |
|---|---|---|---|---|---|---|---|---|---|
| 1 | 1 1.0079 H Hydrogen | | | | | | | | |
| 2 | 3 6.941 Li Lithium | 4 9.0122 Be Beryllium | | | | | | | |
| 3 | 11 22.990 Na Sodium | 12 24.305 Mg Magnesium | | | | | | | |
| 4 | 19 39.098 K Potassium | 20 40.078 Ca Calcium | 21 44.956 Sc Scandium | 22 47.867 Ti Titanium | 23 50.942 V Vanadium | 24 51.996 Cr Chromium | 25 54.938 Mn Manganese | 26 55.845 Fe Iron | 27 58.933 Co Cobalt |
| 5 | 37 85.468 Rb Rubidium | 38 87.62 Sr Strontium | 39 88.906 Y Yttrium | 40 91.224 Zr Zirconium | 41 92.906 Nb Niobium | 42 95.94 Mo Molybdenum | 43 96 Tc Technetium | 44 101.07 Ru Ruthenium | 45 102.91 Rh Rhodium |
| 6 | 55 132.91 Cs Caesium | 56 137.33 Ba Barium | 57-71 La-Lu Lanthanide | 72 178.49 Hf Halfnium | 73 180.95 Ta Tantalum | 74 183.84 W Tungsten | 75 186.21 Re Rhenium | 76 190.23 Os Osmium | 77 192.22 Ir Iridium |
| 7 | 87 223 Fr Francium | 88 226 Ra Radium | 89-103 Ac-Lr Actinide | 104 261 Rutherfordium | 105 262 Dubnium | 106 266 Seaborgium | 107 264 Bohrium | 108 277 Hassium | 109 268 Meitnerium |

| | | | | | |
|---|---|---|---|---|---|
| 57 138.91 La Lanthanum | 58 140.12 Ce Cerium | 59 140.91 Pr Praseodymium | 60 144.24 Nd Neodymium | 61 145 Pm Promethium | 62 150.36 Sm Samarium |
| 89 227 Ac Actinium | 90 232.04 Th Thorium | 91 231.04 Pa Protactinium | 92 238.03 U Uranium | 93 237 Neptunium | 94 244 Plutonium |

- AlkaliMetal
- AlkalineEarthMetal
- Metal
- Metalloid
- Non-metal
- Halogen
- NobleGas
- Lanthanide/Actinide

| Z mass Symbol Name | |
|---|---|

| | | | Group 3 | Group 4 | Group 5 | Group 6 | Group 7 | Group 8 |
|---|---|---|---|---|---|---|---|---|
| | | | | | | | | 2 4.0025 He Helium |
| | | | 5 10.811 B Boron | 6 12.011 C Carbon | 7 14.007 N Nitrogen | 8 15.999 O Oxygen | 9 18.998 F Flourine | 10 20.180 Ne Neon |
| | | | 13 26.982 Al Aluminium | 14 28.086 Si Silicon | 15 30.974 P Phosphorus | 16 32.065 S Sulphur | 17 35.453 Cl Chlorine | 18 39.948 Ar Argon |
| 28 58.693 Ni Nickel | 29 63.546 Cu Copper | 30 65.39 Zn Zinc | 31 69.723 Ga Gallium | 32 72.64 Ge Germanium | 33 74.922 As Arsenic | 34 78.96 Se Selenium | 35 79.904 Br Bromine | 36 83.8 Kr Krypton |
| 46 106.42 Pd Palladium | 47 107.87 Ag Silver | 48 112.41 Cd Cadmium | 49 114.82 In Indium | 50 118.71 Sn Tin | 51 121.76 Sb Antimony | 52 127.6 Te Tellurium | 53 126.9 I Iodine | 54 131.29 Xe Xenon |
| 78 195.08 Pt Platinum | 79 196.97 Au Gold | 80 200.59 Hg Mercury | 81 204.38 Tl Thallium | 82 207.2 Pb Lead | 83 208.98 Bi Bismuth | 84 209 Po Polonium | 85 210 At Astatine | 86 222 Rn Radon |
| 110 281 Darmstadtium | 111 280 Roentgenium | 112 285 Copernicum | 113 284 Nihonium | 114 289 Flerovium | 115 288 Moscovium | 116 293 Livermorium | 117 292 Tennessine | 118 294 Oganesson |

| | | | | | | | | |
|---|---|---|---|---|---|---|---|---|
| 63 151.96 Eu Europium | 64 157.25 Gd Gadolinium | 65 158.93 Tb Terbium | 66 162.50 Dy Dysprosium | 67 164.93 Ho Holmium | 68 167.26 Er Erbium | 69 168.93 Tm Thulium | 70 173.04 Yb Ytterbium | 71 174.97 Lu Lutetium |
| 95 243 Americium | 96 247 Curium | 97 247 Berkelium | 98 251 Californium | 99 252 Einsteinium | 100 257 Fermium | 101 258 Mendelevium | 102 259 Nobelium | 103 262 Lawrencium |

# APPENDIX 2: NATURAL SOURCES OF CERAMIC MATERIALS

| ROCK/MINERAL SOURCE | CHEMICAL COMPOSITION | ELEMENTS | | | | | | |
|---|---|---|---|---|---|---|---|---|
| | | Al | Ba | B | Ca | Pb | Li | |
| Albite | $Na_2O\ Al_2O_3\ 6SiO_2$ | x | | | | | | |
| Anorthite | $CaO\ Al_2O_3\ 2SiO_2$ | x | | | x | | | |
| Barite | $BaSO_4$ (f) | | x | | | | | |
| Bauxite | $Al_2O_3\ 3H_2O$ | x | | | | | | |
| Borax | $Na_2O\ 2B_2O_3\ 10H_2O$ | | | x | | | | |
| Calcite | $CaCO_3$ | | | | x | | | |
| Colemanite | $2CaO\ 3B_2O_3\ 5H_2O$ | | | x | x | | | |
| Corundum | $Al_2O_3$ | x | | | | | | |
| Diaspore | $Al_2O_3\ H_2O$ | x | | | | | | |
| Dolomite | $CaCO_3\ MgCO_3$ | | | | x | | | |
| Flint | $SiO_2$ | | | | | | | |
| Galena | PbS (f) | | | | | x | | |
| Gibbsite | $Al_2O_3\ 3H_2O$ | x | | | | | | |
| Halite | NaCl | | | | | | | |
| Lepidolite | $Li_2O\ Al_2O_3\ K_2O\ 3SiO_2$ | x | | | | | x | |
| Leucite | $K_2O\ Al_2O_3\ 6SiO_2$ | x | | | | | | |
| Magnesite | $MgCO_3$ | | | | | | | |
| Microcline | $K_2O\ Al_2O_3\ 6SiO_2$ | x | | | | | | |
| Nepheline syenite | $KNaO\ Al_2O_3\ 4.65\ SiO_2$ | x | | | | | | |
| Orthoclase | $K_2O\ Al_2O_3\ 6SiO_2$ | x | | | | | | |
| Perthite | Mixed feldspars | x | | | x | | | |
| Petalite | $Li_2O\ Al_2O_3\ 8SiO_2$ | x | | | | | x | |
| Quartz | $SiO_2$ | | | | | | | |
| Smithsonite | $ZnCO_3$ (f) | | | | | | | |
| Sodium carbonate | $Na_2CO_3$ (s) | | | | | | | |
| Spodumene | $Li_2O\ Al_2O_3\ 4SiO_2$ | x | | | | | x | |
| Steatite | $3MgO\ 4SiO_2\ H_2O$ | | | | | | | |
| Ulexite | $NaCaB_5O_9\ 8H_2O$ (f) | | | x | x | | | |
| Witherite | $BaCO_3$ | | x | | | | | |
| Zincite | ZnO (c) | | | | | | | |
| Most sulphide and sulphate minerals | (c) | | | x | x | | | |

Key: f= fritting required, s = soluble, c= must be calcined Adapted from Ralph Mason, Native Clays and Glazes

| ELEMENTS | | | | | | CERAMIC OXIDES | CERAMIC MATERIALS |
|---|---|---|---|---|---|---|---|
| Mg | K | Si | Na | Sr | Zn | | |
| | | x | x | | | Sodium oxide, alumina, silica | Sodium feldspar |
| | | x | | | | Calcium oxide with alumina and silica | Cornwall stone |
| | | | | | | Barium oxide | Barium carbonate |
| | | | | | | Alumina | Alumina hydrate |
| | | | x | | | Boric oxide | Borax |
| | | | | | | Calcium oxide | Calcium carbonate |
| | | | | | | Boric oxide | Colemanite |
| | | | | | | Alumina | Alumina |
| | | | | | | Alumina hydroxide | Alumina hydrate |
| x | | | | | | Magnesium oxide, calcium oxide | Dolomite |
| | | x | | | | Silica | Flint/Silica |
| | | | | | | Lead oxide | Lead oxide |
| | | | | | | Alumina hydroxide | Alumina hydrate |
| | | | x | | | Sodium oxide | Salt |
| | x | x | | | | Lithium oxide | Lepidolite |
| | x | x | | | | Potassium oxide | Potassium feldspar |
| x | | | | | | Magnesium oxide | Magnesium carbonate |
| | x | x | | | | Potassium oxide | Potassium feldspar |
| | x | x | x | | | Sodium, potassium with silica and alumina | Nepheline syenite |
| | x | x | | | | Potassium oxide | Potassium feldspar |
| | x | x | | | | Potassium oxide | Potassium feldspar |
| | | x | | | | Lithium oxide | Petalite |
| | | x | | | | Silica | Silica, silica sand |
| | | | | | x | Zinc oxide | Zinc oxide |
| | | | x | | | Sodium oxide | Soda ash |
| | | x | | | | Lithium oxide with alumina and silica | Spodumene |
| x | | x | | | | Magnesium oxide | Talc |
| | | | x | | | Boron oxide | Gerstley borate |
| | | | | | | Barium oxide | Barium carbonate |
| | | | | | x | Zinc oxide | Zinc oxide |
| x | x | | x | x | x | | |

## APPENDIX 3: DIRECTIONS FOR PROCESSING LOCAL CLAY

### Testing local clay

1. Clay is sticky when moistened with water. Good clay holds together when formed and keeps its shape as it dries.

2. To test for plasticity, moisten a small sample of clay with water. Roll out a coil about the thickness of a pencil and wrap it around a finger. A clay that stretches as it bends, even with small cracks, demonstrates good plasticity. A clay that cracks completely or crumbles lacks plasticity and is not usable clay.

3. To test clay for the presence of soluble alkalis, allow a small piece of clay to dry. If a scum appears on the surface, this indicates the presence of soluble salts or sulphur which can be undesirable. Alkaline clays can be washed after drying. Soak dry clay in water for several days, decant the water and dry to a workable consistency.

4. To test clay for the presence of lime, fire a small sample. After a few days, submerge the fired sample in strong vinegar. A bubbling reaction indicates that lime (CaO) is present. Lime may cause lime pop when it absorbs atmospheric moisture, causing it to swell and crack the clay around it. Extremely small amounts of calcium that are well mixed in the clay will not cause issues because this calcium combines with other oxides during firing.

### Processing local clay

1. Mix local clay with enough water to form a slurry for easier processing that creates less dust. Pour the slip through a well-supported window screen backed with a larger wire grid mesh to remove the larger organic and rocky materials.

2. Pour the screened slurry through a 200 mesh sieve to remove smaller dirt and sand particles.

3. Pour the sieved clay on canvas or plaster to dry to a workable consistency then wedge.

### Calculating dry weight from slurry weight

1. Determining the dry weight equivalent for slurry is useful for calculating percentage additions such as fillers. This eliminates the need to dry out the entire batch to weigh it.

2. Measure out several volumes of slurry by weight such as 10 g, 50 g, 100 g etc. and pour each sample out on a plaster slab to dry.

3. Once the samples are dry, weigh each dry sample to determine its dry weight.

4. Use a simple graph to plot points for the slurry weight in grams on the y axis and the dry weight in grams on the x axis.

5. Draw a line though the points to indicate the correlation between wet slurry and dry weight.

6. Use the dry weight to calculate percentage additions to your clay such as fillers.

# APPENDIX 4: DIRECTIONS FOR TESTING CLAY BODIES

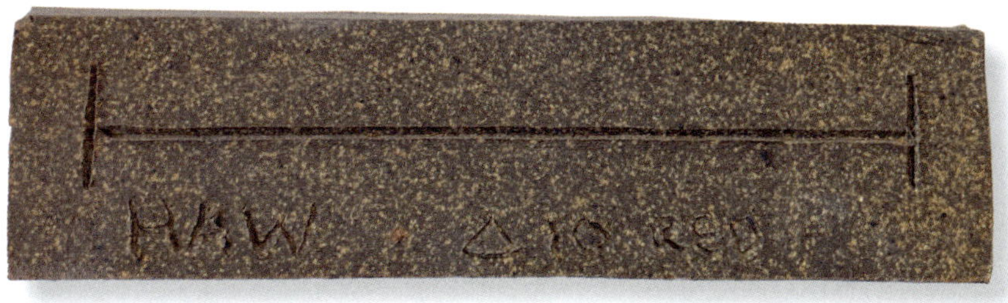

A simple slab test bar from a clay or clay body can provide important information on shrinkage and vitrification.

### Directions for making a test bar

1. Mix a clay or clay body to a workable consistency.
2. Roll the clay into a slab at least 0.5 cm thick.
3. Cut a rectangular bar 4 cm x 12 cm.
4. Draw a straight line down the middle of the bar with a ruler longer than 10 cm.
5. Make 2 crosshatch marks 10 cm apart (wet line length = 100 mm).
6. Dry slowly and evenly. Measure the line prior to firing for drying shrinkage (dry line length).
7. Fire to desired temperature. Measure the line after firing for fired shrinkage (fired line length).

### Determining shrinkage

1. Measure the fired line length in mm.
2. Subtract the fired line length from the wet line length (100mm).
3. The difference in length equals the % fired shrinkage.
   For example, if the fired line length measures 90mm after firing:
   100mm (wet line length) – 90mm (fired line length) = 10 mm (10% fired shrinkage)

### Testing vitrification and absorption

1. Weigh the test bar after firing to obtain the dry weight in grams.
2. Place test bar in boiling water for at least 15 minutes, or in bowl of water at room temperature for 24 hours.
3. Remove the test bar and pat dry to remove all surface water.
4. Weigh the wet test bar.
5. Determine the percentage absorption using the following formula:

$$\%\text{ Absorption} = \frac{\text{wet weight} - \text{dry weight}}{\text{Dry weight}} \times 100$$

### Adding glaze on the test bar

Glazes can be initially tested on the same clay test bar by placing a dab of glaze on the middle of the bar where it will not interfere with other necessary measurements. This glaze sample can help to identify glaze fit issues, crazing, bubbling and colour reaction.

## APPENDIX 5: FILLERS

**Silica based fillers:**

Calcined clay
Feldspar chips
Fibreglass
Grog
Perlite
Pyrophyllite
Silica or flint
Silica sand or quartz sand
Vermiculite
Wollastonite

**Alumina based fillers:**

Alumina hydrate
Alumina oxide
High alumina grog
Kyanite
Mullite

**Organic or burnable fillers:**

Cotton fibres
Flax fibres
Nylon fibres
Paper fibres
Rice
Sawdust
Small grains

## APPENDIX 6: THERMAL EXPANSION

The following list orders the relative levels of thermal expansion in common ceramic materials from high to low.

HIGH THERMAL EXPANSION

$Na_2O$ Sodium
$K_2O$ Potassium
$SiO_2$ Crystalline cristobalite, includes inversion
CaO Calcium / SrO Strontium
$Cr_2O_3$ Chromium
BaO Barium
CoO Cobalt
$SiO_2$ Silica as crystalline quartz, no inversion
PbO Lead / NiO Nickel / $Fe_2O_3$ Iron / $TiO_2$ Titanium
Typical soda glass
Typical white earthenware body
$Li_2O$ Lithium / $P_2O_5$ Phosphorus
ZnO Zinc
Typical Earthenware glaze
Typical Stoneware clay body
CuO Copper / MnO Manganese
Typical Grog
MgO Magnesia
Porcelain clay body
Top quality fire brick
$SnO_2$ Tin / $ZrO_2$ Zirconium
$Al_2O_3$ Alumina
$SiO_2$ Silica fused in clay body or glaze
$B_2O_3$ Boron

LOW THERMAL EXPANSION

Adapted from notes of the Raw Materials course at Alfred University, 2003.

# APPENDIX 7: MIXING TERRA SIGILLATA & CASTING SLIP

**Directions for mixing terra sigillata**

Terra sigillata proportions by weight: 80 water, 20 clay, 1 Darvan #7

1. Use a transparent container with a lid.
2. Measure water and deflocculant and mix together.
3. Weigh the clay and add to the water mixture.
4. Mix well by shaking the closed container with the lid on.
5. Wait for clear separation of layers to appear, may take several hours to a few days. If there is little sedimentation, add more deflocculant and wait again for layers to appear.
6. Slowly pour out the top liquid layer or decant by siphoning without mixing with the darker sludge at the bottom. Discard the sludge.
7. Test the terra sigillata by brushing a thin layer or two on leather hard clay. Some shine on the surface should appear.
8. If the test dries matt, the mixture may be a burnishing slip that require burnishing to shine. Burnishing can be accomplished by rubbing the surface with something smooth like back of a spoon, a smooth pebble or even a finger wrapped in plastic.
9. To make coloured terra sigillata, use white clay for lighter colours and red clay for darker colours. After decanting the terra sigillata, add stains or oxides then ball mill the mixture after adding the colourants to reduce their particle size for best results.
10. Terra sigillata is best applied to greenware from leather hard to bone dry in thin watery layers by brushing, spraying or dipping.

**Directions for mixing casting slip**

1. Weigh the minimum amount of water (usually 40% of dry weight of recipe) and remove 1 cup and set aside.
2. Weigh the minimum amount of deflocculant (usually 0.25% of dry weight of recipe) and add to water.
3. Weigh the clay in the recipe and add to the water mixture.
4. Weigh the remaining ingredients, adding one at a time and mix.
5. At any point during mixing if the mixture becomes too thick to stir, weigh small amounts of deflocculant and record the amount, dilute with water from 1 cup reserve that was set aside, pour into the batch and mix. Continue until slip is moving.
6. Mix slip very well. Because slip may become thinner as it continues to mix, hold back on additions of deflocculant and water until absolutely needed for a fluid mixture.
7. Allow the casting slip to sit for 24 hours prior to casting to allow for thorough hydration and deflocculation.
8. Before casting, test the slip to ensure proper density. Testing by feel is quick way to judge the properties of slip. Dip a hand in the slip and spread the fingers. Slip should web between them for a moment, run off fairly quickly and leave a thin coating of slip on the hand. If it does not web, there is too much water in the mixture. If it clumps on the hand, there is not enough deflocculant. Testing the specific gravity is described in the text in Chapter 4.

### Directions for recycling casting slip

1. Dry the scraps of casting slip completely.
2. Casting slip scraps can be remixed into a new batch of slip at a rate of 25%. Use 25% weight of new batch of dry scraps and an additional 25% of water. For example: 100 lbs dry material/40 lbs water, add max 25 lbs scrap/10 lbs water.
3. No additional deflocculant is required because deflocculant remains in the dry scraps.
4. Add the scraps to the water and let soak for one hour before adding the remaining dry ingredients for the new batch. Please note that mixing casting slip from 100% casting scraps may lead to a slip that is thixotropic, difficult to dry and may lead to other casting problems.

### Casting slip: common problems and remedies

| PROBLEM | CAUSE | REMEDY |
|---|---|---|
| Casting slow | Slip retains water | Reduce plasticity by replacing some ball clay with less plastic clay |
| Collapsing | Draining creates air vacuum | Drain slower |
| Cracking | Slip lacks plasticity | Increase plasticity by adding ball clay or reduce deflocculant if brittle |
| Deforming | Casting not set | Dry mould between castings and wait until mould easily separates before opening |
| Draining issues | Slip has low fluidity | Increase deflocculant or water content based on specific gravity |
| Pinholing | Air bubbles in slip | Tap mould to release air bubbles and increase fluidity of slip with water or deflocculant based on specific gravity |
| Tearing | Casting sticks to mould | Increase plasticity add ball clay |

## APPENDIX 8: COLOURANT GUIDE FOR CLAYS AND ENGOBES

| COLOURANT | FIRED COLOUR |
|---|---|
| Ceramic stains 1–15% | Stain colour light to dark |
| Black stain 3–12% | Light grey to black |
| Chrome oxide 1–5% | Green to grey green |
| Cobalt oxide 0.25–5% | Light blue to dark blue |
| Cobalt carbonate 2%, rutile 4%, chrome oxide 1 % | Green |
| Cobalt carbonate 0.5%, iron chromate 4% | Blue grey |
| Copper oxide 2–4% | Pale green to dark green |
| Ilmenite 6% | Dark speckled |
| Iron chromate 6% | Grey-brown |
| Iron oxide 2–6% | Light tan to dark brown |
| Iron oxide 4%, cobalt oxide 3%, copper oxide 2% | Black |
| Manganese dioxide 3–6% | Light brown to speckled brown |
| Nickel oxide 2–5% | Medium tan to dark tan |
| Nickel oxide 4%, cobalt carbonate 5%, manganese carbonate 2% | Slate blue |
| Rutile 3–6 % | Tan |
| Rutile 4%, red iron oxide 4%, zinc oxide 6% | Rusty tan |
| Titanium dioxide 9% | Pearly |

Adapted from *Advanced Ceramic Manual* by John W. Conrad

### Engobe recipes

| | LOW FIRE (cone 04–3) | MID FIRE (cone 4–6) | HIGH FIRE (cone 7–10) |
|---|---|---|---|
| Kaolin | 20 | 15 | 15 |
| Calcined kaolin | 10 | 20 | 35 |
| Silica | 15 | 20 | 15 |
| Feldspar or nepheline syenite | - | 10 | 15 |
| Frit (3110) | 15 | 10 | - |
| Talc | 25 | 10 | 5 |
| Borax | 5 | 5 | 5 |
| Zircopax | 10 | 10 | 10 |
| Total | 100 | 100 | 100 |

Adapted from *The Ceramic Spectrum* by Robin Hopper

# APPENDIX 9: COLOURANT GUIDE FOR GLAZES

| | **Sodium, potassium lithium** dominated glazes | **Calcium** dominated glazes | **Barium strontium** dominated glazes | **Magnesium** dominated glazes | **Zinc** dominated glazes | **Boron** dominated glazes |
|---|---|---|---|---|---|---|
| $Cr_2O_3$<br>chromium oxide<br>(chrome) | 1–5% bright yellow/green with increasing intensity | 1–5% green with increasing intensity with tin pink | 1–5% yellow-green with increasing intensity (chartreuse) | 1–5% green with increasing intensity (dense and opaque) | 1–5% tan/brown | 1–5% green with increasing intensity |
| | 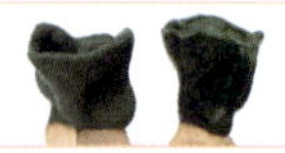 | 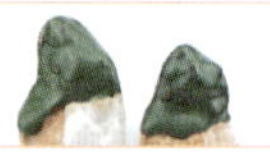 | 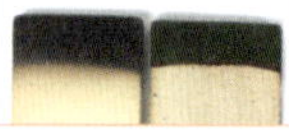 | 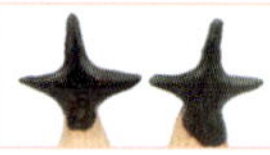 | 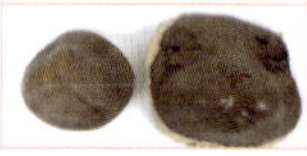 | 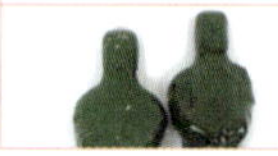 |
| $CoCO_3$<br>cobalt carbonate<br>$CoO$<br>cobalt oxide<br>(1.5 g $CoCO_3$ = 1 g CoO) | 0.25–2% blue 3%+ blackish | 0.25–2% blue 3%+ blackish | 0.25–2% blue 3%+ blackish | 0.25–2% pink or violet, streaks at high temp | 0.25–2% brilliant blue 3%+ blackish | 0.25–2% blue 3%+ blackish |
| | 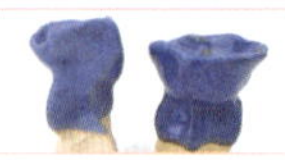 | 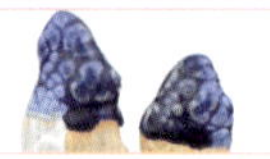 | 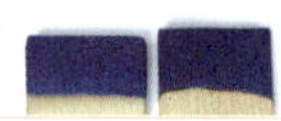 | 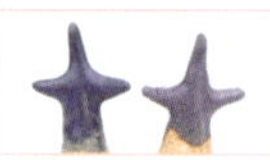 | 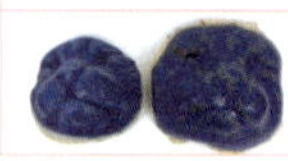 | 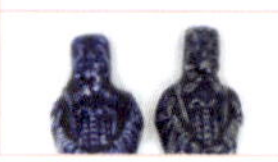 |
| $CuCO_3$<br>copper carbonate<br>$CuO$<br>copper oxide | 0.5% clear 3–5% blue (Na) green (K) | 1–3% light to bluish green 3–5% dark green 5%+ metallic black high fire grey green | 3–5% strong turquoise matt blue | 1–5% green | 1–3% turquoise green (zinc influences copper to go green, not ideal for copper red) | 1–3% blue green |
| | 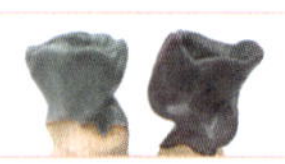 | 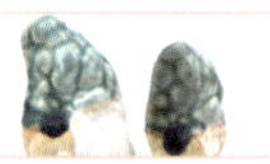 | 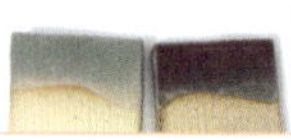 | 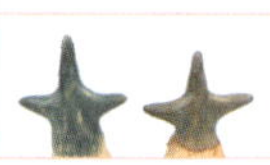 | 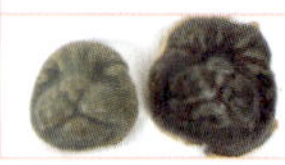 | 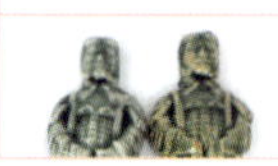 |
| $MnO_2$<br>manganese dioxide<br>Firing fumes toxic | 1–5% range of purple (K) pink, plum, burgundy | 1–5% brownish pink | 1–5% pink, violet, burgundy | 1–5% pinkish brown | 1–5% pinkish | 1–5% reddish purple (in reduction) |
| | 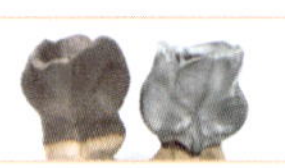 | 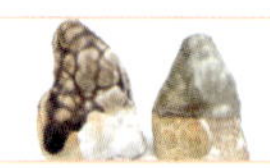 | 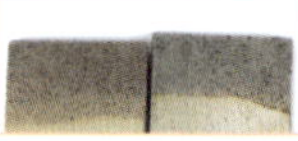 | 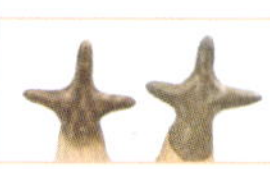 | 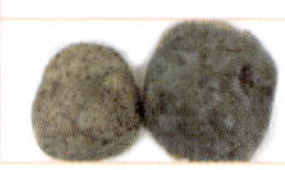 | 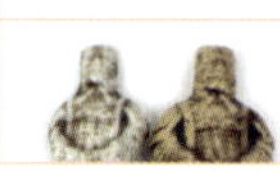 |
| $NiCO_3$<br>nickel carbonate | 1–3% violet to brown pink (K) yellow (Li) | 2% tan brown | range of pink plum red, purple, brown | 1–3% green | 1–3% steel blue to purple | 2–4% tan |
| | 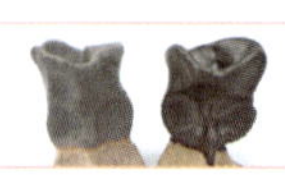 | 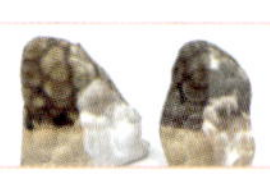 | 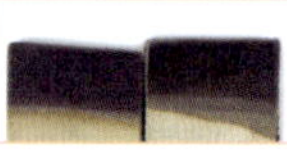 | 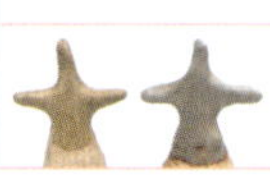 | 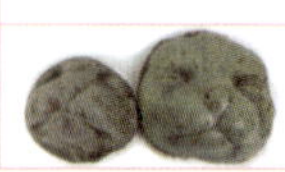 | 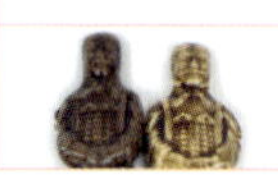 |

| | **Sodium, potassium lithium** dominated glazes | **Calcium** dominated glazes | **Barium strontium** dominated glazes | **Magnesium** dominated glazes | **Zinc** dominated glazes | **Boron** dominated glazes |
|---|---|---|---|---|---|---|
| $Fe_2O_3$ red iron oxide<br>$Fe_2O_3 + FeO$ black iron oxide<br>$Fe_2O_3.H_2O$ yellow iron oxide | 0.5–2% amber or honey<br><br>0.5–2% celadon green in reduction | 1–2% pale yellow (bleaching effect)<br>2–4 % brown amber<br>7–10% saturated | 0.5–2% amber or honey (in reduction iron blues and celadons) | 1–2% yellow/ brown | 1–2% dull muddy brownish | 1–2% yellow amber<br>5–10% red brown<br>12% blackish |
| In oxidation refractory<br>In reduction acts as flux | 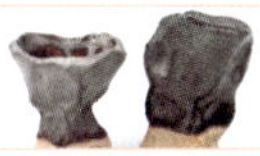 | 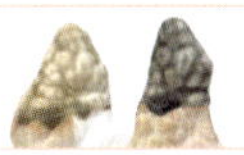 | 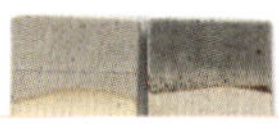 | 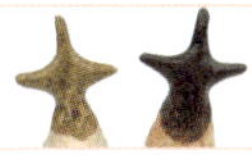 | 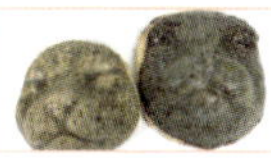 | 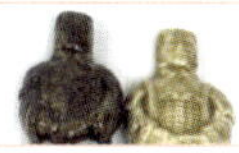 |

| | | | | | | |
|---|---|---|---|---|---|---|
| $TiO_2$ + $Fe_2O_3$ rutile<br><br>(titanium dioxide + iron) | 1–3% pale straw to cinnamon brown<br>3%+ gradually more brown | tan yellows<br>8–10% creates crystals (more in oxidation) | 1–3% yellow | 1–3% pale blue to tan with titanium dioxide will be matt due to tiny white crystals | 1–3% yellowish bluish in reduction can form crystals | 2% yellow |
| | 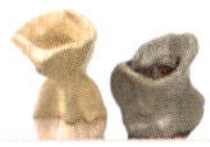 | 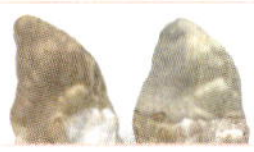 | 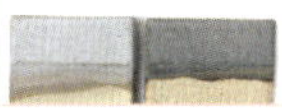 | 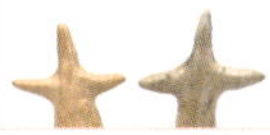 | 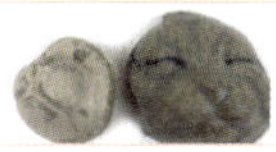 | 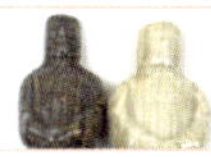 |

## Opacifiers

| | |
|---|---|
| $SnO_2$<br>tin oxide | 5 % required to opacify a glaze, strongest opacifier. Gives softer whites, creamy or bluish milk white.<br>Opacity lost in high fire reduction causing grey cloudiness or scumming.<br>tin + iron = warm oranges/reds<br>tin + chrome = pinks |
| $TiO_2$<br>titanium dioxide | 5–10% required to opacify a glaze. Gives a crystalline soft matt surface and provides creamy opacity to glaze.<br>20% gives drier crystalline surface. Hardens a glaze and promotes formation of crystals.<br>titanium + cobalt = soft greens<br>titanium + iron = bright yellow to orange<br>titanium + iron in reduction = green<br>titanium + copper = bluish streaks<br>titanium + copper in reduction = copper reds and purplish blues |
| $ZrO_2$<br>zirconium oxide | 10–15% required to opacify a glaze.<br>Gives bright even harsh whites.<br>Brands: Zircopax, Superpax, Ultrox |

This Colourant Guide serves as a general overview of how colour reactions correlate to the dominant flux oxide in a glaze. Variations in glaze composition, clay body colour and firing atmosphere can contribute to the colour of a glaze. These glaze samples made of stoneware fired in oxidation on the left and in reduction on the right.

# APPENDIX 10: USA TO UK MATERIAL CONVERSION

| USA | UK |
|---|---|
| Bentonite, Macaloid, Veegum | Bentonite |
| Cornwall stone | Cornish stone |
| Custer feldspar, G-200, K-200 | Potash feldspar |
| EPK Edgar Plastic Kaolin | China clay |
| Ferro frit 3110 | High-alkaline frit |
| Ferro frit 3124 | Borax frit |
| Ferro frit 3134 | Calcium borate frit |
| Ferro frit 3249 | Low-expansion frit |
| Foundry Hill Creme | Buff stoneware |
| Georgia kaolin, Tile 6 kaolin | China clay |
| Gold Art | Fire clay |
| Helmer kaolin | China clay |
| Kentucky OM-4 ball clay | Hymod AT ball clay (high iron from Dorset) |
| Kentucky Stone | Hyplas 71 ball clay (high silica from Devon) |
| Kona F-4, NC-4, Minspar 200 | Soda feldspar |
| Red Art | Etruria Marl, Keuper Marl |
| Silica | Flint, quartz |
| Tennessee ball clay | HVAR ball clay (Devon) |
| Zircopax, Superpax, Ultrox | Zirconium silicate |

## APPENDIX 11: CERAMIC MATERIALS LIST

Ceramic materials, chemical formulae and molecular or equivalent* weights (containing one molecule).

| MATERIAL | FORMULA | MOLECULAR WEIGHT |
|---|---|---|
| Alumina | $Al_2O_3$ | 102 |
| Alumina hydrate | $Al(OH)_3$ | 78* |
| Barium carbonate | $BaCO_3$ | 197.3 |
| Bentonite | $Al_2O_3.4SiO_2.H_2O$ | 360.3 |
| Bone ash (calcium phosphate) | $Ca_3(PO_4)2$ | 103* |
| Borax | $Na_2O.B_2O_3.10\ H_2O$ | 381.4 |
| Boric oxide | $B_2O_3$ | 69.6 |
| Calcium borate | $Ca(BO_2)_2$ | 125.7 |
| Cerium oxide | $CeO_2$ | 172.1 |
| China clay | $Al_2O_3.2SiO_2.2H_2O$ | 258.2 |
| Colemanite | $2CaO.3B_2O_3.5H_2O$ | 206* |
| Cornwall stone | $K_2O.Al_2O_3.8SiO_2$ | 676.8 |
| Chromium oxide | $Cr_2O_3$ | 152 |
| Cobalt oxide | CoO | 74.9 |
| Copper oxide | CuO | 79.5 |
| Cryolite | $Na_3AlF_6$ | 210 |
| Dolomite | $CaCO_3.MgCO_3$ | 184.4 |
| Erbium oxide | $Er_2O_3$ | 382.5 |
| Feldspar soda (albite) | $Na_2O.Al_2O_3.6SiO_2$ | 524.4 |
| Feldspar potash (orthoclase) | $K_2O.Al_2O_3.6SiO_2$ | 556.4 |
| Feldspar lime (anorthite) | $CaO.Al_2O_3.2SiO_2$ | 278.2 |
| Fluorspar | $CaF_2$ | 78.1 |
| Ilmenite | $FeO.TiO_2$ | 151.7 |
| Iron (ferric) oxide (red) | $Fe_2O_3$ | 159.7 |
| Iron (ferrous) oxide (black) | FeO | 71.8 |
| Kaolin | $Al_2O_3.2SiO_2.2H_2O$ | 258.2 |
| Kyanite | $Al_2O_3.SiO_2$ | 162 |
| Lead bisilicate | PbO.2SiO2 | 343.4 |
| Lead oxide (litharge) | PbO | 223.2 |
| Lead sulphide (galena) | PbS | 239.2 |
| Lead sesquisilicate | $2PbO.3SiO_2$ | 313.3* |
| Lepidolite | $LiFKF.Al_2O_3.3SiO_2$ | 366.3 |

| MATERIAL | FORMULA | MOLECULAR WEIGHT |
|---|---|---|
| Lithium carbonate | $Li_2CO_3$ | 73.9 |
| Magnesia | $MgO$ | 40.3 |
| Magnesium carbonate | $MgCO_3$ | 84.3 |
| Manganese dioxide | $MnO_2$ | 87 |
| Mullite | $3Al_2O_3.2SiO_2$ | 426.1 |
| Neodymium oxide | $Nd_2O_3$ | 336.5 |
| Nepheline syenite | $K_2O.3Na_2O.4Al_2O_3.8SiO_2$ | 389.6* |
| Nickel oxide | $NiO$ | 74.7 |
| Petalite | $Li_2O.Al_2O_3.8SiO_2$ | 612.5 |
| Praseodymium oxide | $Pr_2O_3$ | 329.8 |
| Quartz | $SiO_2$ | 60.1 |
| Rutile | $TiO_2$ | 79.9 |
| Silicon carbide | $SiC$ | 40.1 |
| Silica | $SiO_2$ | 60.1 |
| Spinel | $MgAl_2O_4$ | 142.3 |
| Spodumene | $Li_2O.Al_2O_3.4SiO_2$ | 372.2 |
| Strontium carbonate | $SrCO_3$ | 147.6 |
| Talc (magnesium silicate) | $3MgO.4SiO_2.H_2O$ | 126.4* |
| Tin oxide | $SnO_2$ | 150.7 |
| Titanium dioxide (anatase) | $TiO_2$ | 79.9 |
| Vanadium pentoxide | $V_2O_5$ | 181.9 |
| Whiting (calcium carbonate) | $CaCO_3$ | 100.1 |
| Wollastonite (calcium silicate) | $CaSiO_3$ | 116.2 |
| Zinc oxide | $ZnO$ | 81.4 |
| Zirconium silicate | $ZrSiO_4$ | 183.3 |

*Equivalent weights (containing one molecule) are sometimes given where the chemical formula contains a multiple number of molecules, for example, talc contains three molecules of magnesia, so the total molecular weight is divided by three to give the equivalent weight of talc containing one molecule of magnesia.

### Molecular formula calculation table

Divide the glaze recipe weight percent for each material by the molecular weight to get the molecular equivalent.

Add together the total silica, alumina and fluxes. Divide 1 by the total fluxes. Multiply the total silica, alumina, potassium and calcium by this number to get the unity molecular formula.

| MATERIAL | WEIGHT % | MOL WT | MOL EQUIV | $SiO_2$ | $Al_2O_3$ | $K_2O$ | CAO |
|---|---|---|---|---|---|---|---|
| Potash feldspar | 30 | 556 | 0.054 | 0.32 | 0.054 | 0.054 | |
| Silica | 30 | 60 | 0.5 | 0.5 | | | |
| Kaolin | 20 | 258 | 0.078 | 0.16 | 0.078 | | |
| Whiting | 20 | 100 | 0.2 | | | | 0.2 |
| Total | 100 | | | 0.98 | 0.132 | 0.054 | 0.2 |
| UMF | | | | 3.86 | 0.52 | 0.21 | 0.79 |

**Limits for stable glazes in the unity molecular formula**

Stable glazes (those that are dishwasher safe and food safe) will generally have a ratio of alkali metal (potassium, sodium) to alkaline earth (calcium, magnesium) oxides of around 0.3:0.7 (±0.1). The amounts of silica and alumina will increase with firing temperature. For glazes fired below cone 9 (1280°C (2336°F)), enough boron is needed to fully melt the glaze (0.1 B2O3 for cone 8 glazes to 0.5 B2O3 for cone 04 earthenware glazes).

Alumina and silica limits (Cooper and Royle, 1984)

Cone number and temperature. Number of molecules in unity formula.

| | | |
|---|---|---|
| Cone 04 1060°C (1940°F) | $Al_2O_3$ 0.1–0.45 | $SiO_2$ 1.375–3.15 |
| Cone 5 1200°C (2192°F) | $Al_2O_3$ 0.275–0.65 | $SiO_2$ 2.4–4.7 |
| Cone 6 1225°C (2237°F) | $Al_2O_3$ 0.325–0.70 | $SiO_2$ 2.6–5.15 |
| Cone 8 1250°C (2282°F) | $Al_2O_3$ 0.375–0.75 | $SiO_2$ 3.0–5.75 |
| Cone 9 1275°C (2327°F) | $Al_2O_3$ 0.45–0.825 | $SiO_2$ 3.5–6.4 |
| Cone 10 1300°C (2372°F) | $Al_2O_3$ 0.50–0.90 | $SiO_2$ 4.0–7.2 |

Recommended maximum flux in molecular unity formula (Cooper and Royle, 1984)

| CONE | TEMP °C | MgO | BaO | ZnO | CaO | $B_2O_3$ | K+Na |
|---|---|---|---|---|---|---|---|
| 5 | 1200 | 0.325 | 0.40 | 0.30 | 0.55 | 0.35 | 0.375 |
| 6 | 1225 | 0.330 | 0.425 | 0.32 | 0.60 | 0.30 | 0.35 |
| 8 | 1250 | 0.335 | 0.45 | 0.34 | 0.65 | 0.25 | 0.325 |
| 9 | 1275 | 0.340 | 0.475 | 0.36 | 0.70 | 0.225 | 0.30 |
| 10 | 1300 | 0.345 | 0.50 | 0.38 | 0.75 | 0.21 | 0.275 |

## APPENDIX 12: ORTON PYROMETRIC CONE TEMPERATURES

Pyrometric cones measure heat work, and so they depend on the heating rate. A slower temperature rise will cause the cone to bend at a lower temperature.

| CONE NO. | 60°C/HOUR | 108°F/HOUR | 150°C/HOUR | 270°F/HOUR |
|---|---|---|---|---|
| 014 | 807 | 1485 | 836 | 1537 |
| 013 | 837 | 1539 | 859 | 1578 |
| 012 | 858 | 1576 | 880 | 1616 |
| 011 | 873 | 1603 | 892 | 1638 |
| 010 | 898 | 1648 | 913 | 1675 |
| 09 | 917 | 1683 | 928 | 1702 |
| 08 | 942 | 1728 | 954 | 1749 |
| 07 | 973 | 1783 | 984 | 1805 |
| 06 | 995 | 1823 | 985 | 1852 |
| 05 | 1030 | 1886 | 1046 | 1915 |
| 04 | 1060 | 1940 | 1070 | 1958 |
| 03 | 1086 | 1987 | 1101 | 2014 |
| 02 | 1101 | 2014 | 1120 | 2048 |
| 01 | 1117 | 2043 | 1137 | 2079 |
| 1 | 1136 | 2077 | 1154 | 2109 |
| 2 | 1142 | 2088 | 1162 | 2124 |
| 3 | 1152 | 2106 | 1168 | 2134 |
| 4 | 1160 | 2120 | 1181 | 2158 |
| 5 | 1184 | 2163 | 1205 | 2201 |
| 6 | 1220 | 2228 | 1241 | 2266 |
| 7 | 1237 | 2259 | 1255 | 2291 |
| 8 | 1247 | 2277 | 1269 | 2316 |
| 9 | 1257 | 2295 | 1278 | 2332 |
| 10 | 1282 | 2340 | 1303 | 2377 |
| 11 | 1293 | 2359 | 1312 | 2394 |
| 12 | 1304 | 2379 | 1324 | 2415 |
| 13 | 1321 | 2410 | 1346 | 2455 |
| 14 | 1388 | 2530 | 1366 | 2491 |

# INDEX

*References to images are in italics*

## M



## N



## O



## P



## R



## S

## ACKNOWLEDGEMENTS

We are grateful to all of the artists, designers and Concordia students who contributed images of their work to help illustrate the qualities, properties and potential of ceramics.

Thank you to Mette Blum Marcher, Director of Guldagergaard International Research Center, and to the staff for providing us with a fertile meeting ground. Thank you to Matthew P Loocke, Assistant Professor of Research, Department of Geology, Louisiana State University for providing access to the mineral and rock samples. Thank you to Paul Holmquist and Daniel Gillberg for the photography of student work and mineral samples and thank you to Chloe Holmquist for editing. A huge thank you to Henry Bloomfield for editing and creating the illustrations. Thank you to Andy Chapman and the 525 team for their work on the layouts.